U.S. Marine Corps Physical Readiness Training for Combat

Plus

Technical Manual for 5.56MM, M16A2, M4 and M4A1 Carbine

Department of Defense

Pentagon
Publishing

UNITED STATES MARINE CORPS
Marine Corps Combat Development Command
Quantico, Virginia 22134-5001

29 January 1988

FLEET MARINE FORCE REFERENCE PUBLICATION............0-1B

MARINE PHYSICAL READINESS TRAINING FOR COMBAT

1. PURPOSE

Fleet Marine Force Reference Publication (FMFRP) 0-1Marine Physical Readiness Training for Combat is intended for use by all Marines. It provides the information and references necessary to establish and conduct physical conditioning programs to prepare Marines for the physical demands of combat.

2. APPLICABILITY

Although not directive, this handbook is intended for use as a reference by all Marines in developing unit physical readiness training for combat. This FMFRP does not establish policy for the conduct of physical fitness training. Refer to Marine Corps Order 6300.3_ for applicable Marine Corps policy on physical fitness. It may also be used as a basis for instruction conducted by unit leaders or as a reference for instruction or training.

3. RECOMMENDATIONS

To improve this manual, the MAGTF Warfighting Center seeks comments and recommendations from the user. Officers and noncommissioned officers with constructive comments are encouraged to respond. Lack of response will be considered as concurrence. The User Suggestion Form which follows can be detached and forwarded to - -

Commanding General
Marine Corps Combat Development Command (WF 06)
Quantico, Virginia 22134-5001

4. DISTRIBUTION

Initial distribution of this FMFRP 0-1B is shown below. Addressees on this list should ensure that appropriate subordinate organizations are provided copies.

JOHN I. HOPKINS
Brigadier General, U.S. Marine Corps
Assistant Chief of Staff, Warfighting
Marine Corps Combat Development Command
Quantico, Virginia

DISTRIBUTION: TDJ

From:

To: Commanding General, Marine Corps Combat Development Command (WF 06), Quantico, Virginia 22134-5001

Subj: FMFRP 0-1B, MARINE PHYSICAL READINESS TRAINING FOR COMBAT, RECOMMENDATIONS CONCERNING

1. In accordance with the Promulgation Page to FMFRP 0-1B, which invites individuals to submit suggestions concerning this FMFRP directly to the above addressee, the following unclassified recommendation is forwarded:

Page_________Article/Para No._________Line No.________Figure/Table No.________

Nature of Change:_____Add_____Delete_____Change_____Correct

2. Proposed New Verbatim Text: (Verbatim, double spaced; continue on additional pages as necessary.)

3. Justification/Source: (Need not be double spaced.)

NOTE: Only one recommendation per page.

MARINE PHYSICAL READINESS TRAINING FOR COMBAT

Chapter 1
PHYSICAL READINESS LEADERSHIP
Section I. Role of Physical Fitness in Combat

1101. PURPOSE OF PHYSICAL FITNESS TRAINING

Physical fitness training in the Marine Corps has one purpose: to prepare Marines to physically withstand the rigors of combat. All other goals of physical fitness training are subordinate to and must support attainment of this goal. The idea that only infantry or reconnaissance units and their attachments normally face physically demanding combat is wrong. This error must not influence the priority commanders of combat support, combat service support, aviation, and headquarters units give to physical fitness for combat. Physical fitness for combat has a high priority for all Marines. A sound, effective unit program, requiring limited time and material, offers a greater payoff in combat than many more expensive and time-consuming training programs. This handbook describes the concept and provides the ingredients of a program for commanders to use in carrying out one of their most serious responsibilities: ensuring that their Marines are physically ready for combat.

1102. SCOPE OF THE MANUAL

This handbook provides guidance for all leaders, trainers, and planners of physical training programs. It describes unit physical fitness for combat training in the following chapters:

a. Physical Readiness Leadership. Chapter 1 provides guidance to leaders in the conduct of physical combat readiness training.

b. Physical Readiness Training Programs. Chapter 2 discusses how to structure programs to meet physical readiness goals in a variety of combat training situations.

c. Physical Conditioning Activities. Chapter 3 discusses the primary physical conditioning activities which commanders may use to attain readiness for combat goals. Sections in this chapter address foot marches under load, strength-building activities, and activities which build confidence and the aggressive spirit needed in combat.

d. Combat Water Survival. As an amphibious force, Marines cannot ignore the importance of combat water survival training. Chapter 4 describes a program for commanders to conduct this training.

e. Competitive Conditioning Activities. Chapter 5 describes competitive activities which stress the relationship to combat actions, featuring team-building types of competition.

f. Evaluation of Performance During Training. Chapter 6 describes tests which measure individual and unit physical fitness for combat.

g. The Human Body. Chapter 7 covers the structure and functioning of the body.

1103. COMMANDER'S ROLE

Major General Commandant John A. Lejeune in the 1921 edition of Marine Corps Manual expressed our philosophy of leadership. Among other things, General Lejeune stated that "it will be necessary for officers ... to endeavor to enlist the interest of their men in building up and maintaining their bodies in the finest physical condition." It is significant that the Commandant prefaced these writings in 1921 with the statement that his thoughts were greatly influenced by the experience of World War I. We have never varied as a Corps from this belief that one of a commander's most serious responsibilities is to have Marines physically prepared for combat. Senior commanders and their staffs carry out this role primarily by providing command emphasis, including planning, support, and supervision. Commanders at the battalion/squadron and company/battery level execute a program of physical fitness for combat by close supervision and, most importantly, leadership by example.

1104. HISTORY OF MILITARY PHYSICAL READINESS

Every war has revealed our military physical deficiencies during the initial periods of mobilization. This realization followed the Civil War and has recurred regularly with each national emergency.

a. Training programs in each war were geared to the physical need of the era. Success was dependent upon the amount of time available during training to physically prepare Marines for battle conditions. Frequently, casualties in initial engagements were attributed to the inability of our Marines to physically withstand the rigors of combat over rugged terrain and under unfavorable climatic conditions. With adequate preparation, Marines have always handled the test of battle.

b. During World War II, the first physical conditioning doctrine that could be scientifically justified by testing was introduced. As the war progressed, this program was effective in the physical conditioning of millions of Marines for combat.

c. Postwar periods have traditionally been a time of consolidation. Unfortunately, some leaders considered the conditioning phase of training to be a wartime tool. With such a philosophy prevalent between wars, physical readiness was relegated to secondary importance resulting in a serious lowering of combat effectiveness. The initial commitment of Marines in Korea dramatically displayed this failure to recognize the extreme physical demands of warfare. Call-up of large numbers of reserve Marines and pressing them into the war within weeks after call-up mainly caused the loss in effectiveness.

d. Costly lessons learned from our military experiences over a period of years and the course of several wars led to an increasing interest in the physical conditioning of the individual Marine. No longer can we afford to emphasize physical fitness during wartime and de-emphasize it during peacetime. It is evident that, in spite of increased mechanization and modern weapons, physical readiness retains a vital place in the life of each individual Marine and in every unit within the Marine Corps regardless of mission.

e. Commanders are well aware of the need for rugged and well conditioned Marines. The daily demands of housekeeping, maintenance, support, training, operations, and other time-consuming tasks make it necessary for commanders to set aside time for frequent, regular, and vigorous exercise periods.

1105. PHYSICAL DEMANDS OF COMBAT

There are three primary elements of effective physical fitness for combat: lower body strength and stamina; upper body strength and stamina; and a competitive, combative spirit. Cardiovascular functioning is not treated as a separate element since it is a necessary foundation for the other three. Exercises and activities designed to improve Marines' condition in the three primary elements will foster good cardiovascular and respiratory condition.

a. Lower Body Strength and Stamina. Some common demands that combat places on lower body strength and endurance are marching long distances under load and functioning effectively at the destination; moving quickly and evasively under fire; and carrying wounded Marines to safety.

b. Upper Body Strength and Stamina. Some common demands that combat places on upper body strength and stamina are rapidly emplacing crew-served weapons; handling large-caliber ammunition for extended periods; climbing walls, cliffs, and other high obstacles; and performing field maintenance on air craft or heavy machinery.

c. Competitive, Combative Spirit. While part of this critical aspect of combat readiness is mental and emotional, robust physical condition and a training program which promotes physical aggressiveness greatly enhance a competitive and combative spirit. Activities which force Marines to overcome natural physical fear, which directly leads to fatigue, are particularly valuable.

1106. PHYSICAL EFFECTS OF COMBAT STRESS

It has been thoroughly documented that the added uncertainty and stress of combat have a major physical effect on Marines as well as the acknowledged psychological effect. In other words, in combat, fear equals fatigue. In training, we cannot easily reproduce this combat stress which reduces the effectiveness of individuals and units. However, we can produce fatigue and psychological doubt in training situations by developing a challenging physical training program which stresses the participants. This compels Marines to exercise their ability to continue to function under difficult and trying conditions. Such activities as long foot marches under load and difficult confidence courses train Marines to overcome their own fear and fatigue. Rugged competitive activities such as martial arts and pugil stick fighting prepare Marines psychologically to overcome an opponent as well as their own fear and fatigue. This sort of physically demanding training builds self-confidence and unit morale. The adage that "the unit that sweats more in peacetime will bleed less in battle" certainly applies to its physical training program. Athletic coaches have long followed this approach by such practices as having teams shoot foul shots after wind sprints in order to strengthen their mental conditioning and toughness. Although pushing Marines to their limit is necessary, commanders often set lower standards initially than ultimately desired. The program should then call for steadily increasing the difficulty. A good guide for a starting point is a standard that will be physically and psychologically demanding for no less than three-fourths of the unit. Expressed another way, no more than one-fourth of the unit should be working at less than their limit. The program should progress quickly, augmented, if needed, by remedial training for those unable to meet the unit standards, until the training is challenging for all but the most exceptional Marines in the command. Commanders must never forget that Marines expect to be challenged and thrive on being pushed to their limit.

1107. LIMITS OF PHYSICAL READINESS

A well-conditioned Marine unit has a significant advantage in combat, but this does not mean that physical conditioning can substitute for technical knowledge or good planning. For instance, units must, as discussed in chapter 3, pursue a continuing and challenging program of marches under loads. This does not disprove the axiom that fighting men, if required to carry over one-third of body weight, will generally become ineffective in battle. Leaders and planners must decide what is essential to be carried and must use an effective logistic distribution system, such as unit trains, to make remaining items available when they are needed. In a similar vein, the valid requirement to foster a competitive and combative spirit in Marines does not excuse leaders from the responsibility to practice tactics which will not needlessly risk lives. Physical readiness for combat is a vital goal, but it is only one part of a unit's overall readiness for the lest of combat.

Section II. Fundamentals of Physical Fitness

1201. FITNESS FOR MARINES

Total combat readiness includes technical proficiency and mental, emotional, and physical fitness. If any one of these attributes is lacking, combat effectiveness suffers proportionately. Without technical fitness, Marines lack the knowledge and skill to fight; without mental and emotional fitness, they lack the incentive and will to fight; and without physical fitness, they lack the physical ability and confidence to fight. Physical fitness in a Marine means a whole and healthy body, the capacity for skillful and sustained performance, the ability to recover from exertion rapidly, the desire to complete a designated task, and the confidence to face any eventuality.

1202. COMPONENTS OF PHYSICAL FITNESS

A sound body, free of disease and defect, does not in itself constitute physical fitness. Before an untrained Marine can be considered physically fit for combat, the following traits must be developed:

a. Strength. Every Marine needs enough strength to perform the heaviest task encountered in routine and emergency activities. The basic areas where strength is required are in the arm and shoulder girdle, abdomen, back, and legs. Muscles increase in size, strength, and firmness with regular and strenuous exercises. Without work, they grow flabby and weak.

b. Endurance. Each Marine needs sustaining power to maintain the maximum ability without undue fatigue. There are two types of endurance:

(1) Muscular Endurance. The Marine needs muscular endurance to fight the enemy under the most tiring combat conditions. Muscular endurance is the ability to perform continuous work over long periods of time. Endurance depends on the bloodstream's ability to deliver large amounts of oxygen and nutrition to the muscle masses and then carry away the waste products quickly.

(2) Cardiovascular-Respiratory Endurance. The development of cardiovascular-respiratory endurance ("wind") is necessary to maintain muscular endurance. Cardiovascular-respiratory endurance depends on the efficiency of the blood vessels, lungs, and heart. The maximum effort a Marine can exert over a period of time is limited by the capacity to absorb oxygen and expel carbon dioxide. The average Marine's cardiovascular-respiratory capacity can be greatly increased by exercise.

c. Agility. A Marine must be able to change direction quickly and as faultlessly as possible. The ability to react instantly and to maintain orientation during rapid changes of body position is important to survival. This important ability may be developed by conditioning exercises that require varied and rapid changes of body position on the ground and in the air, such as obstacle courses.

d. Coordination. Coordination is the ability to move all parts of the body in a smooth, efficient, and concerted effort (commonly called timing). A well-coordinated individual does not make useless movements. An individual who moves with precision and accuracy saves energy. Coordination is best developed by practicing diversified muscular activities and skills affecting all body parts.

1203. TYPES OF EXERCISE

Basically, there are two forms of exercise: isotonic and isometric. Both forms are contained in the activities found in this manual.

a. Isotonic. Isotonic exercises are those in which the expenditure of energy is regulated and released during concentrated efforts. The regulated expenditure of energy is controlled by both the mode of exercise and the individual's effort. This type of effort is common to most exercise and sports. To develop endurance, coordination, and flexibility, isotonic exercise must be used. Strength can also be developed or increased through this type of exercise.

b. Isometric. Isometric exercises are those in which maximum effort is applied and held until the engaged muscle energy is depleted during a single contraction effort. The individual exerts full force against an immovable object for a relatively short period of time (6 to 10 seconds) and then repeats the contraction several times with a short period of rest between each contraction. This type of exercise develops only strength; therefore, it has limited application.

1204. PRINCIPLES OF PHYSICAL CONDITIONING

Men and women vary in their physical makeup. Physiological (body) function and reaction also vary in proportion to heavy demands placed upon the body. To attain the maximum benefit without sacrificing Marines' welfare, fitness must grow with a careful program of conditioning. To allow for adjustments in body functioning as the conditioning program progresses and to ensure attainment of objectives, the following principles of physical conditioning must be applied:

a. Overload. As strength and endurance increase, the physical load must be increased until the desired level of fitness is reached.

b. Progression. In beginning stages, the load must be moderate. Gradual progression from this low state of fitness to a higher state is possible through application of a progressive program.

c. Balance. An effective program utilizes various types of activities and provides for the concurrent development of strength, endurance, and coordination as well as basic physical skills.

d. Variety. Some programs fail because the routine becomes boring. The most successful programs always include conditioning activities, competitive events, and military physical skill development.

e. Regularity. There is no easy or occasional way to develop physical fitness. Regularity of exercise is a must, with daily exercise preferred.

1205. THREE STAGES OF PHYSICAL CONDITIONING

Unconditioned or poorly conditioned Marines pass through the following stages in reaching the desired state of physical condition.

a. Toughening Stage. This stage is approximately 2 weeks long and is usually characterized by muscular stiffness and soreness followed by recovery.

b. Slow Improvement Stage. This stage is approximately 6 to 10 weeks long and is characterized by slow and steady improvement until the desired level, or a high level, of fitness is attained.

c. Sustaining Stage. This stage goes on indefinitely in order to maintain the level of conditioning achieved by passage through the previous stages.

1206. EFFECTS OF CLIMATIC CONDITIONS

Temperature, both atmospheric and body, affects the physical performance of Marines. Proper maintenance of body temperature through warm up exercise, proper dress in cold weather, and removal or adjustment of clothing in hot weather is necessary for effective performance and health. Climatic factors to be considered are the following:

a. Exercise In High Temperatures and High Humidity. Marines can endure strenuous physical activity in extremely hot temperatures if they are given an opportunity to become acclimated and if they take enough salt and water. It is essential to continue physical training programs in hot climates. Marines can better withstand high temperatures when they are well-conditioned. High humidity combined with high temperatures presents a serious danger. These conditions prevent the natural cooling of the body by the evaporation of perspiration. Training schedules should conform to the provisions of the base commander in respect to wet bulb conditions. Those who conduct training under these conditions should monitor weight loss and be careful to make fluid replacement adjustments.

b. Exercise at High Altitudes. Certain problems are encountered in conditioning Marines stationed in high altitudes because the heart undergoes greater exertion during exercise. It is particularly important that only light exercise be given initially at such altitudes. A Marine's body gradually adjusts to high altitudes within a few weeks. After this adjustment, progressively greater amounts of exercise are possible.

c. Exercise in Arctic Regions. Military duty in the arctic is so arduous that a high level of physical conditioning is essential. Because of the difficulties of carrying on physical conditioning exercises in extreme cold, Marines should be conditioned to the highest level possible before they arrive. A sustaining program will then maintain that level. When exercising in cold weather, Marines should be required to remove excess clothing to prevent them from becoming damp with perspiration.

1207. WARMING UP AND COOLING OFF

It is a fundamental physiological principle that Marines should warm up gradually before taking strenuous exercise. Such action speeds up the circulation to prepare the body to take an overload and helps to prevent injury to muscles and joints after exercising. Marines should be kept mildly active, walking, stretching, or performing some other mild muscular activity until breathing and temperature have returned to normal. Marines should never be allowed to cool off too rapidly; in cool or cold weather, they should put on additional clothing during the cooling-off period.

1208. PHYSICAL ACTIVITY AS AGE INCREASES

Combat makes severe physical demands on Marines. All Marines, regardless of age, must be physically ready to meet these demands. There is no physiological reason to cease exercise or exertion with age. Increased age usually brings increased responsibility which, in many instances, leads to a routine that can become almost devoid of physical activity. The key to fitness with increased age is to continue exercising at a reasonable level and to include exercise of a vigorous type in the daily routine. Older persons who have not regularly maintained a reasonable state of physical fitness require more time than younger persons to become fit. Such individuals usually require a longer period of time to recover from physical effort than younger Marines. If general health is good, evidence shows that older Marines can develop and maintain a degree of fitness which permits vigorous activity and proper performance of duties. It is both the individual's and the commander's responsibility that all Marines continue a daily sustaining exercise program.

Section III. Goals of Physical Readiness Training

1301. NECESSITY FOR PHYSICAL READINESS TRAINING

A very important objective of training is attainment and maintenance of operational readiness. Marines must be physically ready for operations at any time and under all conditions of climate and environment. A combination of training to develop proficiency in physical skills and conditioning to improve strength and endurance results in physical readiness for combat. The degree of physical fitness required of Marines can be acquired through physical exercises in a normal well-coordinated and closely monitored training routine. The performance of purely military duties, such as drills and marching, is not enough to build all the desired areas of fitness. Few recruits are physically fit for the arduous duties ahead of them. The softening influences of our mechanized civilization and the difficulties of conditioning Marines make physical fitness more important than ever before. If Marines are to be fully developed to and maintained at the desired standard of physical fitness, a well-conceived plan of physical readiness training must be a basic part of every training program. Marines cannot be adequately prepared in any other way for the hard work and arduous demands required on the battlefield.

1302. OBJECTIVE OF PHYSICAL READINESS TRAINING

The overall objective of the physical readiness training program is to develop individuals and units who are physically able and ready to perform their duty assignments or missions during training and in combat. To attain the objective of physical readiness, exercise activities must develop the following areas:

a. Strength and Endurance. Developing adequate strength to perform required duties and adequate endurance to sustain activity over a long period of time.

b. Muscle Tone. Developing muscle tone adequate to maintain proper posture and reasonable weight control.

c. Skills. Developing military physical skills which are essential to personal safety and effective combat performance. As skill is developed, agility and coordination are attained. The essential skills are-

(1) Marching Under Load. Marching with individual and unit weapons and equipment.

(2) Running. Distance and sprint running on roads and cross-country.

(3) Jumping. Broad jumping and vertical jumping downward from a height.

(4) Dodging. Changing body direction rapidly while running.

(5) Climbing and Traversing. Vertical climbing of ropes, poles, walls, and cargo nets. Traversing horizontal objects such as ropes, pipes, and ladders.

(6) Crawling. High crawling and low crawling for speed and stealth.

(7) Throwing. Propelling objects, such as grenades, for a distance with accuracy.

(8) Vaulting. Surmounting low objects, such as fences and barriers, by use of hand assists.

(9) Carrying. Carrying objects and employing man-carries.

(10) Balancing. Maintaining proper body balance on narrow walkways and at heights above normal.

(11) Falling. Contacting the ground from standing, running, and jumping postures.

(12) Surviving in Water (Or Other Specialized Situations) Using water survival techniques.

d. Character Traits. Jnstilling character traits which help accomplish military missions to include-

(1) Confidence. Confidence develops through achieving progressively more difficult tasks as physical ability grows.

(2) Aggressiveness. Participation in combative activities and contests develops desire and willingness to overcome an opponent.

(3) Reaction Under Pressure. Competitive contests and game situations train Marines to think and to act quickly while under pressure.

(4) Teamwork. Teamwork develops through competitive events in which a number of Marines must coordinate their efforts to accomplish a physical task.

1303. BENEFITS OF EXERCISE

The benefits of exercise are not always understood. Some of the more important results of exercise are listed below:

a. Improved Muscle Tone. Muscular tone improves and, at the same time, muscular strength and endurance are built up.

b. Cardiovascular-respiratory Endurance. Cardiovascular-respiratory endurance, or wind, improves through a process of opening up dormant lung capacity to absorb greater amounts of oxygen.

c. Circulation. Circulation of the blood speeds up and extends to a greater portion of the body as exercise forces the blood to service all parts of the body. The efficiency and effectiveness of the heart, lungs, and blood vessels improve.

d. Flexibility. A wider range of muscular movement is possible and rapidity in physical skills grows.

e. Elimination of Body Waste. Bending and twisting the body and the general speedup of body processes caused by exercise regulate and help eliminate body wastes.

f. Tension. Working off excess nervous energy and relief from daily worries and cares relieve tension. Participation in exercises leaves little time for worry.

g. Sleep. Sleep improves because muscles are healthfully tired after about of exercise. A by-product of sound sleep is relief of tension.

h. Obesity Control.. Control of obesity (fat) is made possible by using up excessive amounts of fat-producing food elements.

i. Injury Susceptibility. Susceptibility to injury is reduced through exercise. Muscles, tendons, and joints are strengthened. Injuries such as hernia, back strain, and joint sprains are less likely to occur if muscles are maintained in proper tone.

Section IV. Leadership Roles

1401. PSYCHOLOGICAL LEADERSHIP

The full development of a Marine's resources is not all physical. To be effective in developing physical readiness, leaders must realize that mind and attitude are also important to success. The more important psychological considerations are to—

a. Promote Understanding of the Value of Physical Readiness. A desire to be physically ready should be created in all Marines. Motivation is increased and Marines take greater interest in their individual physical fitness if they understand the value and benefits of vigorous exercise. When Marines realize their efforts are an investment in their own personal welfare, it should not be difficult to obtain their cooperation. Marines should understand the objectives, the benefits, and the value of each type of exercise activity in their program. They should also understand the relation of physical readiness to survival in combat.

b. Maintain a Positive Approach. Physical readiness training for combat is strenuous and demanding. It is a responsibility of leadership to create an atmosphere where all desire to participate fully. This attitude should be fostered. A negative approach must not be identified with physical readiness training, even with those having difficulty. Only in unusual cases should fear of punishment be the motivating factor behind good performance. For those few who cannot keep up or attempt to malinger, an effective remedial program is essential.

c. Seek Cooperation and Develop Morale. In a program placing maximum physical stress upon individuals, it is necessary to gain their cooperation. Favorable reaction is enhanced by proper planning and organization, challenging requirements, use of competition, and application of a progressive program resulting in physical fitness. As physical fitness grows, morale also grows.

1402. COMMAND AND SUPERVISORY FUNCTIONS

a. Command Functions. Commanders should take the following actions to support physical readiness training:

(1) Lead by personal example.

(2) Instill command interest and indicate to subordinate personnel the importance of this training to the welfare of the organization.

(3) Allot sufficient time for the achievement of objectives and monitor the use of such allotted time. The substitution of other training or routine duties for scheduled physical readiness training is unsound and unwise.

(4) Assign and properly utilize qualified personnel to supervise and conduct physical readiness training. If leaders are not competent, take action to ensure they become competent quickly. Reassign those who do not meet this standard to other duties.

(5) Make necessary facilities and funds available to support a program to develop physical readiness within all personnel.

(6) Measure the physical fitness of individuals and units in order to evaluate progress and to determine if the program is successful.

b. Supervisory Functions. Leaders responsible for planning, conducting, and supervising physical readiness training should take the following actions:

(1) Prepare physical readiness training schedules which apply the principles of physical conditioning and which aim for a particular type of program plan.

(2) Provide for wide participation of as many Marines as possible. All Marines, regardless of position or age, will benefit from regular exercise. In some instances, special efforts are necessary to overcome obstacles to regular and frequent training. Special effort is also necessary to ensure remedial conditioning. Such conditioning should occur for those who are physically substandard and after extended absence due to leave, sickness, injury, and travel.

(3) Prevent waste or unwise use of time allotted for physical readiness training. Time-wasters include unprepared instructors; assignment of one instructor to a group larger than a platoon; progression which does not keep pace with the physical development of the Marines; extreme formality; inadequate equipment or facilities which require waiting turns to exercise; and lengthy rest periods between exercises which interfere with the application of overload.

(4) Ensure that the program contains vigorous physical activity. Such activity places progressively greater demands upon the body during each exercise session and also over the duration of the training program. To be of benefit, exercise must tire the muscles and cause the heart to increase its rate of beat.

(5) Set an overall objective for each physical fitness program.Observe the training as necessary to ensure that the established objectives are being achieved.

(6) Observe physical readiness training to insure the use of a positive approach. To implement a positive attitude, small-unit leaders and instructors should personally set the example; have an understanding, fair, and sympathetic attitude; recognize individual differences; and motivate Marines toward their best effort.

(7) Guide and inform small-unit leaders and instructors concerning approved techniques, directives, and literature. As necessary, arrange for local training of instructors to include clinics, conferences, schools, and demonstrations.

(8) Determine the effectiveness of physical readiness training by personal participation in and observation of training, analysis of field inspection reports, and analysis of individual physical fitness test scores. Scores may be combined to reflect the fitness of the unit.

1403. SMALL-UNIT LEADERS AND INSTRUCTORS

a. Responsibility. The instruction and conduct of physical readiness activities are the function of company/battery commanders, platoonleaders and persons assigned as instructors. Experience has proved the effectiveness of physical fitness development when conducted in company-and platoon-size units under direct control of the leader with overall supervision by the parent-unit commander. For example, all the platoons of a company may exercise at the same time under the general supervision of the company commander, with each platoon conducting the assignment separately and under its own leadership.

b. Leader's Assignment. A small-unit leader or an instructor in a school or training activity is assigned to a combat unit or to a support unit. In this assignment, the leader is responsible for all training to include physical readiness training. In a different situation, the noncommissioned officer (NCO) or officer may beassigned as a full-time physical readiness instructor. This contrasts to a unit leader assignment where only part of the time is devoted to such training. In either case, the leader will hold an important and vital position for the physical fitness of Marines.

c. Leader's Training. Leaders may come to the assignment either fully or partially trained. It may be their first responsibility for the development of physical fitness. If they have had previous training through experience, make certain that their information is supplemented with study of this manual. If they have had professional training in physical education during civilian life, but no military experience, they should also use this manual to learn the methods used by the Marine Corps. A new leader should take advantage of various ways to learn including attendance at leader training courses, self-study, practice, and discussion with more experienced leaders.

d.Leader's Objective. As a physical readiness training instructor, the leader has two general objectives. The first is to motivate Marines to want to be physically fit. The second is to conduct a program that will develop a high degree of physical fitnss. Motivated Marines will react enthusiastically to such a program. It aids greatly in achieving local program objectives.

e. Leader's Personal Fitnes. A unit leader who must instruct and demonstrate physical activities must be in physical condition to do the job without undue physical stress.The leader should be able to do those things that must be demonstrated. The leader's strength, endurance, posture, and skill should set the example. This does not mean that the leader must excel, as other Marines do not expect championship performance. However, they do expect, and deserve, a creditable showing of fitness for the job.

f. Leader's Knowledge. The leader must have three types of knowledge to properly administer physical readiness training. Theyare- -

(1) Knowledge of Marines. The leader must understand Marines, know how to lead and motivate them, understand how they learn, and apply this knowledge wisely in the day-to-day training situation.

(2) Understanding of Body Functioning. A more intelligent exercise program results from understanding and applying the principles which govern physical conditioning of the body. The leader with such knowledge can better prescribe, adjust, and regulate exercise types, amounts, and progression to attain fitness.

(3) Understanding Exercise Activities. The leader needs to understand the contribution each type of physical activity makes to physical fitness, and how to use each activity to develop fitness. Skill to demonstrate and lead the various activities is a necessary part of technique and is invaluable to the instructor or small-unit leader.

Chapter 2
PHYSICAL READINESS TRAINING PROGRAMS
Section I. Development of a Program

2101.GUIDANCE FOR PLANNERS

This chapter instructs planners on physical readiness training procedures. It contains program planning guidance including factors to consider when developing programs; steps in assembling a program; definitions of activity packages and systems of exercise; selection of activity packages; and selection of systems used in implementing those packages.

2102. MARINE CORPS MISSION

The mission of the Marine Corps is to seize and defend advanced naval bases and to perform other missions as the President of the United States may direct. Traditionally, to perform this mission, Marines have been projected into the area of operations by Navy vessels and have attacked hostile beaches over the shore in landing craft, assault amphibian vehicles, and helicopters. Recently, performance of this task has been somewhat complicated by the introduction of the concepts of maritime prepositioning and the air-landed Marine expeditionary-brigade. Under these concepts, Marines, landed at airfields in foreign countries, will join their equipment at commercial ports. The distance between these airfields and ports is often 10 or more miles. In time of international crisis, it can be expected that transportation assets will not be adequate. Marines from all elements of a Marine Air-Ground Task Force may have to march from the airfield to the site of their equipment. All Marines must be physically conditioned to rigors of conducting foot marches with individual weapons and equipment.

2103. EVALUATION OF FITNESS

The Marine Corps mission is such that all Marines must constantly and consistently achieve a high level of fitness that prepares them for the demands of combat. While Marines have many different MOS's, all Marines must be prepared for the demands of marching under load and performing basic infantry tasks, such as rear security and patrolling. While the physical fitness test remains our universal measure of individual fitness, it is the commander's responsibility to observe and evaluate the unit's ability to perform effectively in combat. Often the commander's most effective evaluation tool is weekly physical training sessions. The commander should participate in these sessions and see that they are fully integrated into the unit physical conditioning program.

2104. INTEGRATION OF TRAINING

Training time must be used efficiently and wisely. Every opportunity to integrate physical training into other training activities should be seized. For example, foot marches under load can be integrated into the normal daily activities as a way to move from one training or work site to another. This integration reflects realism in training and should be used to maintain overall proficiency. To accomplish this requires imagination and enthusiasm on the part of the commander. The area of operations in which the training is conducted determines what can be integrated and how. If possible, training should also consider the terrain and climate of the area in which the unit will subsequently conduct training or operations. Training should include familiarization with special equipment and the application of specialized techniques to tactical principles. Activities such as gun drills and command post emplacement drills also have physical conditioning value, particularly if supervised and done competitively or measured against time standards.

2105. ACTIVITIES AT THE UNIT LEVEL

Physical readiness training is a command responsibility and is generally conducted in Marine units at the company and platoon levels. Battalion/squadron commanders must constantly supervise, making sure that all Marines are physically prepared for combat.

a. Physical readiness training should usually be conducted weekly by each battalion/squadron-level unit, using events designed for all personnel (e.g., marches under load, battalion/squadron runs, military field days, etc.).These events enable the commanders to demonstrate personal leadership, and observe and evaluate the physical condition and combat physical readiness of their Marines. This exercise period should be designed to elevate the unit's morale and emphasize unit identity through the wearing of uniforms and the carrying of battalion and company guidons. The battalion/squadron commander should make every effort to be a visible participant in the unit's physical readiness program on a weekly basis. These activities could be conditioning marches under load or a weekly calisthenics/unit run.The suggested times for these activities are early on Monday morning or late on Friday afternoon after the training week has been completed. The uniform may be physical training gear orutilities and boots.

b. Commanders should be cautious of long runs in boots and utilities. This is not to suggest that units should not run in utilities, but to remind commanders that boots are designed for marching, not running. If commanders choose boots and utilities, they should select off-road routes which will provide a cushioning effect.

c. Units should be made well aware of the commander's intent before the unit falls out for exercise. If the exercise period is conducted early in the morning, it need not necessitate an extensive shower/clean-up period afterward, so long as all Marines from the commander on down remain dirty. Commanders should inform their Marines not to fall out for these sessions in clean, pressed uniforms.

d. The physical readiness training of headquarters units is often difficult to manage. Often physical training is left for individuals or sections to develop and manage on their own. This approach can be effective, but if not supervised, it can result in a "paper" program. Headquarters unit commanders should endeavor to conduct weekly integrated training which will enable them to evaluate the physical readiness of their personnel. Often this training has to be scheduled before or after normal working hours to maximize participation of the entire unit.

2106. STEPS IN PLANNING

To implement workable and effective programs (as directed by CMC ALMAR 261/87) the planner must-

a. Determine the Type of Program Needed. Marine units are inherently different in organization and mission. The physical readiness program must be tailored to the mission and to the current physical condition of most unit personnel. Programs to meet this need are of the following types:

(1) **Developmental Programs.** Marines in a beginning or poor state of physical readiness need a program which will develop strength, endurance, physical skills, and character traits which are beneficial to successful accomplishment of military missions. Such programs should be applied progressively to rise gradually to a peak of fitness and skill.

(2) Maintenance Programs. Once Marines reach the sustaining stage of conditioning, their goal is then to maintain this level by participation in a maintenance program.

(3) Remedial Programs. The term "remedial" is usually applied to those individuals or groups who possess substandard physical fitness. For example, a remedial physical conditioning program could be applied to persons who are overweight, who fail to reach physical fitness test standards, or who have missed extended periods of conditioning due to illness, injury, extended hospitalization, or other absence. Saturday training sessions are designed as remedial sessions for all individuals who either miss the daily training or have failed to meet adequate standards. Note that personnel in light duty or no-duty status should be expectedto attend all training sessions, observing, supervising, or assisting as needed.

b. Determine the Time Required. The amount of time for training operations varies considerably. However, every unit can find time to conduct physical readiness training. Frequent (i.e.,daily) physical training of short duration-30 to 60 minutes-is preferred over occasional longer periods. Other demands for training time are so urgent that every minute of time allotted for physical readinesstraining should be used. Determine the time required per week anddivide it into daily blocks.

c. Organize for Various Group Size. It is essential to stress exercise rather than formality. Marines must complete the program where they are- -on the training field, in the motor pool, on the range, next to the classroom, in the office area, in the shop, aboard ship, or elsewhere. Although desirable, it is not always possible to assemble company-size units for physical training. Platoon-size groups are appropriate for the proper conduct of physical conditioning activities. Certain situations may require exercise programs for section- or squad-size units. This manual outlines programs for all situations and types of organizations.

d. Allow for Weather and Exercise Area. In programing and scheduling, the climate and terrain often govern the selection of activities.

(1) Weather changes cause differences in temperature, rainfall, wind chill, and snow. These changes should be anticipated as they dictate the type of program. Alternate plans should be part of the schedule. Nevertheless, normal weather changes and rain should not drive training inside.

(2) Local terrain and available exercise areas may also influence the selection of activities and the type of program which it is possible to support. Some activities can be completed in nearly any area.

e. Plan for Seasonal Change. As most physical readiness training is conducted outdoors, it is necessary to recognize seasonal change. A program should be divided into fall, winter, spring, and summer parts. Seasonal change also causes change in light. For example, an early morning program started in the summer will have ideal lightconditions, yet in the fall or winter, darkness will occur at that same hour and interfere with the conduct of the program. Develop programs in seasonal blocks, and make provisions for anticipated changes.

f. Consider Needed Facilities. An excellent program can be conducted with practically no facilities since there are exercises which require no equipment. However, a better program can be developed when supported by certain facilities and items of equipment. Proper command support, plus ingenuity, will solve this problem. (Items of equipment, when necessary to support the recommended exercises, are included in the chapters on exercise activities.)

g. Specify Appropriate Uniform. The uniform worn for exercising depends upon the season of the year, the state of the weather, and local regulations. Whenever possible, Marines should be dressed alike. Undershirts are preferred as the upper garment when the weather permits. A uniform that restricts the free movement of the body should not be worn when exercising.

h. Consider Availability of Instructors. Leaders who can lead and direct the scheduled activity must be available. Organizational units should train junior officers and noncommissioned officers down to squad or section leaders to instruct and lead the various activities.

i. Select Activity and System. With the type of program needed and the objectives in mind, select an activity package or an integrated training package for each day's scheduled physical training. At this same time, the system to be used in employment of the selected activity or activities must be determined. This selection affects equipment, areas, instructors, transportation, and other support requirements.

j. Secure Command Participation and Support. Prepare and brief the commander to assure full understanding of the objectives and administration of the program. The full participation and support of the commander will greatly improve the level of success.

k. Supervise Execution. Determine needs, publish the program, and supervise its execution as a necessary part of developing the unit's training schedule.

2107. SELECTION OF ACTIVITIES AND SYSTEMS

a. Exercise Activities. The planner must determine exercise activities which will be appropriate to include in the program. An exercise activity is a single means of exercise usually identified by the name applied; for example, running, log exercises, and obstacle course. Many exercise activi ties can become part of activity packages.

b. Activity Packages. Many of the physical activities described in this manual are arranged in prescribed sequences and are known as activity packages. An activity package is a number of exercises of the same type, assembled as a group or a set, and arranged in a specific sequence. Exercise packages are organized in such manner that not more than 15 minutes will be required to complete the execution of any package. Each type of activity is explained in laterchapters. The number of available drills, tables, or circuits; the manner of organization; and the contribution each makes to the total program are covered. Full understanding of this information will greatly assist in developing effective programs. Various designations are used to identify exercise packages; for example, conditioning exercises when arranged in a set order are known as drills, and other packaged activities are designated as tables or circuits. The following are activity packages:

(1) Activity packages for groups:

o Conditioning Drill Two
o Conditioning Drill Three
o Rifle Drill
o Log Drill
o Grass Drill
o Running Tables
o Guerrilla Tables
o Circuit-Interval Table
o Combatives Tables
o Relay Tables

(2) Strength Circuits:

o Fixed Circuit
o Movable Circuit
o Simplified Circuit (Circuit Interval Table)

(3) Activity packages for individuals:

o The 6-12 Plan
o Weight Training
o Isometric Exercise

c. Advantages of Using Packages. These of exercise packages simplifies scheduling and conducting of exercise and results in the following benefits:

(1) Schedule development is simplified as the planner assembles packages which will satisfy the training objective. There is no need to deal with selection of individual activities or to be concerned about the amount of time to be expended on each.

(2) Any 15-minute period, and in some cases less time, can be scheduled or used to perform an activity.

(3) If longer periods of time are available or if the objective demands, several packages can be assembled to provide a more complete period of activity.

(4) Marines are assured a balanced set of exercises or activities as each package is carefully arranged to reach all muscle groups.

(5) The instructor can concentrate on the conduct of a vigorous workout as the type and duration of the activity have already been determined.

d. Nonpackaged Activities. Several types of activity are not packaged. Activities in this category are conditioning marches, unit runs, obstacle courses, combat water survival swimming, teamcontests, and team sports. These activities can be scheduled in combination with packaged activities, or they may be scheduled separately. Often they require a longer period of time. Most nonpackaged activities require a 50-minute period to satisfactorily complete their objective. For conditioning marches, a few hours are required. The benefits of these activities should not be overlooked as some desirable objectives cannot be attained without them.

e. Systems of Exercise. Several methods or systems of organizing exercise and activity packages can be used. Each system is based upon a specific organization as follows:

(1) Single Activity System. The unit leader immediately assumes command of the unit at the beginning of the exercise period. The leader moves the unit to a predesignated exercise site at double time, forms the unit in a circle around the leader, groundsclothing and equipment as appropriate, and quickly moves into the exercise routine. There is usually no time to teach; therefore,the Marines must know the activity to be used. At the conclusion of a 5- to 15-minute period, the unit leader returns the unit to the instructional area at double time and releases the unit for the next scheduled activity.

(2) Progressive Activity System. All Marines (company or platoon) complete activities in the same order during the period. For example, Drill One is followed by dual combatives, and finally a 1-mile run. This system is usually progressive from a warm up activity, such as Drill One, to an activity which contributes in a major way to one of the objectives, such as aggressiveness development through combatives or cardiovascular development, such as running.

(3) Rotating Activity System. Set up the same number of activities or stations as there are platoons in the company. Each platoon rotates through each station in turn. With four platoons in a 50-minute period, about 10 minutes can be devoted to each station. With three platoons, approximately 15 minutes may be spent at each station. Activities must be of a type that can be covered in the time allotted. For example, with three platoons, Station 1 could be Conditioning Drill One; Station 2, running; and Station 3, a team contest.

(4) Circuit System. Set up a number of stations to provide various types of exercise equipment and/or items of apparatus. The idea is to keep all Marines busy and exercising vigorously for a short period of time at each station. The fixed strength circuits and the movable strength circuits are examples of this system. Station changes must be rapid, and the exercise must be started quickly after each change. Since the objective is to exercise at top speed, the motivation comes from frequent change of activity by moving to another station. Rotation by station groups continues until all Marines have covered all stations.

(5) Interval System. This system stresses the development of strength and endurance. It involves heavy work for a given distance within a specified time, alternated with lighter work and recovery, but never stopping during the workout. This procedure is repeated, and the intensity is increased gradually as physical condition improves, but always with adequate recovery. The important concept is stress, recover, stress, recover, and so on. This system is often applied through running, but other activities of a continuous nature may also be used. An example is the Circuit-Interval Table.

f. Steps in Selecting Activities and Systems. The planner should follow these steps.

(1) Evaluate the needs of one unit with emphasis upon unit mission, objective, and time available.

(2) For each day, select an activity package or a combination of activities which will contribute to the objective.

(3) Then determine the system to be used in implementing the selected activities. There are several possible choices and much opportunity for flexibility in program development. For example, the weekly program may contain various exercise packages, systems, and time periods. Additional guidance is contained in following chapters.

2108. SAMPLE WEEKLY PROGRAMS

The variety of weekly programs given shows available alternatives. Programs and schedules other than those illustrated may be assembled to provide schedules to fit any situation. Many other scheduling combinations are possible through use of the packaged activities, varied time periods, and different types of assigned duty.

a. Sample A. Figure 2-1 illustrates a 60-minute time allotment in which the single activity system is used with activity packages that vary from day to day. In this program, 5 hours of training are scheduled for the week. This program needs some type of integrated unit training to supplement it each week. A bimonthly conditioning march of 10 miles in 3 hours is suggested to round out this program. The Saturday session is designed for remedial physical training.

NOTE: A physical training activity must be conducted at a minimum of once every 72 hours. After 72 hours, muscles begin to atrophy.

Figure 2-1 is available in the illustration section at the end of the book

b. Sample B. A more comprehensive weekly program (fig. 2-2) illustrates the scheduling of various-length periods, a variety of activities, and the use of three systems during the week. In this schedule, 3.5 hours are included for the week. The following should be noted concerning this schedule:

Figure 2-2 is available in the illustration section at the end of the book

(1) On Monday, a 60-minute period is available for physical readiness training. Three 20-minute packages are scheduled. Each platoon, under its own leadership, will progress through each scheduled activity in turn.

(2) On Tuesday, Wednesday, and Thursday, only 30-minute periods are available. Here each platoon leader supervises each platoon in a single activity system.

(3) On Friday, the company commander directs training to exhibit personal leadership example and supervision. In this example, only one strength circuit and one cross-country course are available, and each will accommodate only one platoon. For this reason, Conditioning Drill One station is included. A platoon rotates to each station within the 60-minute period.

(4) The Saturday session once again is designed for remedial physical training.
(5) Bimonthly, a 10-mile/3-hour march under load will be necessary to round out this program.
NOTE: Cardiovascular/respiratory exercise to improve endurance must be of long enough duration (20 minutes or more) to induce and maintain oxygen debt.

c. Sample C. A third sample schedule (fig. 2-3) illustrates the use of the single activity system for 5.25 hours of training. This schedule includes sustaining-type activities for a unit that has passed through both the toughening and slow improvement stages of conditioning. The use of competitive activities is featured to holdinterest and provide self-motivation. A bimonthly 10-mile/3-hour march under load will be necessary to round out this program. The Saturday session once again is designed forremedial physical training.

Figure 2-3 is available in the illustration section at the end of the book

NOTE: Thc body will respond to hard work or physical training without soreness and stiffness provided the work or physical training is conducted regularly.
NOTE: Excrciscs of short duration increase strength and bulk but do little to effect endurance.
NOTE: Thc times shown in the charts arc participation times.

Section II. Remedial Physical Training

2201. NEED FOR ATTENTION

Remedial physical training is a process by which physically substandard individuals are conditioned to meet standard levels of performance. To achieve a full degree of operational readiness, it is necessary to bring all Marines up to the prescribed standard. Experience has demonstrated that some Marines have difficulty due to a poor state of fitness, obesity, or lack of motivation. Attention to these individual deficiencies will help improve unit combat readiness.

2202. IDENTIFICATION OF INDIVIDUALS

The company or battery commander identifies persons who cannot achieve the prescribed level of proficiency through the physical fitness test or as a result of observation during training. The commander notes particular weakness of body parts as indicated by failure of certain test events. These Marines are then placed in a special remedial program either at platoon, company/battery, or battalion/squadron level. Extra time is devoted to overcoming the weakness. These extra periods of conditioning may be during or after duty hours, as determined by the commanders. Other persons need reconditioning as result of hospitalization. Restoring physical fitness to damaged areas of the body is accomplished through progressively graded physical activities under professional supervision, not by the physical fitness trainers.

a. Types of Deficiencies. Physical deficiencies which can be corrected by exercise fall into several categories.

(1) Lack of Strength in One or More Body Parts. The major muscle areas concerned are the arms and shoulder girdle, back, abdomen, and legs.

(2) Lack of Overall Endurance. Usually, there is deficient muscular and cardiovascular endurance.

(3) Deficiency in Coordination and Agility. In these cases, physical skill is not developed to a satisfactory degree in activities such as crawling, running, jumping, climbing, traversing, vaulting, pushing, pulling, lifting, and carrying.

(4) Overweight or Underweight. Either condition may interfere with physical fitness and mission accomplishment. Lack of exercise is not always the cause. The cause may be malfunction of normal physiological functions or it may be poor health habits such as over- or under-eating, lack of adequate rest, or overconsumption of alcohol.

(5) Lack of Motivation. Not all persons are motivated to attain or maintain **a** desirable state of fitness. Some Marines do not understand the importance of physical fitness, some find proper exercise too difficult, and others find it inconvenient.

b. Causes of Deficiencies. There are several causes for Marines being deficient in physical fitness:

(1) Absence of exercise.

(2) Exercise which fails to develop all muscle groups and components of fitness.

(3) Exercise which is not vigorous enough or which lacks progression.
(4) Injury or illness which depletes fitness.
(5) Inadequate amounts of sleep or rest.

c. Methods of Detection. The commander has several means by which to detect physical deficiencies:
(1)
from those who pass. Make an analysis to determine the cause of failure as revealed by test scores.
(2) Observe Marines as they perform physical tasks-
o Marines who have difficulty during training or in physical types of work,
o Marines who have become obese and therefore experience difficulty.
(3) Be alert to those Marines who are often on sick call, returnees from hospitalization, or newly assigned.

2203. ADMINISTRATION OF REMEDIAL ACTION

a. Group. Attitude Marines who are singled out as being physically deficient are self-conscious and are not always convinced they need extra help. Within a deficient group, motivation may be low. These Marines must be convinced that a remedial program which is tailored to their needs will help them and will eliminate their deficiencies. Encouragement is often needed and desirable.

b. Leadership. The leader of this group must study individuals and know their deficiencies. The unit leader should counsel Marines individually, observe them closely as they progress through the remedial program, maintain records, and adjust the program as required.

c. Measurement.. Whenever improvement in performance is noted, Marines should be measured by the physical fitness test either individually or by group. When testing reveals an individual to be satisfactory, that individual should be released from the remedial group. There may be exceptions to this policy in the case of Marines who are overweight or in the execution of an exercise program prescribed for some special purpose.

d. Organization of a Remedial Group
(1) A remedial group is usually a small group of Marines within a company or battalion. In some situations, it may be a larger group numbering 50 to 75 Marines. In the case of a larger group, divide the Marines initially into subgroups according to ability. Prescribe exercise loads commensurate with their ability. General conditioning is sufficient in the early part of the program to qualify Marines who are on the borderline. Qualification will reduce the size of the group, permitting more individual and specialized attention to those who remain.
(2) As the program progresses, regroup individuals who have like deficiencies in order that they can concentrate on their weakness. For example—

(a) One group may be weak in the arms and shoulders as revealed by failure in the chin up event. This group, with an assistant instruction charge, can work on pull-ups, rope climb, pushups, rifle or logexercises, horizontal ladder, and similar types of exercise.

(b) Another group may be weak in general endurance as revealed by the3-mile run. This group could profit from participation in conditioning drills, running, grass drill, and strength circuit.

(3) It may be necessary to form some groups to overcome weaknesses in skills, such as an inability to throw, to quickly change direction while running, to crawl rapidly, or to carry a load. Lack of coordination or lack of practice may cause these deficiencies. In this instance, instructors must provide an opportunity to practice and correct poor form and other errors as they are noted.

Chapter 3
PHYSICAL CONDITIONING ACTIVITIES
Section I. Marching Under Load

3101. GENERAL

Few physical fitness activities are as directly related to readiness for combat as foot marches under load. In addition to the obvious physical conditioning and unit cohesiveness benefits, marches under load prepare most Marines for numerous foreseeable tasks in combat. The idea that only infantry and reconnaissance units actually require training marches to prepare them for combat is clearly a mistake. In addition to the very real possibility of Marines from all types of units being used to fill combat-depleted infantry units, it is likely in maritime prepositioning force operations that aviation, combat service support, and command element personnel will have to march with weapons and equipment, from the arrival airfield to the equipment marshalling site. In American coastal cities, the harbor is nearly always 10 or more miles from the airport. It is difficult to imagine any different condition in a lesser developed country .A further combat-related benefit is that foot marching under load is the most effective way to develop leg strength in the context of a unit training program. In his classic study of troop performance in World War II, Men Against Fire, S.L.A. Marshall noted that the greatest single weakness of replacements of all occupational fields was leg strength. Truck drivers must push trucks which are stuck in mud, and headquarters communicators must climb hills to emplace antennas. The demands on the lower body in combat apply to all Marines, and foot marching is one activity which can help build these muscles while being conveniently integrated with the rest of the training program.

3102.TRAINING GOALS

The standard for success of a foot march is very simple to measure: did the unit arrive at the destination at the prescribed time with Marines in condition and required equipment present to accomplish the mission?A progressive program can increase a unit's readiness. It can instill pride by increasing distances and rates of march and by selecting routes over increasingly challenging terrain as the Marines become better conditioned. However, the ability to execute the mission at the conclusion of the march must remain the standard of success.

3103.MOVEMENT PLANNING

The basic considerations in planning a foot march are the mission, tactical situation, terrain and weather, and the units to participate. The success of the march will depend largely upon the thoroughness with which it is planned. A successful march is characterized by adherence to prescribed routes and time schedules, the efficient employment of the means available, and the ability of the unit to accomplish its assigned mission upon arrival at the destination.

a. Movement Orders. Movement planning culminates in the preparation and issuance of an operation order prepared in the standard, five-paragraph format. Necessary annexes are attached to furnish detailed information required for the movement. Written movement orders are rarely prepared at company level. A discussion of the annexes is contained in this chapter.

b.March Planning. March planning, as discussed here, is the planning conducted at battalion level. March planning may be organized into the following steps:

(1) **Preparation and Issuance of the Warning Order.** In order to afford subordinate units the maximum possible time to prepare for a pending move, a warning order containing all available information about the march is issued. The amount of planning time available will determine the time of issuance and the content of the warning order.

(2) Estimate of the Situation. In his estimate, the commander considers the mission, terrain, weather, time, and space factors, available routes, available transportation for the movement of equipment and/or shuttling of Marines, enemy capabilities, disposition of own forces, physical condition and training of Marines, and courses of action available to the command.

(3) Organization and Dispatch of a Reconnaissance. Party very march plan is based on as thorough and complete a ground reconnaissance as time and the situation will permit. Map and aerial reconnaissance are valuable in formulating a plan, but are not a substitute for ground reconnaissance. Route reconnaissance is accomplished by a reconnaissance party which usually consists of a reconnaissance element, an engineer element from the attached or supporting engineer unit, and a traffic control element. Unit standing operating procedures generally establish the basic composition of the reconnaissance party. It is modified as necessary to meet the requirements of a particular march. A recommended method of reporting information obtained by the route reconnaissance party is shown in figure 3-1.The minimum information required from the reconnaissance party is—

(a) Available routes and conditions.(Routes may be specified by higher headquarters.)

(b) Recommended rate of march.

(c) Selection of start point and release point, or confirmation of the suitability of start point and release point previously selected by map reconnaissance.

(d) Confirmation of location of the assembly or bivouac area.

(e) Location of critical points on the route.

(f) Distance between critical points on the route and total distance from start point to release point.

Figure 3-1 is available in the illustration section at the end of the book

(g) Location of obstacles and estimation of necessary Marines and equip- ment needed to repair and maintain routes.

(h) Number of guides required and their location on the route.

(4) Development of Detailed Movement Plan. Some of the elements included in a detailed movement plan are-

(a) Organizing the Column. To facilitate control and scheduling, units will be organized into serials and march units and given an order of march. In determining the order of march, the march planner must consider the enemy situation and the desirable order of arrival of the units at the destination. Where dispersion is required, a unit may be organized into two or more columns, each assigned a different route.

(b) Using Reconnaissance Information. Results of the route reconnaissance will be used to select the route(s); determine the start point, critical points along the route(s), and the release point; and select the rate of march.

(c) Determining March Computations. March unit pass time is based on the strength, formation, and rate of march. The pass time of the marching columns, plus necessary time distance computations, will be used to determine the completion time of the march.

(d) Drafting of Road Movement Table. Using the completed march computations, a draft road movement tabic is compiled.

(e) Checking the Plan. Using the draft road movement table and a road movement graph, the movement plan is checked to ensure that it conforms to the directive of the higher headquarters and the battalion commander's instructions.

(5) Preparation and Issuance of the Road Movement Order. After the plan has been checked and approved by the commander, an operation order is prepared and issued. The operation order may be in written form or issued orally, and is accompanied by a road movement table, overlay and/or strip map, and appropriate administrative details.

(6) Road Movement Table. A road movement table, prepared as an annex to the operation order, provides serial commanders with arrival and clearance times at critical points along the route of march. It also provides the column commander with information as to the proposed location of elements of the column at various times.

(7) Overlay/Strip Map. An overlay serves the normal purpose and should show, as a minimum, the present location of units, route of march, critical points, and the new location of units at the destination. A strip map is a schematic diagram of the route of march and shows landmarks and critical points with the distances between them. A strip map may be issued as an annex to the road movement order, in addition to or in lieu of an overlay.

c. March Computations. Prior to issuance of the operation order, the S-3 must verify time and space computations as they provide him with the necessary data for the preparation of a road movement table.

(1) Time-Distance. Time distance (TD) is determined by dividing the distance to be traveled (D) by the rate of inarch (R):

TD = D/RWhere TD = hours D = distance in kilometers R = kilometers per hour

(2) Length of ColumnThe length of column (LC) is used to determine the pass time of a column. The sum total of the following two parts determines the length of column. (See pars, (a) and (b)).The two parts are the space occupied by the Marines alone (including the distance between Marines) and the sum of the distances between the units of the column (column gap).

(a) The length of column of Marines alone is determined by multiplying the number of Marines by the appropriate factor selected from the table below. The length of column does not include distances between units. LC Marines = No. of Marines x factor.

(b) The total distance in meters between units is determined as follows:

Step One. Determine the number of gaps between serials (total serials minus one).

Step Two. Multiply the number of serial gaps from Step One by the length (in meters) between respective units.

Step Three. Determine the number of gaps between march units (total march units minus one, minus the number of serial distances).

Step Four. Multiply the number of march unit gaps from Step Three by the length (in meters) between respective units.

Step Five. Add the totals from Step Two and Step Four to get the total meters for the column gap.

EXAMPLE:A battalion foot column is organized into 12 platoon-sized march units and 3 company-sized serials. REQUIRED: total column gap distances when there are 100 meters between serials and 50 meters between march units.

(3)Pass-Time.Pass-time (PT) is the time a unit takes to pass a specified point. For foot columns, the pass-time is determined by applying the following formula:PT (minutes) = LC x FACTOR (for appropriate rate of march).

PASS-TIME FACTORS - FOOT TROOPS

.0150 for 4.0 km/h
.0187 for 3.2 km/h
.0250 for 2.4 km/h
.0375 for 1.6 km/h

EXAMPLE: Determine the pass-time of a unit whose length of column is 1,500meters and is marching at a rate of 4 km/h.PT (min) = 1,500 x .0150 (the factor for 4.0 km/h) = 22.5 mm.

(4) Completion Time. Completion time is the time of day that a march will be completed. Completion time is determined by using the following formula: Completion Time = SP time + TD + PT + Scheduled Halts (other than normal hourly halts).

EXAMPLE: A column's starting point (SP) time is 0700.The time-distance is6 hours and 40 minutes. Pass-time of the column is 30 minutes. A 35-minutelunch halt has been scheduled. What is the completion time of the march?Employing the 24-hour clock system, the formula can be applied as follows for simplified addition of the times:

HrMin SP time................0700
TD......................0640
PT......................0030
Lunch halt..............0035

Completion time.........13 105or 1445

The march will be completed at1445.

(5) Experience Tables. Based on previous movements made by a unit, data is accumulated to facilitate march planning. Such data includes approximate pass-times lor various elements of the battalion. The S-3 can utilize these data rather than computing them each time a march is scheduled. Such experience tables serve to reduce the time required to complete the computation phase of march planning. Matter appropriate to the unit standing operating procedure should be integrated therein.

d. Road Movement Graph

(1) A road movement graph is a time-distance diagram used in planning, preparing, or checking road movement tables, and for controlling marches. The graph provides a visual representation of a march plan so that conflicts and discrepancies may be prevented in the planning stage, before congestion occurs on the route. It is not usually issued as a part of the order. Road movement graphs may be applied to small units, to movements of a single column, or to a large organization scheduling separate elements, moving by various means, with different rates of march, over one or more routes.

(2) To construct a road movement graph, use the following steps: (See fig. 3-2.)

Figure 3-2 is available in the illustration section at the end of the book

Step One. Determine the number of hours available for the march. Designate the lower left corner of a sheet of graph paper as the SP time or the earliest even hour before the march is to begin. Select a scale and plot the hours available in sequence from the left on the horizontal axis.

Step Two. Determine the distance to be moved in kilometers. Indicate the SP at the lower left corner of the graph sheet and, using an appropriate scale, plot the number of kilometers on the vertical scale from the SP to the release point. Indicate the location of critical points on the vertical scale.

Step Three. At the proper distance from the SP, draw a horizontal line indicating the location of the RP. Indicate by a vertical line the hour when the movement must be completed. Plot lines representing route restriction, if any, at the proper distances and times on the graph.

Step Four. Determine the pass-times of foot and motor elements of the column.

Step Five. Starting at the SP at the specified hour, plot the movement of the head of the leading element. If the rate of march is 3 kilometers per hour, the lead element will have moved 1.5 kilometers in 30 minutes, 3 kilometers in 1 hour, etc. Plot the trace of the lead man to the RP. The last Marine will cross the SP in pass-times after the lead man. Measure this time on the graph and plot the trace of the last Marine of the column. The lines describing the head and tail of a march serial are parallel. Indicate the time subsequent serials reach the SP and plot the trace of the head and tail. Check to see that the plan complies with all restrictions and orders. If there are any violations or conflicts, the plan may be made to conform by changing the order of march, the starting time of the serials, the planned rate of march, or the organization of the column.

e. Conversion to the Metric System. The adoption of the metric system for expressing linear distances requires a simple means of conversion.

(1) To accomplish this, use the following tables:

o Multiply inches x 2.54 to obtain centimeters. Multiply centimeters x .39 to obtain inches,

o Multiply yards x .91 to obtain meters. Multiply meters x 1.1 to obtain yards,

o Multiply miles x 1.6 (or 8/5) to obtain kilometers. Multiply kilometers x .62 (or 5/8) to obtain miles.

(2) Linear distances can be accurately measured on maps which have been converted to the metric system by using a straightedge graduated in centimeters (cm).The centimeter scale can be used in lieu of the graphic scale when the scale of the map is shown, as follows:

Map Scale:	One cm Equals:
1:500,000	5,000 meters
1:250,000	2,500 meters
1:100,000	1,000 meters
1:50,000	500 meters
1:25,000	250 meters

3104. INDIVIDUAL LOAD

Since the backbone of Marine ground combat forces is the infantry, tactical mobility is largely dependent on the marching ability of the individual Marine. The load that each Marine must carry is the greatest impediment to mobility. In general, the commander must make every effort to reduce the individual load of his Marines to the absolute minimum. While the load has the greatest effect on Marines marching on foot, the value of carrying the minimum essential supplies and equipment applies for all Marine forces.

a. Fighting Load. The fighting load consists of items of clothing, equipment, weapons, and ammunition that are carried by, and essential to, the effectiveness of the combat Marine and the accomplishment of the immediate mission. The fighting load should not weigh more than 40 pounds. The commander must ensure that his Marines' loads- -and his own- -be stripped to the bare minimum. The addition of small, light-weight items in excess of the limit should not be tolerated, since cumulatively, these items will take a toll in energy. Every extra pound a Marine must carry decreases his combat effectiveness.

b. Existence Load. The existence load consists of items other than those in the fighting load that are required to sustain or protect the combat Marine, that may be necessary for increased personal and environmental protection, and that are not normally carried by the individual. The existence load is normally carried by the Marine's unit. Marching, but not engaged in combat, a Marine may be expected to carry a maximum of 50 pounds of supplies and equipment.

c. Considerations

(1) The primary consideration is not how much a Marine can carry, but how much he can carry without impaired combat effectiveness-moral or physical. The combat strength of a unit is not counted simply in numbers of Marines, but in the numbers of willing and physically able Marines.

(2) It is generally better to risk temporary inconvenience from lack of health and comfort items than to exhaust Marines due to overloading.

(3) It is a fundamental truth that men become physically exhausted more quickly when under the stress of combat. Marines must be required to carry less into battle than they are conditioned to carry in training. Marines should be conditioned for carrying weight, but should be equipped in combat for fleetness of foot.

(4) A common mistake is to base the fighting load on the gear and supplies necessary to meet every contingency. The commander cannot reasonably expect the Marines to carry enough gear for every possible eventuality. The items to be carried must be based on reasonable expectations.

(5) It is the commander's responsibility to procure the transport to carry any additional gear. As a rule of thumb, a rifle company, or unit of similar size, requires one 5-ton truck and trailer in support to carry additional gear. In cold weather, or other conditions in which the necessary personal gear increases, this requirement will increase.

(6) The commander must ensure that the supply system provides, on a dependable and timely basis, the balance of essential supplies and equipment not carried by the unit. Marines must feel confident that they will be supported with the necessary supplies and equipment.

(7) In training, commanders must implant in their Marines the pride in operating under austere conditions. To effectively demonstrate the need for this spartan mentality, the commander must personally set the standard. Marines should be trained in field craft, foraging techniques, and the use of caches and field expedients. Maximum use should be made of captured stores.

3105. MARCH TECHNIQUES AND PROCEDURES

a. Organization. A command executing a march is organized into march units, march serials, and march columns, as necessary for control.

(1) March Units. A march unit is a unit of command which moves and halts at the command of a single commander. The march unit normally corresponds to one of the smaller Marine units such as a platoon or company.

(2) March Serials. A march serial consists of one or more march units organized under the senior officer and given a numerical or alphabetical designation to facilitate control. A serial is usually a battalion, but may be a company if the battalion is marching alone.

(3) March Columns. A march column is composed of elements of a command moving over the same route. It may be composed of one or more serials. To facilitate control, a column commander is designated. A column is normally a regiment or larger unit, but may be a battalion if marching alone.

b. Order of March. The order of march is determined by the mission, terrain, and the commander's desired order of commitment of units into action.

c. Control and Coordination Measures. The commander establishes initial control of the march by designating control measures in his march order. The most frequently used measures include:

o Start point and release point.
o Time at which head or tail of column passes the start point.
o Rate of march.
o Order of march.
o Route of march.
o Location of command post.
o March communications.

d. The Pacesetter. The pace-setter is an experienced individual carrying the same load as the majority of the Marines and marching from 4 to 10 meters at the head of the column. (See fig. 3-3.)The pace-setter's primary duty is to maintain the rate of march ordered by the column commander.

(l) Setting the Pace. The pacesetter does this by establishing the pace (length of step) and cadence (steps per minute) to obtain the prescribed rate of march. He should be of medium height so normal strides will be taken. Overstriding or understriding tends to tire leg muscles quickly and affects the combat efficiency of marching Marines. The officer marching at the head of the column supervises the pacesetter to ensure that he takes normal strides and maintains a uniform cadence.

Figure 3-3 is available in the illustration section at the end of the book

(2) Reducing Accordion Effect. Accordion effect occurs when the column alternately lengthens and shortens, causing the rear elements of the column to double time to maintain distance. It is caused by changes in the rate of march by the lead elements, usually after passing a slowing obstacle. To reduce this effect as much as possible, lead elements should slow their rate of march for a sufficient time after crossing obstacles to allow the rear elements to maintain the prescribed distance without running. The rear elements can signal the head when the last man has cleared the obstacle and the rate can be resumed. Since some accordion effect is nearly inevitable, the order of march should be periodically rotated to prevent units in the rear from becoming physically exhausted at the conclusion of the march.

e. Length, Rate and Pace

(1) **Length of March.** The length of march varies depending on terrain and weather, enemy situation, and physical and mental condition of Marines. The normal length of march for a 24-hour period is from 20 to 32 kilometers (12 to 20 miles) marching from 5 to 8 hours at a rate of 4 kilometers (2.5 miles) per hour (km/hmi/h).A march in excess of 32 kilometers in a 24-hour period is considered a forced march. Well-trained units that have been progressively trained will be able to achieve a unit goal of 60 miles in a 3-day period.

(2) Rate of March. The same factors of terrain, weather, load to be carried, and condition of Marines affect the rate of march. The commander considers these and selects a rate which will place his unit at the destination in the shortest possible time in condition to accomplish the mission. Unit standing operating procedure should state the standard rate of march for that unit over normal terrain for both road and cross-country movement and in both day and night. The commander then modifies this rate if the situation requires. Normally prescribed standards are as follows:

Figure is available in the illustration section at the end of the book

(3) Pace and Cadence. The normal pace is 30 inches. A pace of 30 inches and a cadence of 106 steps per minute result in a speed of 4.8 kilometers per hour or 3 miles per hour and a rate of 4 kilometers per hour or 2.5 miles per hour if a 10-minute rest halt per hour is taken. Since the pace of each man may vary, the cadence may have to be adjusted to maintain the prescribed rate. Slope of ground and surface footing affect pace. A normal cadence is relatively simple to maintain on level or gently rolling terrain on firm footing. Mud, sand, loose gravel, and hills all greatly affect pace.

f. Halts. Halts during a day's march are taken at regular intervals to rest personnel and adjust loads .Halts are specified by standing operating procedures or march order. Such factors as midday heat or enemy action may force the adoption of long midday halts or night marches. At long halts, each unit disperses to a previously selected location near the route of march.

(1) Normal Time and Frequency. Under normal conditions, a 15-minute halt is made after the first 45 minutes of marching. Following the first halt, a 10-minute halt is made after each 50 minutes of marching. Variations of this schedule are made when a scheduled halt time occurs when passing through a built-up area or when cover and concealment are required by the tactical situation and none is available. Observation posts may be established, if required for security of the unit during halts.

(2) Actions at Halts. All units in the column should be halted at the same time. At the halt signal, Marines should move to the side of the road, staying within the immediate vicinity of their unit. Marines should remove or loosen gear and sit or lie with feet elevated. Commanders inspect Marines and equipment, and corpsmen administer medical treatment, as required.

g. Communications. The four primary means of communications used in controlling foot marches are radio, visual, sound, and messenger.

(1) Radio. Radios are normally used for communications between platoons and higher headquarters in the march column. Radio transmissions should be held to the absolute minimum necessary for control and should be at minimum power required.

(2) Visual.. Visual communications may include flashlights, luminous markers, panel sets, pyrotechnics, and hand and arm signals. When contact with the enemy is not expected, the loss of light discipline from some of these signals may be a lesser security risk than the risk of disclosure from radio transmissions.

(3) Sound. Sound communications include voice and such signaling devices as horns and whistles. Some of their uses include assembling Marines at the conclusions of halts and providing warnings of air or chemical attacks.

(4) Messengers. Messengers are particularly effective during periods of radio silence and reduced visibility. Messages should be simple and brief to preclude mistakes by the messenger.

h. Security. Both active and passive measures are used against attack by enemy aircraft and indirect fire weapons during movement. Active measures include the use of air sentries and organic and attached weapons in accordance with the unit air defense standing operating procedure. Passive measures include the use of concealed routes and assembly areas, night marches, and extended interval between elements of the column. With imagination and planning, all of these measures can be practiced in training without detracting from the physical conditioning benefit of the march. In fact, such drills as rapidly taking cover during simulated attacks and then quickly resuming the march can prove physically challenging.

i. Reconnaissance. Reconnaissance determines in advance critical points along the march route such as bridges, fords, and obstacles in order that congestion or delay may be held to a minimum and local security provided. Paragraph 3103.C describes organization of the reconnaissance party.

3106.TACTICAL MARCHES

These marches are movements of Marines and equipment not in direct ground contact with the enemy but expecting early ground contact either en route or upon arrival at the objective. Some characteristics of this type of movement include security elements to the front, flanks, and, if necessary, to the rear of the march column. Additional considerations include the selection of concealed areas on defendable terrain for start and release points and rest stops. In training, infantry units of regimental size or smaller will normally derive the greatest benefit from tactical marches. The security requirements usually dictate a slower rate of march and, therefore, reduce the physical conditioning value for other types of units which would seldom move in combat with the expectation of early ground contact with the enemy. Two key points should be remembered regarding tactical marches in a training program.

a. Security. Although security requirements often slow the rate of march, it is a valid and important training objective for infantry units to work toward reducing march times while still practicing sound tactical security. March standing operating procedures, good land navigation skills, and simple but complete march orders contribute as much to this goal as physical conditioning. The idea that proceeding slowly enhances security is seldom true. Keeping the enemy off balance by rapid movement frequently is the best guarantee of security.

b. Training. In peacetime training, it is critical that support personnel who normally are located with infantry units in combat, participate completely with infantry units in forced march training. Not only must they not slow the infantry rate of march because of inferior conditioning, but they must be thoroughly familiar with the infantry unit's standing operating procedure for security on the march. Personnel from within the ground combat element such as artillery forward observer teams and combat engineer units, can be easily identified for this training and should be habitually associated with the same infantry units. Personnel from other elements, such as low-altitude air defense teams and radio battalion detachments, are more difficult to identify and earmark for support of the same infantry unit. Their normal mode of employment is in vehicles and their units frequently provide general support vice direct support. This does not, however, lessen their requirement to be prepared to provide support to dismounted infantry over terrain or in a tactical situation not conducive to vehicles. Such training needs to take place, and planning and cooperation among the elements of the MAGTF can overcome the obstacles.

3107.ADMINISTRATIVE MARCHES

These marches are movements of Marines and equipment made when no enemy interference is expected except from aircraft or indirect fire. During these marches, units are administratively grouped for ease of control and speed of movement. The term administrative foot marches should not be taken to imply that tactical considerations are disregarded. There are simply less stringent security measures used than when ground contact with the enemy is a distinct possibility. The normal formation is the route column with one file moving on either side of the road and with negligible security. Air sentries are used and dispersion between individuals and units is practiced. Although roads usually provide the fastest route, there may be reasons to conduct these marches cross-country (particularly as part of a training program). Administrative marches are the type of foot marches most practical for training programs for non-infantry units. These marches may also play a key role in the physical conditioning program of infantry units, because of the greater speeds possible due to lessened security requirements.

3108. FORCED MARCHES

A forced march is a march which covers a greater distance than 32 kilometer (20 miles) in a 24-hour period. Normally the extra required distance should be achieved by increasing the number of hours marched in a day beyond the norm (8 hours) rather than increasing the rate of march beyond the norm (4 kilometers per hour/2.5 miles per hour). However, there will be occasions when the situation or mission demands an increase in the rate of march.

a. Maximum Recommended Distances. The maximum recommended distances for forced marches are-56 kilometers (35 miles) in 24 hours; 96 kilometers (60 miles) in 48 hours; or 128 kilometers (80 miles) in 72 hours.

b. Sample Time Breakdown. A sample time breakdown for a forced march of 52 kilometers, beginning at first light is as follows:

Figure is available in the illustration section at the end of the book

3109. NIGHT MARCHES

Night marches are characterized by closed formations, more difficult control and reconnaissance, and a slower rate of march than day marches.

a. Control. Control is increased by reducing the distance between individuals and units, and by using connecting files to maintain contact between platoons and companies. Connecting files are normally constituted from personnel from the rear most march unit.

b. Safety on Roads. Because of reduced visibility, night marches on roads used by vehicles require attention to safety procedures to prevent accidents. Commanders should not think these safety measures apply only to peacetime training. In combat, nearly all vehicles operate with blackout lights at night, making visibility even more difficult than in garrison driving. If consistent with the tactical situation, the following measures can reduce the possibility of marchers being struck by vehicles:

(1) Use off-road trails and routes as much as possible.

(2) Guards to the front and rear of march units should be marked with strips of luminous tape and carry red-filtered flashlights. These measures are generally consistent with light discipline, while still affording reasonable warning to approaching drivers.

Section II. Conditioning Drills (One, Two, and Three)

3201. GENERAL

a. Description and Objective. Conditioning drills are calisthenic exercises. Each drill contains seven exercises organized and numbered in a set pattern. Each drill takes 15 minutes to complete. The objective of conditioning drills is to exercise all major muscle areas in order to develop strength, endurance, coordination, and flexibility.

b. Area and Equipment. Any level area is satisfactory for conduct of the drills. Drills One and Two contain ground exercises. If ground conditions are unsatisfactory, Drill Three should be used as it contains no ground positions. Usually, no equipment is required;however, if the group exceeds a platoon in size, an instructor's stand is necessary.

c. Formation. The extended rectangular formation is used in this drill. (See app. A.)

d. Starting Level and Progression. The starting level is six repetitions of each exercise. An increase of one repetition for each three periods of exercise in which the drill is performed is an acceptable rate of progression. This rate is continued until 12 repetitions can be completed. To maintain, continue the drill at 12 repetitions. To progress, move to 6 repetitions at a more difficult drill which exercises the same muscle group. Progression can also be gained by moving from Drill One to Drill Two, as Drill Two is more demanding.

e. Starting Positions. Starting positions vary with the exercise and are explained in each exercise.

f. Leadership. A principal instructor demonstrates and leads the drills. The instructor must be familiar with leadership techniques peculiar to conditioning drills to include the exercises, commands, counting cadence, cumulative count, formation, method of teaching the exercises, and utilization of assistant leaders.

g. Use With Other Programs. Conditioning Drills One, Two, and Three reach all major muscles of the body. They are easy to learn and to perform, and they are simple to administer and supervise. These features, coupled with the short time required for completion, the fact that no equipment is necessary, and adaptability to most areas of execution, make these drills possible in any programs.

3202. CONDITIONING DRILL ONE

This conditioning drill is similar to the calisthenics drill commonly known throughout the Marine Corps as the "Daily Seven."The "Daily Seven" may be used as a substitute for Conditioning Drill One.

a. Exercise l:High Jumper

(1) Starting Position. Feet separated shoulder width, knees flexed, body bent forward at the waist, arms aligned with the trunk and hips, elbows locked, palms facing, fingers extended and joined, head and eyes to the front.(See fig. 3-4, A.)(Elbows remain locked throughout the exercise.)

Figure 3-4 is available in the illustration section at the end of the book.

(2) Cadence. Moderate.

(3) Movement. A four-count exercise: at the count of—

(a) ONE -- Take a slight jump into the air, swinging the arms forward and up to shoulder level.

(b) TWO -- Take a slight jump into the air and swing the arms downward and back, returning to the starting position.

(c) THREE -- Take a vigorous leap into the air, swinging the arms forward and up to an overhead position, momentarily looking skyward, on returning to the ground the knees are flexed, head and eyes return to the front.

(d) FOUR--Repeat the action of count two.

b. Exercise 2: Bend and Reach

(1) Starting Position. Feet spread more than shoulder width, arms overhead, elbows locked, palms facing, fingers extended and joined, head and eyes to the front.(See fig. 3-4, B.)

(2) Cadence. Moderate.

(3) Movement. A four-count exercise: at the count of—

(a) ONE-Bend at the knees and waist, swing the arms straight downward and reach between the legs. Touch the ground as far to the rear as possible and look to the rear. (Elbows remain locked throughout the exercise).

(b) TWO-- Recover sharply to the starting position.

(c) THREE--Repeat the action of count ONE.

(d) FOUR-- Repeat the action of count TWO.

c. Exercise 3: Pushup

(1) Starting Position. Front leaning rest position. To assume this position there is a silent one-two count:on the silent count of one, assume the squatting position, heels together, elbows locked inside the knees, hands flat on the ground directly beneath the shoulders; on the silent count of two, thrust the legs to the rear, toes and heels together, body straight from head to heels.(Seefig. 3-4, C.)

(2) Cadence. Moderate.

(3) Movement. A four-count exercise: at the count of—

(a) ONE -- Flex the elbows lowering the body until the thick portion of chest touches the ground.

(b) TWO -- Raise the body until elbows are straight and locked.

(c) THREE--Repeat the action of count ONE.

(d) FOUR -- Repeat the action of count TWO. (On returning to position of attention, the silent one-two count is used in reverse).

d. Exercise 4: Trunk Twister

(1) Starting Position. Feet are spread more than shoulder width apart, fingers placed behind neck, thumbs pointing downward, elbows back. (See fig. 3-4, D.)(Elbows remain well back throughout the exercise).

(2) Cadence. Slow.

(3) Movement. A four-count exercise: at the count of-

(a) ONE-- Keeping the knees locked and the back straight, bend forward at the waist sharply, with a slight recovery.

(b) TWO-- Twist the trunk to the left vigorously at the waist, keeping the elbow back. The left elbow is higher than the right.

(c) THREE-- Twist vigorously to the right, so the left elbow comes under the right.

(d) FOUR-- Straighten sharply to the starting position. NOTE: Do not attempt to touch the elbows to the knees on counts two and three.

e. Exercise 5: Squat Bender

(1) Starting Position. Feet are spread less than shoulder width apart, hands on hips, thumbs in small of back, elbows back. (See fig. 3-4, E.)

(2) Cadence. Moderate.

(3) Movement. A four-count exercise: at the count of—

(a) ONE-- Assume the squatting position, maintain balance on the balls of the feet, with trunk erect thrust arms forward to shoulder level, elbows locked, palms down.

(b) TWO-- Recover to starting position. Elbows are well back.

(c) THREE-- Keeping the knees locked, bend forward at the waist, touching the ground in front of the toes.

(d) FOUR-- Vigorously recover to the starting position.

f. Exercise 6: Body Twist

(1) Starting Position. On the back, arms extended sideward on the ground, palms down. The legs are raised to a near vertical position, feet together, knees locked.(See fig. 3-4, F.)

(2) Cadence. Slow-fast.

(3) Movement. A four-count exercise: at the count of-

(a) ONE-- Lower legs slowly to the left until they touch the ground near the left hand, keeping the knees straight and shoulders on the ground.

(b) TWO-- Recover the starting position by quickly raising the legs, keep knees straight and feet together.

(c) THREE-- Repeat movement of count ONE, except the movement is to the right side.

(d) FOUR-- Recover sharply to the starting position.

g. Exercise 7: Stationary Run

(1) Starting Position. Position of attention. (See fig, 3-4, G.)

(2) Cadence. Fast.

(3) Movement.

(a) At the command of execution, start running in place at double-time cadence, lifting the left foot first time cadence. Follow the instructor as he counts two repetitions of cadence;

e.g., 1, 2, 3, 4--1, 2, 3, 4. The instructor then gives informal commands such as FOLLOW ME, running on the toes and balls of the feet, keeping the back straight, speeding up the cadence to a sprint, raising the knees high, leaning forward at the waist, and pumping the arms vigorously.

(b) To halt the exercise, the instructor will count two repetitions of cadence as the left foot strikes the ground: 1, 2, 3, 4 - 1, 2, 3, HALT.

NOTE: When counting cadence, the instructor counts only as the left foot strikes the ground. The duration of the exercise is approximately 1 1/2 minutes.

3203. CONDITIONING DRILL TWO

a. Exercise 1: Jumping Jack

(1) Starting Position. Feet separated more than shoulder width, arms overhead. (See fig. 3-5, A.)

(2) Cadence. Moderate.

(3) Movement. A four-count exercise: at the count of~

(a) ONE-- Jump to position with the feet together and assume the squatting position, swinging the arms sideward and downward, placing the hands palms down on the ground, elbows locked inside the knees.

(b) TWO-- Recover to the starting position by jumping to the side straddle and swinging the arms sideward overhead.

(c) THREE-- Repeat the action of count ONE.

(d) FOUR-- Recover to the starting position.

b. Exercise 2: Turn and Bend

(1) Starting Position. Side straddle, arms overhead. (See fig. 3-5, B.)

(2) Cadence. Moderate.

(3) Movement. A four-count exercise: at the count of-

(a) ONE-- Turn the trunk to the left and bend forward over the left thigh, attempting to touch the fingertips to the ground outside the left foot. Keep the left knee straight. On successive repetitions attempt to touch farther and farther to the side.

(b) TWO-- Recover to the starting position.

(c) THREE-- Turn the trunk to the right and bend forward over the right thigh, trying to touch the hands to the ground outside the right foot. Keep the right knee straight.

Figure 3-5 is available in the illustration section at the end of the book.

(d) FOUR-- Recover to the starting position.

c. Exercise 3: Eight-Count Pushup

(1) Starting Position. Position of attention. (See fig. 3-5, C.)

(2) Cadence. Moderate.

(3) Movement. An eight-count exercise: at the count of—

(a) ONE-- Assume the squatting position, palms on the ground directly beneath the shoulders, elbows locked inside the knees.

(b) TWO-- Thrust the legs to the rear assuming the front leaning rest position.

(c) **THREE**-- Flex the elbows until the thick portion of the chest touches the ground.

(d) **FOUR**-- Raise the body on a straight plane until the elbows are locked.

(e) **FIVE**-- Repeat the action of count THREE.

(f) **SIX**-- Repeat the action of count FOUR.

(g) **SEVEN**-- Recover to the squatting position as in count ONE (elbows locked inside the knees).

(h) **EIGHT**—Return sharply to the position of attention.

d. Exercise 4: Turn and Bounce

(1) Starting Position. Feet spread more than shoulder width apart, arms sideward at shoulder level, palms up. (See fig. 3-5, D.)

(2) Cadence. Slow.

(3) Movement. An eight-count exercise: at the count of—

(a) **ONE**-- Turn sharply to the left as far as possible, then recover slightly.

(b) **TWO**-- Again turn to the left as far as possible and recover as in ONE.

(c) **THREE**-- Repeat the action of count TWO.

(d) **FOUR**-- Recover sharply to the starting position.

(e) **FIVE**—Turn sharply to the right as far as possible, then recover slightly.

(f) **SIX**—Again turn to the right as far as possible and recover as in FIVE.

(g) **SEVEN**-Repeat the action of count SIX.

(h) **EIGHT**-Return to the starting position.

NOTE: The head and hips remain to the front throughout the exercise and the knees and elbows are locked at all times.

e. Exercise 5: Squat Stretch

(1) Starting Position. Attention. (See fig. 3-5, E.)

(2) Cadence. Moderate.

(3) Movement. A four-count exercise: at the count of—

(a) **ONE**-- Squat, placing the hands on the ground about 12 inches in front of the feet.

(b) **TWO**— Keeping the fingertips on the ground, straighten the knees completely and raise the hips.

(c) **THREE**-- Recover to position ONE.

(d) **FOUR**-- Recover to the starting position.

f. Exercise 6: Leg Circular

(1) Starting Position. On the back, arms stretched sideward, palms down, feet raised foot from ground, knees straight. (See fig 3-5,F.)

(2) Cadence. Slow.

(3) Movement A four-count exercise: at the count of—

(a) **ONE--** Swing the legs as far as possible to the left, keeping the knees straight and the legs together.

(b) **TWO--** Swing the extended legs over head with the thighs as close as possible to the trunk.

(c) **THREE--** Swing the legs as far as possible to the right.

(d) **FOUR--** Recover to the starting position.

g. Exercise 7: Stationary Run

(1) Starting Position. Position of attention. (See fig. 3-5, G.)

(2) Cadence. Fast.

(3) Movement

(a) At the command of execution, start running in place at double-time cadence, lifting the left foot first. Follow the instructor as he counts two repetitions of cadence; for example: 1, 2, 3, 4--1, 2, 3, 4. The instructor then gives informal commands such as FOLLOW ME. Running on the toes and balls of the feet, keeping the back straight, speeding up the cadence to a sprint, raising the knees high, leaning forward at the waist, and pumping the arm vigorously.

(b) To halt the exercise, the instructor will count two repetitions of cadence as the foot strikes the ground: 1, 2, 3, 4-1,2, 3, HALT.

NOTE: When counting cadence, the instructor counts only as the left foot strikes the ground. The duration of the exercise is approximately 1 1/2 minutes.

3204. CONDITIONING DRILL THREE

a. Exercise 1: Side Straddle Hop

(1) Starting Position. Position of attention. (See fig. 3-6, A.)

(2) Cadence. Moderate.

(3) Movement. A four-count exercise: at count of-

(a) ONE-- Take a slight jump into the air, moving the legs sideward (more than shoulder width apart); at the same time, swing the arms overhead (to an overhead position) clapping the palms together.

Figure 3-6 is available in the illustration section at the end of the book.

(b) TWO-- Take a slight jump into the air, swing the arms sideward and downward returning to the starting position.

(c) THREE-- Repeat the action of count ONE.

(d) FOUR-- Repeat the action of count TWO.

b. Exercise 2: Back Bender

(1) Starting Position. Standing, feet 12 inches apart, fingers placed behind the head. (See fig. 3-6, B.)

(2)Cadence. Slow.

(3) Movement. A four-count exercise: at the count of-

(a) ONE-- Bend the upper trunk backward, raising the chest high, pulling the elbows back, and looking upward. Keep the knees straight.

(b) TWO-- Recover to the starting position.

(c) THREE-- Repeat the action of count ONE.

(d) FOUR-- Recover to the starting position.

c. Exercise 3: Squat Thrust

(1) Starting Position. Position of attention. (See fig. 3-6, C.)

(2) Cadence Moderate.

(3) Movement. A four-count exercise: at the count of—

(a) ONE-- Assume the squatting position; heels together, placing the hands flat on the ground, shoulder width apart, elbows locked and inside the knees.

(b) TWO-- Thrust the legs to the rear, assuming the front leaning rest position, body in line from head to toe, heels and toes together.

(c) THREE-- Return to the squatting position as in ONE.

(d) FOUR-- Return to position of attention.

d. Exercise 4: Side Bender

(1) Starting Position. Feet are spread more than shoulder width apart, arms are raised sideward and overhead, thumbs interlocked palms to front, fingers extended and joined, elbows locked. (See fig. 3-6, D.)

(2) Cadence. Slow.

(3) Movement. An eight-count exercise: at the count of—

(a) ONE-- Bend to left as far as possible, then recover slightly.

(b) TWO-- Again bend to the left as far possible, then recover slightly.

(c) THREE-Repeat the action of count TWO.

(d) FOUR-- Recover sharply to the starting position.

(e) FIVE-- Bend to the right as far as possible, then recover slightly.

(f) SIX-- Again bend to the right as far as possible, then recover slightly.

(g) SEVEN-- Repeat the action of count SIX.

(h) EIGHT-- Recover sharply to the starting position.

NOTE:Keep the elbows and knees locked throughout the exercise. The bend should occur to the side and not the front.

e. Exercise 5: Knee Bender

(1) Starting Position. Feet are spread less than shoulder-width apart, hands on hips, thumbs in small of back, elbows back. (See fig 3-6, E.)

(2) Cadence. Moderate.

(3) Movement. A four-count exercise: on the count of—

(a) ONE-Do a knee bend, lean trunk forward at the waist, thrust arms between legs until the extended fingers touch the ground palms to the ground, hands 6 inches apart.

(b) TWO~Recover sharply to the starting position.

(c) THREE-Repeat the action of count ONE.

(d) FOUR-Repeat the action of count TWO.

f. Exercise 6: Bottoms Up

(1) Starting Position. Front leaning rest position. A silent one-two count is used as in the pushups. (See fig. 3-6, F.)

(2) Cadence. Moderate.

(3) Movement A four-count exercise: at the count of-

(a) ONE-- With the weight on the hands, and knees locked, jump forward bringing the feet as close to the hands as possible; look to the rear.

(b) TWO--Keeping the knees locked, thrust the legs backward assuming the front leaning rest position.

(c) THREE-- Repeat the action of count ONE.

(d) FOUR-- Repeat the action of count TWO.

g. Exercise 7: Stationary Run

(1) Starting Position. Position of attention. (Sec fig. 3-6, G.)

(2) Cadence. Fast.

(3) Movement

(a) At the command of execution, start running in place at double time, lifting the left foot first. Follow the instructor as he counts two repetitions of cadence; for example: 1, 2, 3, 4--1, 2,3, 4. The instructor then gives informal commands such as FOLLOW ME, running on the toes and balls of the feet, keeping the back straight, speeding up the cadence to a sprint, raising the knees high, leaning forward at the waist, and pumping the arms vigorously.

(b) To halt the exercise, the instructor will count two repetitions of cadence as the left foot strikes the ground: 1, 2, 3, 4 - 1, 2, 3, HALT.

NOTE: When counting cadence, the instructor counts only as the left foot strikes the ground. The duration of the exercise is approximately 1 1/2 minutes.

Section III. Rifle and Log Drills

3301. RIFLE DRILL

a. Description and Objective. Rifle exercises are conditioning exercises performed with a rifle. Each drill contains six exercises (fig. 3-7) and they are numbered in a set pattern. The drill takes 15 minutes to complete. The objective of rifle drills is to exercise the arms, shoulders, and back muscles in order to develop strength and endurance, particularly in the upper body. In units without rifles, log drills may be substituted.

b. Area and Equipment. Any level area is satisfactory for conducting this drill. Each Marine completes these exercises from a standing position and no ground contact is required. Each Marine will need a rifle and, if the group exceeds a platoon in size, then the instructor will need an instructor's stand.

c. Formation. The extended rectangular formation is used in this drill. (See app. A.)

d. Starting Positions. Starting positions vary with the exercises and are explained in each exercise. As in all set conditioning drills, the command used to start the exercise is STARTING POSITION, MOVE. The following directions apply to rifle drill.

(1) In those exercises which start from the rifle downward position, on the command MOVE, execute port arms as prescribed in NAVMC 2691, Drill and Ceremonies Manual, and then assume the starting position. The command to return to the men to the position of attention at the conclusion of the exercise is POSITION OF ATTENTION, MOVE.

(2) In exercises which terminate in the rifle downward position, on the command MOVE, execute the position followed by order arms as prescribed in NAVMC 2691.

(3) In the exercises which terminate in a position other than the rifle downward position, Marines first assume the rifle downward position before executing port arms and order arms.

(4) These movements are executed without command. This procedure promotes uniformity, but precision is not expected. To be effective, rifle exercises must be strenuous enough to tire the arms, but not to the point where the arms cannot move with precision.

e. Leadership. A principal instructor demonstrates and leads the drill. He must be familiar with leadership techniques for conditioning exercises and the peculiar techniques for rifle drill.

3302. EXERCISES PERFORMED WITH RIFLES

The exercises of rifle drill are outlined in the following paragraphs,

a. Exercise 1: Foreup, Behind Back

(1) Starting Position. Rifle downward, feet together.(See fig. 3-7, A.)

Figure 3-7 is available in the illustration section at the end of the book.

(3) Movement. A four-count exercise: at the count of—

(a) ONE-- Swing the arms forward and upward to the overhead position. Inhale.

(b) **TWO**-- Lower the rifle to the back of the shoulders. Exhale.
(c) **THREE**-- Recover to position ONE and inhale.
(d) **FOUR**-- Recover to the starting position and exhale.

b. Exercise 2: Lunge Side, Turn and Bend
(1) Starting Position. Rifle downward, feet together.(See fig. 3-7, B.)
(2) Cadence. Moderate.
(3) Movement. An eight-count exercise:at the count of—
(a) **ONE**-- Lunge sidewards to the left, swing the rifle forward and upward to the overhead position.
(b) **TWO**-- Turn the trunk to the left and bend forward over the left hip. At the same time, swing the rifle to a **low** horizontal in front of the left ankle.
(c) **THREE**-- Recover to position ONE.
(d) **FOUR**-- Recover to the starting position.
(e) **FIVE, SIX, SEVEN, and EIGHT**-- Repeat on the right side.

c. Exercise 3: Foreup, Back Bend
(1) Starting Position. Rifle downward, feet together.(See fig. 3-7, C.)
(2) Cadence. Moderate.
(3) Movement. A four-count exercise:at the count of—
(a) **ONE--** Swing the arms forward and upward to the overhead position.
(b) **TWO**-- Bend backward, emphasizing the bend in the upper back. The face is up.Keep the knees straight.
(c) **THREE**-- Recover to position ONE.
(d) **FOUR**-- Recover to the starting position.

d. Exercise 4:Up and Forward
(1) Starting PositionRifle downward, feet together.(See fig. 3-7, D.)
(2) Cadence. Fast.
(3) Movement. A four-count exercise:at the count of—
(a) **ONE**-Swing the arms forward and upward to the overhead position.
(b) **TWO**- Swing the arms forward to shoulder level.
(c) **THREE**-Recover to position ONE.
(d) **FOUR**-Recover to the starting position.

e. Exercise 5: Foreup, Full Squat
(1) Starting Position. Rifle downward, feet in narrow stance. (See **fig.** 3-7, E.)
(2) Cadence. Moderate.
(3) Movement. A four-count exercise: at the count of—
(a) **ONE**-- Swing the arms forward and upward to the overhead position.
(b) **TWO**-- Swing the arms down to shoulder level and assume the squatting position.
(c) **THREE**-- Recover to position ONE.
(d) **FOUR**-- Recover to the starting position.

f. Exercise 6: Arms Forward, Side Bend

(1) **Starting Position.** Sidestraddle, regular stance, rifle forward.(See fig. 3-7, F.)

(2) **Cadence.** Moderate.

(3) **Movement.** A four-count exercise:at the count of-

(a) **ONE--** Bend the trunk to the left. Keep the knees straight.

(b) **TWO**-- Recover to the starting position.

(c) **THREE**-- Bend the trunk to the right. Keep the knees straight.

(d) **FOUR**-- Recover to the starting position.

NOTE:Keep the rifle on the same level as the shoulders throughout the exercise.

3303. LOG DRILL

a. Description and Objective. Log exercises are conditioning exercises performed with a log. Each drill contains six exercises and they are numbered in a set pattern. The drill takes 15 minutes to complete. The objective of log drills is to develop strength and muscular endurance and, in this instance, under maximum loads. Log exercises also develop teamwork. Log may be used in lieu of conditioning drills after the Marines have become somewhat conditioned.

b. Area and Equipment. Any level area is satisfactory for conducting this drill. Each Marine completes these exercises from a standing position and no ground contact is required. If the group exceeds a platoon in size, then the instructor will need instructor's stand. Each six-person group or eight-person group will need a log. The logs should be from 6 to 8 inches in diameter. They may vary in length from 14 feet (for 6 people) to 18 feet (for 8 people).The logs should be skinned, smoother, and dried. The 14-foot logs should weigh approximately 300 pounds and the 18-foot logs, approximately 400 pounds. Rings should be painted on the logs to indicate each person's position. When not in use, the logs should be stored on a rack to keep them off the ground.

c. Formation. All the Marines assigned to the same log team should be about the same height at the shoulders. The recommended method of dividing the platoon is to have the Marines form a single file or column with short people to the front and tall people to the rear. Have the Marine assume their positions in the column according to shoulder height, not head height. When they are in position, they are given the command COUNT OFF BY SIXES (OR EIGHTS), COUNT OFF, to divide them into six- or eight-person log teams. Each team in turn, can then proceed to the log rack, shoulder a log, and carry it to the designated exercise area. The log teams form in columns in front of the instructor. With the Marine holding the log in the chest position, have them face the instructor and ground the log at least 10 yards between columns and 10 yards between log teams within the columns.

d. Starting Positions. The Marines fall in, facing the log, with their toes about 4 inches from it. The basic starting positions and commands are as follows:(See fig. 3-8.)

(1) **RIGHT-HAND STARTING POSITION, MOVE.** It the command MOVE, move the left foot 12 inches to the left, and lower the body into a flat foot squat. Keep the back straight, head up, and arms between the legs. Encircle the far side of the log with the left hand. Place the right hand underneath the log.(See fig. 3-8, A.)

(2) LEFT-HANDED STARTING POSITION, MOVE. These commands are executed in the same manner as in paragraph a. except that the left hand is underneath the log and the right hand encircles its far side. (See fig. 3-8, B.)

Figure 3-8 is available in the illustration section at the end of the book.

(3) RIGHT SHOULDER POSITION, MOVE. At the command MOVE, pull the log upward in one continuous motion to the right shoulder. At the same time, move the left foot to the rear and stand up, facing left. Balance the log on the right shoulder with both hands.(See fig. 3-8, C.)This movement cannot be performed from the left-hand starting position because of the position of the hands.

(4) LEFT SHOULDER POSITION, MOVE. These commands should be given from the left hand starting position. At the command MOVE, pull the log upward in one continuous motion, to the left shoulder. At the same time, move the right foot to the rear and stand facing right. Balance the log on the left shoulder with both hands.(See fig. 3-8, D.)This movement cannot be performed from the right-hand starting position.

(5) WAIST POSITION, MOVE. From the right hand position pull the log waist high. Keep the arms straight and fingers laced underneath the log. The body is inclined slightly to the rear and the chest is lifted and arched. (See fig. 3-8, E.)

(6) CHEST POSITION, MOVE. This command should be given after the waist position has been assumed .On the command MOVE, shift the log to a position high on the chest, bring the left arm under the log and hold the log in the bend of the arms.(See fig. 3-8, F.)Keep the upper arms parallel to the ground.

(7) To move the log from the right shoulder to the left shoulder, the command is: LEFT SHOULDER POSITION, MOVE. On the command MOVE, push the log overhead and lower it to the opposite shoulder.

(8) To return the log to the ground from any of the above positions, the command is: STARTING POSITION, MOVE. At the command MOVE, slowly lower the log to the ground. The hands and fingers must be kept from under the log.

e. Leadership. A principal instructor demonstrates and leads the drill. He must be familiar with the leadership techniques for conditioning exercises and the peculiar techniques for log drill.

3304. EXERCISE PERFORMED WITH LOGS

The exercises of log drill are outlined in the following paragraphs. Figure 3-9 graphically explains log drill.

a. Exercise l:Two-Arm Pushup

(1) Starting Position. Right or left shoulder position. Regular stance. (See fig. 3-9, A.)

(2) Cadence. Moderate.

(3) Movement. A four-count exercise:at the count of—

- **(a) ONE--** Push the log overhead until the elbows lock.
- **(b) TWO--** Lower the log to the opposite shoulder.
- **(c) THREE--** Repeat the action of count ONE.
- **(d) FOUR--** Recover to the starting position.

b. Exercise 2: Forward Bender

(1) Starting Position. Chest position. Regular stance.(See fig. 3-9, B.)

(2) Cadence. Moderate.

(3) Movement. A four-count exercise:at the count of-

(a) ONE-- Bend forward at the waist, keeping the back and legs straight.

(b) TWO-- Recover to the starting position.

(c) THREE-- Repeat the action of count ONE.

(d) FOUR-- Recover to the starting position.

c. Exercise 3: Straddle Jump

(1) Starting Position. Right or left shoulder position, feet together, fingers interlaced on top of the log.(See fig. 3-9, **C.)**

(2) Cadence. Moderate.

(3) Movement. A four-count exercise:at the count of—

(a) ONE-- Jump to a side-straddle. Pull down on the log with both hands to keep it from bouncing on the shoulder.

(b) TWO-- Recover to the starting position.

(c) THREE-- Repeat the action of count ONE.

(d) FOUR-- Recover to the starting position.

Figure 3-9 is available in the illustration section at the end of the book.

d. Exercise 4: Side Bender

(1) Starting Position. Right shoulder position, feet regular stance.(See fig. 3-9, D.)

(2) Cadence. Moderate.

(3) Movement. A four-count exercise:at the count of—

(a) ONE-- Bend sideward to the left as far as possible, bending the left knee.

(b) TWO-- Recover to the starting position.

(c) THREE-- Repeat the action of count ONE.

(d) FOUR-- Recover to the starting position.

(4) After completing the required number of repetitions, change shoulders and execute an equal number of repetitions to the other side.

e. Exercise 5: Knee Bend

(1) Starting Position. Right or left shoulder position. Narrow stance. Fingers interlocked on top of the log.(See fig. 3-9, E.)

(2) Cadence. Slow.

(3) Movement. A four-count exercise:at the count of—

(a) ONE-- Flex the knees to a quarter-squat position.

(b) TWO-- Flex the knees to a half-squat position.

(c) THREE-- Lower the body to a three-quarter squat position.(Lean slightly forward.)

(d) FOUR-- Recover to the starting position.

NOTE:Pull forward and downward on the log throughout the exercise.

f. Exercise 6: Overhead Toss

(1) Starting Position. Right or left shoulder position, regular stance. The knees are bent to a quarter-squat.(See fig. 3-9, F.)

(2) Cadence. Moderate.

(3) Movement. A four-count exercisc:at the count of-

(a) ONE-Straighten the knees and toss the log into the air approximately 12 inches overhead. Catch the log with both hands and lower it toward the opposite shoulder. As the log is caught, lower the body into a quarter-squat.

(b) TWO-- Again toss the log into the air and when caught return it to the original shoulder.

(c) THREE-- Repeat the action of count ONE.

Section IV. Grass Drills

3401. GENERAL

a. Description and Objective. Grass drills are extremely strenuous exercises and are performed at top speed for only short periods of time. No cadence is counted but the Marines continue to execute the multiple repetitions of the command until the next command is given. The grass drill consists of two drills:Drill One and Drill Two. Each drill contains six exercises. The objective of the drills is to decrease reaction time, to develop cardiovascular endurance, and to provide a vigorous workout for all major muscles.

b. Area and Equipment. Any level area suitable for ground contact and of a size to accommodate the group is adequate. No equipment is needed.

c. Formation. All movements are executed in place. The extended, rectangular formation is recommended for a platoon- or company-size unit. The circle formation is suitable for groups of squad or section size. At the beginning of an exercise program, 2 to 3 minutes of grass drills will insure a good workout.

d. Progression. Progression is gained by gradually increasing the length of time devoted to the drills. As the physical condition of the Marines improves, the periods should be gradually lengthened to 5 minutes. As the second drill is more difficult than the first, some progression can be attained by initially executing Drill One; then as the program and the Marines progress, introduce Drill Two. To extend the duration of the drill, it may be necessary to repeat the drill.

e. Starting Position

(1) The drills are started from the GO position. Other basic positions are FRONT, BACK, and STOP. (See fig. 3-10, A.)

(a) GO. Running in place (top speed): on the toes and balls of feet, knees raised high, arms pumping, body bent forward at waist.

(b) FRONT. Prone position: elbows bent (along body), palms flat on ground directly under the shoulders, legs together and straight.

(c) BACK. Supine position (flat on back): arms extended near side on ground with palms down, legs together and straight, feet toward the stand or instructor.

(d) STOP. Football lineman stance: feet spread and staggered, left arm across left thigh, right arm straight, knuckles on ground, head up, back parallel with ground.

(2) To assume the FRONT or BACK position from the STANDING, GO, or STOP position, vigorously get into the prescribed position as quickly as possible. (See fig. 3-10, B.)

(3) To change from the FRONT to the BACK position, quickly do a pushup, move the feet several short steps to the right or left, lift the arm on the side toward which the feet move, and thrust the legs vigorously to the front. (See fig. 3-10, C.)

Figure 3-10 is available in the illustration section at the end of the book.

(4) To move from the BACK to the FRONT position, sit up quickly, place both hands on the ground to the right or the left of the legs. Move the feet several short steps to the rear on the side opposite the hands. When the feet are opposite the hands, thrust the legs vigorously to the rear and lower the body to the ground.(See fig. 3-10, D.)

f. Leadership. A warm up activity of lesser intensity should proceed grass drill. During the instructional phase and conduct of these drills, the following points should be applied.

(1) The instructor executes only GO and STOP with the Marines. This allows the instructor to supervise the drill.

(2) The commands for grass drills are given in rapid succession without the usual preparatory command.

(3) To prevent confusion, the instructor should give the commands sharply to distinguish them from comments or encouragement.

(4) As soon as the Marines know the drill, they should respond to the instructor's commands and perform all exercises vigorously and as rapidly as possible. All exercises are executed continuously until the next command is given. Insist on top speed performance; anything less is not effective.

(5) The commands peculiar to each exercise are identical to the name of the exercise.

(6) Marines are not to be required to assume the position of attention once the drills are started. To halt the drill for instructions or for rest, the command UP is used. At this command, the Marines assume a relaxed standing position. Do not demand formality. At the conclusion of a fast and vigorous 5-minute grassdrill, it is physically impossible for people to stand at attention.

(7) The sequence of commands for the execution of grass drills should occur as follows:Drill One. GO, FRONT, Bouncing Ball; GO, BACK, Bicycle; GO, Full Squatter; GO, BACK, Situps; GO, FRONT, Mountain Climber; GO, FRONT, Roll Left; GO, STOP, UP.

g. Use With Other Programs. Since grass drills can be executed in a short period of time, they may be executed where only a few minutes are available for exercise or in conjunction with another type of activity. Grass drills are an excellent substitute for running when time is a factor.

3402. GRASS DRILL ONE AND TWO

a.Grass Drill One

(1) Bouncing Ball. From the FRONT position, push up, supporting the body on the hands (shoulder-width apart) and feet. (See fig. 3-11.)Keep the back and legs in line and the knees straight. Bounce up and down by a series of short, upward springs from the hands, hips, and feet simultaneously,

(2) Bicycle. From the BACK position, raise the legs and hips. Keep the elbows on the ground and support the hips with the hands. Move the legs vigorously as if pedaling a bicycle.

(3) Full Squatter. From the STOP position, assume a full knee bend, the feet on line, hands on hips. Bounce up and down in place by short, bouncing jumps.

Figure 3-11 is available in the illustration section at the end of the book.

(4) Situps. From the BACK position and with arms stretched overhead, sit up, reach forward, and touch toes. Return to the supine position.

(5) Mountain Climber. From the STOP position, place both hands on the ground directly under the shoulders. Thrust the right leg to the rear, knee straight. The left foot should be close to the left hand, the left knee outside the left arm. Shift the weight to the hands, thrust off with the rear (right) foot and bring that foot up close to the right hand, the right knee outside the right arm. At the same time, thrust the left leg vigorously to the rear, knee straight. Continue at a fast cadence, alternating the legs.

(6) Roll Left. From the BACK or FRONT position, make one complete roll in the direction commanded. On completing the roll, return to the FRONT or BACK position.

b. Grass Drill Two

(1) Legs Over. .From the BACK position and with arms stretched overhead, palms up, raise the legs upward and then swing them backward over the head until the toes touch the ground behind the head. Return legs to the starting position. (See fig. 3-12.)

(2) V-Up and Touch Toes. .From the BACK position, raise the legs with the knees straight, sit up until the trunk and legs form a V, and touch the toes with the hands. Return to the BACK position.

(3) Rocker. .In the FRONT position, clasp the hands behind the back, arch the body, holding the head back. Start rocking, using the front part of the trunk as a rocker.

(4) Bounce and **Clap Hands**. The procedure is the same as for bouncing ball, but while in the air, clap the hands. This requires a more vigorous bounce or spring. (See fig. 3-12.)

(5) Leg Spreader. .From the BACK position, raise the legs so that the heels are 10 to 12 inches from the ground, spread them apart as far as possible, then close them together. Open and close legs as rapidly as possible.

(6) Forward Roll. .For forward roll from the STOP position, place both hands on the ground, tuck the head, and do one complete forward roll, keeping the legs tucked as you roll, and come back to the STOP position.

Figure 3-12 is available in the illustration section at the end of the book.

Section V.Guerrilla Exercise

3501.GENERAL

a. Description and Objective. Guerrilla exercises are individual exercises of various basic skills that are performed rapidly while moving forward in a circle formation. There are two tables of guerrilla exercises. Each table takes 15 minutes to complete. The objective of this exercise is to increase strength and endurance, aid flexibility, and develop coordination.

b. Area and Equipment. Any level area is suitable for conducting guerrilla exercises. No ground contact is required other than the hands. There is no equipment requirement.

c. Formation

(1) The circle formation (app. A) is used for guerrilla exercises. Each platoon forms its own circle and engages in guerrilla exercises under a platoon instructor. If the platoon exceeds 30 Marines, double or concentric circles may be used.

(2) When the circle is formed, the instructor steps into the center of the circle and moves clockwise in a small circle. He commands:QUICK TIME, MARCH, 1-2-3-4. (Rapid cadence of approximately 130 counts per minute. Cadence and step are maintained between exercises.)

(3) To reform the platoon after completing guerrilla exercises, the instructor halts the Marines and places the base man or platoon guide where he wishes and commands:

(a) BASE MAN (or platoon guide),POST.

(b) FALL OUT AND FALL IN ON THE BASE MAN (or platoon guide).

d. Progression. Progression may be attained by moving from table I to table II or by shortening the quick time marching periods between exercises and performing all exercises a second time.

e. Leadership

(1) To execute the exercises, the Marines continue at quick time while the instructor simultaneously explains and demonstrates the exercise to be performed, and then commands the Marines accordingly. In each instance, the preparatory command will be the name of the exercise and, in all instances, the command of execution will ***MARCH***.. *To* terminate each exercise, the command *is* ***QUICK TIME, MARCH.*** *The* Marines immediately pick up the step as the instructor counts cadence.

(2) Unless specified differently, each exercise should be continued for 20 to 40 seconds depending upon the vigor of the exercise. The leader can determine the duration of each exercise by observing its effect upon the Marines.

(3) To form for double guerrillas, the commands for pairing the Marines (who are in circle formation) are-

(a) PLATOON, HALT.

(b) FROM (designate an individual), BY TWO'S, COUNT OFF(.Examplel-2; 1-2; 1-2; etc.)

(c) EVEN NUMBERS MOVE UP BEHIND ODD NUMBERS(At this time, adjust pairs according to height and weight.)

(d) YOU ARE NOW PAIRED UP FOR DOUBLE GUERRILLAS. To change the Marines position, merely command CHANGE.)
(e) FORWARD, MARCH.

f. Place in the Program. Many Marines have not had the opportunity to perform the simple skills involved in guerrilla exercises. The conduct of these exercises is a simple matter since they can be performed easily and quickly in almost any situation. The tables of exercise are applicable to all personnel. The tables can constitute a station within a 1-hour period or be completed within a separate 15-minute period.

3502.GUERRILLA TABLES

a. Table I
(1) Double Time.(See fig. 3-13, A.)Hold arms in the thrust position. Execute a double-time run, maintaining the circle formation and the prescribed distance between your fellow Marines. Duration-1 minute.
(2) All Fours.(See fig. 3-13, B.) Face downward. Support the body with the hands and feet. Walk forward hands first.
(3) Crab Walk.(See fig. 3-13, C.) Get in the sitting position, face upward, and lift the hips. Support the body with the hands and feet. Walk forward feet first.
(4) Squat Walk.(See fig. 3-13, D.) Assume a full knee bend position. Grasp the ankles (left ankle with the left hand, right ankle with the right hand). Walk forward.
(5)Broad Jump.(See fig. 3-13, E.) Jump forward on both feet in a series of broad jumps. Swing the arms vigorously to assist the jumps.
(6)Toe-Touch Walk.(See fig. 3-13. F.) Walk forward, bending at the waist and touching one hand to the toe of the opposite foot while it is on the ground. Raise the trunk to the vertical position between steps. Keep the knees straight.
(7)Bottoms-Up Walk.(See fig. 3-13, G.) Assume the front leaning rest position and move the feet toward the hands in short steps, keeping the knees locked. When the feet are as close to the hands as possible, walk forward on the hands to the front leaning rest position.
(8) Straddle Run.(See fig. 3-13, H.) Run forward, leaping to the right from the left foot and to the left from the right foot.
(9) Fireman's Carry .(See fig. 3-13,1.)
(IO)Single-Shoulder Carry(See fig. 3-13, J.) Two men execute the carries as indicated by the diagram. No. 1 man executes one type; No. 2 man executes the other.

Figure 3-13 is available in the illustration section at the end of the book.

b. Table II
(1) Double Time.(See fig. 3-14, A.)Hold arms at the thrust position. Execute a double-time run, maintaining the circle formation and the prescribed distance between your fellow Marine.Duration-1 minute.
(2) Toe-Grasp Walk.(See fig. 3-14, B.) Bend forward and grasp toes. With knees slightly bent, walk forward.

(3) Hand-Kick Walk..(See fig. 3-14, C.) Walk forward, kicking the moving foot upward on every step. At the same time, lean forward and touch the elevated toe with the hand of the opposite arm.

(4) Pike Juinping.(See fig. 3-14, D.) Jump forward and upward from both feet, keeping the knees straight, and at the same time, swing the legs forward and touch the toes with the hands at the top of each jump.

(5) Squat Jump.(See fig. 3-14, E.) Leap forward from the squatting position, with the hands on the ground and the arms between the legs. Land on the ground with hands and legs extended. Bring up the legs to the squatting position.

(6) Steam Engine..(See fig. 3-14, F.) Lace the fingers behind the neck and walk forward in the following manner: as the left leg moves forward, raise the knee high, bend the trunk forward, and touch the outside of the right elbow to the outside of the knee. Then lower the left leg and step forward on the left foot and raise the right leg. Repeat with the right leg and left elbow.

(7) Knee-Touch Walk.(Scc fig. 3-14, G.) Walk forward, bending the knees and touching the ground on each step. The knees are bent and straightened on each step.

(8)Hobble Hopping.(See fig. 3-14, H.) Hold foot behind back with opposite hand and hop forward. On the command CHARGE, grasp the opposite foot with opposite hand and hop **forward.**

(9) Cross Carry.(See fig. 3-14,1.)(10)Saddle Back Carry. (See fig. 3-14, J.) Two Marines execute the carries as indicated in the diagram. No. 1 Marine executes onetype;No. 2 Marine executes the other.

Figure 3-14 is available in the illustration section at the end of the book.

Section VI. Running and Orienteering

3601. RUNNING

a. General

(l)Description. The general form and technique for all types of running are fairly constant. (See fig. 3-15.)The head is erect, body slightly forward without bending at the waist, and the arms are at a loose thrust position alternating from front to rear in straight planes. A cross-body arm movement wastes energy .The movement of the legs and feet will be discussed in subsequent paragraphs dealing with the different types of running. Of primary importance is the fact that in all types of running, the toes should be pointed straight ahead. Toeing out is a common error in both running and walking and should be an item of individual correction.

(2) Objective. The objective of running is to develop cardiovascular endurance. Despite the fact that Marines have developed their muscle structure and the strength of their muscle tissue, unless they have developed cardiovascular endurance to a satisfactory degree, they are not entirely physically fit or combat ready. Cardiovascular endurance (wind) depends on the efficiency of the lungs and heart. The efficiency of the lungs and heart depend on the amount of oxygen the lungs can absorb with each breath inhaled and the amount of carbon dioxide the lungs can expel. The process of absorbing oxygen and expelling carbon dioxide (cardiopulmonary process) is performed by the blood that circulates through the lungs. The condition of this process will determine the amount of effort a person can exert over a period of time. Running is one of the best activities to develop this vitally important endurance.

b. Running Skills. In the development of running skills, individuals may require instruction to improve their proficiency. Some important skills to consider are-

Figure 3-15 is available in the illustration section at the end of the book.

(1) Action of the Arms. Arm action is important. Check to see that arms are held loosely and that the action is relaxed. The faster the run, the more rapid the arm action.

(2) Breathing. Allow the individual to breathe through the mouth as the body demands a large supply of oxygen. Oxygen can be inhaled in greater quantities through the mouth.

3602. TYPES OF RUNNING

a. Double Time

(1) Description and Objective. Double timing is marching at the rate of 180 steps per minute, each step being 36 inches in length. It takes practice to double time with precision in formation. The Marines should keep in step, placing their feet flat on the ground. This, however, should not be a stamping motion, but should be done with as slight a jolt as possible. Double timing is like a jog, the difference being that in a jog the feet are lifted well off the ground and the running motion is bouncy. In double timing, the feet skim the ground and there is no bounce to the ran. Double timing is a vehicle for teaching proper running form and for the development of the cardiovascular system.

(2) Area and Equipment. This type of running can be completed over a variety of surfaces. Usually a Marine uses a field or road. There is no equipment requirement.

(3)Progression. There is no set standard for alternating quick time and double time in the early conditioning of Marines. A general rule is to begin with enough quick-time marching to ensure a thorough warming up, then double time about 100 paces. Change again to quick time until the individuals have made a reasonable recovery from the running, then double time another 100 paces. The amount of double time can be increased and the quick time decreased from week to week, until the individuals are double timing about 1800 yards. This type of training should be given at least twice a week, but by no means is it adequate as the sole means of conditioning.

(4) Leadership

(a) The instructor should be to one side of the column or group and toward the rear so there is a full view of all Marines. Inexperienced instructors have a tendency to supervise from a position too far forward.

(b) Select an individual who can maintain the proper cadence to act as the guide during double-time running.

(c) There are several ways for the instructor and group to count cadence while double timing. If not contrary to local policy, learn several methods and use them for variety.

b. Wind Sprints

(1) Description and Objective. This type of running involves a series of 30- or 40-yard dashes, usually conducted in successive waves of squads. Each squad is in line and the squad leader on the right flank. Wind sprints assist indeveloping speed and cardiovascular endurance. Any flat and level area may be used which will permit the squad to form a line and run the required distance.

(2) Progression. One or two 30-yard sprints will be adequate at the beginning. As time passes, sprints can be lengthened and up to six or seven sprints may be used.

(3) Leadership

(a) At the command READY (given by squad leader), each runner assumes the sprinter's starting position. At the command GO, the squad sprints approximately 30 yards, takes 10 yards to stop, and lines up immediately with the squad leader who repeats, READY, GO, and again the squad sprints. At the conclusion of the third sprint, the squad waits until all the squads of the platoon have made three sprints. Then they all line up and the squad leaders conduct three more wind sprints in the opposite direction.

(b) Valuable time is gained by having each squad ready to go when the preceding squad has moved off its second sprint mark.

c. Cross-Country Running

(1) Description and Objective. Cross-country running is a distance run conducted on a course laid out along roads, across fields, over hills, through woods, and on any irregular ground. The cross-country ran may be utilized as a conditioner or as a competitive event; the objective is to cover the distance in the shortest possible time. The course should be 2 to 2.5 miles in length and be laid out to avoid heavy vehicular traffic. The course should be marked by directional arrows until the runners know the course. These runs build leg muscles, increase lung capacity, and develop endurance. Any local area of varied terrain is suitable.

(2) Progression. In the mass training of a large group, leaders should be stationed at the head and the rear of the column and should make every effort to keep the individuals together .After determining the abilities of the individuals in cross-country running, it is advisable to divide the unit into three groups. The poorest conditioned group starts first, and the best conditioned group, last. The starting time of the groups should be staggered so that all of them finish about the same time. In preliminary training, the running is similar to ordinary road work in that it begins with rather slow jogging, alternating with walking. The speed and distance of the run is gradually increased. As the condition of the individual improves, occasional sprints may be introduced. At first the distance run is from 1/2 to 1 mile. It is gradually increased to 2 or 2.5 miles. Well-conditioned personnel can run 2 to 2.5 miles within a 15-minute period.

(3) Leadership. Marines should not be required to take part in distance running until they have been through a progressively scheduled training program which requires a considerable amount of running. Cross-country runs should be scheduled occasionally to provide variety in the program. Cross-country running has the advantage of allowing mass participation. Interest can be stimulated by putting the runs on a competitive basis. As a single activity, short cross-country runs can be scheduled once a week, gradually increasing the distance as the physical conditioning improves.

(4)Use With Other Programs. Cross-country running can be combined with other activities such as conditioning exercises.

d. Fartlek Training

(l)Description and Function. Fartlek training is various running exercises conducted along an intense combat-like course. The running is conducted to overwork the lungs, allowing only partial recovery which is followed by another intense period of overload. This sequence is repeated for the duration of the workout. Fartlek training is a useful combat training method and general cardiovascular conditioner. The entire unit must be in uniformly good to excellent condition to qualify for this training.

(2) Progression. The following is an example of one fartlek session:

(a) Warm up by stretching 3-5 minutes.

(b) Warm up running easily 5-10 minutes.

(c) Run at a fast, steady pace for 3/4 to 1 3/4 miles (dependent on terrain).

(d) Walk/jog at a moderate pace for 5 minutes (recovery).

(e) Run easily sprinting 15 to 20 meters occasionally.

(f) Run full speed uphill for 175 to 200 meters.

(g) Warm down by running easily for 1/2 to 1 mile.

(h) Warm down by stretching 3 to 5 minutes.

(3)Leadership. To add variety, try incorporating combat movements, terrain association, a series of exercises or carrying table of equipment weapons (hit and roll). This will take some initiative in setting the course, but it allows for variety. In this manner, the fartlek course can be used as a training session or part of one. lt works extremely well with highly motivated, competitive Marines.

3603.ORIENTEERING

Orienteering is land navigation over a prescribed course as a timed, competitive event. It is an excellent way to integrate land navigation training with physical conditioning and requires Marines to be skilled in both areas. Commanders adjust the training as needed to change the degree of difficulty or to emphasize either navigation or conditioning. Taken to the extreme, an orienteering exercise is conducted in full combat gear over rugged terrain against aggressors. Orienteering requires the following physical skills: endurance, speed, strength, and all-around conditioning.

Section VII. Circuit Training

3701. GENERAL

a. Description and Objective. A strength circuit consists of a series of stations where individuals in small groups exercise vigorously for a short period of time and then move (on signal) to the next station where a different form of exercise is conducted. This rotation continues until all groups move through all stations. Strength circuits contain no set or specific types of exercise stations within the circuit. The objective of circuit training is to develop strength. There are three general types of circuits.

(1) Fixed Circuit. This is a circuit in which apparatus of an immovable type (fixed into the ground) is used.

(2) Movable Circuit. This circuit consists of individual exercise apparatus which is portable and can be moved to and from the training area.

(3) Simplified Circuit. This circuit requires no equipment or apparatus.

b. Formation. The exercises are done at will, but rapid, steady, and continuous work is required of all. Each Marine's nervous and muscular system reacts differently to timed vigorous exercises. His performance should be measured on how many movements per exercise he can complete as an individual. For example, one Marine may be able to complete 5 movements, while another may be able to complete 20, and yet each is receiving the maximum benefit. All three circuits contained in this chapter are designed for platoon-sized groups. Expansion beyond this capacity requires a large amount of equipment, as each Marine in the fixed and movable types of circuits must have an item of equipment available for exercise at each station. A group larger than a platoon could be exercised through use of the simplified type of circuit; however, the group would be unwieldy and control could be a problem.

c. Place in the Program. All circuits illustrated can be completed in a 15-minute period. This feature allows the exercise of a platoonor smaller group on the circuit for a single 15-minute period, or the scheduling of the circuit as a 15-minute period within a longer period. A circuit can thus be utilized within a rotating activity system of scheduling. Choice of a circuit by the unit depends upon area, facilities, and other local factors; however, there is a circuit for every need.

3702. FIXED STRENGTH CIRCUIT

a. Description and Objective. The strength circuit is an arrangement of various types of exercise apparatus which are fixed in position. (See fig. 3-16.)Seven basic exercises are used and each exercise requires an apparatus. All apparatus of one type are positioned together to constitute a station. Each station will accommodate 10 Marines. The circuit is designed to be accomplished in 15 minutes when conducted on a time-rotation basis, as normally executed for unit training. For individual use, a Marine may complete the circuit by executing a specific number of repetitions for each exercise. In this case, the time required for completion of the circuit would vary slightly depending on the number of repetitions the Marine completed. The objective of this circuit is to provide a series of exercises which will improve and maintain the strength of the body's major muscle groups.

Figure 3-16 is available in the illustration section at the end of the book.

b. Starting Level. Marines must be thoroughly warmed up prior to participating in the circuit system. If Marines have not engaged in vigorous exercise immediately prior to starting the circuit, then an instructor should execute the following warm up exercises. These exercises should be conducted in the normal formation for set drills. Seven repetitions of each exercise will normally provide sufficient warm up. The exercises are-

(1) High jumper, Conditioning Drill 1.

(2) Bend and reach, Conditioning Drill 1.

(3) Squat bender, Conditioning Drill 1.

c. Progression. The instructor can adjust the circuit system through several methods to accommodate participants of varying physical ability. One method of adjusting an exercise is to change the method of executing the leg lift or to select a heavier weight for the barbell curls. When it becomes apparent during a unit program that the overall fitness of the group has Improved, then the exercise may be made more strenuous in two additional ways. First, the instructor can have one or more of the supplementary stations added. Second, the time spent exercising at each station can be increased in 5-second increments to a maximum of 60 seconds. Also, as Marines learn the circuit, the 45-second period for movement and instruction between stations can be eliminated, thus allowing only enough time to change stations.

d. Starting Positions. The Marine starts on any station, exercises steadily for a certain period (45 seconds initially), then moves on command to the next type of apparatus where he again exercises steadily for an equal period. The Marine continues until the required exercise is completed at each station. If it is desired to expand the number of stations in the circuit to accommodate more participants at one time, the instructor can provide four additional supplementary exercises, requiring no equipment. This will increase the amount of time required to complete the circuit.

e. Leadership. Close control of all Marines is necessary to ensure that a minimum amount of time is spent in moving them to their initial stations and in moving them between stations. One instructor can control the activity on the strength circuit. A stop watch or wrist watch with a second hand is required. When Marines arrive at the strength circuit, the instructor will form them for exercise and conduct the warm up drill. The group is then reassembled and formed into a number of files equal to the number of stations being used in the circuit. Each file is then directed to a station. As soon as all participants have reached an exercise position at a station, the instructor gives the command READY, GO. After 45 seconds of exercise, the instructor gives the command STOP, CHANGE OVER. The instructor allows Marines 45 seconds for moving to the next station and for preparing for the next exercise before the command EXERCISE is again given. In lieu of verbal commands, a whistle may be used to stop and start the exercises. For large groups, a megaphone or loudspeaker is useful.

f. Individual Conditioning Program. For the Marine working alone on the strength circuit, it would be impractical to time the exercise periods. For individual exercise, the Marine should select a number of repetitions of each exercise to complete, then rotate to the next stations after completing these repetitions. The number of repetitions selected should be at or near the maximum that the Marine is capable of doing without halting for rest.

g. Fixed Circuit Stations

(1) Primary Stations (See fig. 3-16.)

(a) Pullups/Chinups. A horizontal bar placed 8 feet above the ground is necessary for each Marine at this station. The Marine will also need a space on the bar that is 45 inches wide. On the command**EXERCISE**,the Marine grasps the bar with both palms facing either forward or to the rear, arms fully extended, and feet free of the ground and executes the exercise as described in MCO 6100.3_.The exercise is repeated as many times as possible until the command **STOP** is given. Then the Marine moves to the next exercise. If a Marine has done his maximum number of pullups/chinups prior to the command **STOP**, he will remain in the "dead" handing position until the command **STOP** and move to the next station.

(b) Twist Grip. The apparatus is a horizontal bar, free to turn, held between uprights placed 30 inches apart. The bar is 52 inches above the ground. A weight of 20 pounds is attached to the center of the bar by a light rope long enough to permit the weight to rest on the ground. The Marine stands at arm's length from the bar and grasps it with his hands on either side of the rope, palms down, thumbs under the bar. On the command **EXERCISE,** the hands are rotated so that the backs of the hands are rotated away from the body, thus winding the rope on the bar. The elbows are kept straight to ensure that the exercise is performed by the hand and forearm. When the weight is drawn up to the bar, the bar is then rotated in the opposite direction to lower the weight to the ground. This exercise is continued until the command **STOP** is given. The Marine then moves to the next station.

(c) Pulley Weights. The apparatus is a T frame with a system of pulleys that suspends a weight of about 90 pounds. The weight is attached to a light steel cable which has a drawbar attached to the other end. The Marine grasps the drawbar and sits down directly under the bar, legs extended to the front and arms extended overhead. The exercise is executed by pulling the drawbar down behind the head, then extending the arms slowly again until they are fully extended overhead. The exercise is repeated as many times as possible until the command STOP is given. The Marine then will move to the next exercise. Upon completion of the exercise, the weight is lowered slowly to the ground.

(d) Barbell Curls. A barbell is necessary for each Marine at this station. The barbell is constructed of 1 1/4-inch pipe 5 feet long, and two concrete-filled No. 10 cans. Each barbell should weigh about 40 pounds. Variance in the weight of the barbells, up to about55 pounds, will allow appropriate overload to be applied to Marines who are above average in strength or weight. The Marine grasps the bar with the palms forward and assumes a standing position with the barbell held in front of the hips, hands approximately shoulder's width apart. On the command **EXERCISE,** the Marine flexes the elbows and draws the barbell up until it touches the upper chest. The elbows remain at the sides. Breath is inhaled with the upward movement and exhaled as the barbell is lowered to the starting position. The exercise is repeated as many times as possible until the command **STOP** is given. The Marine moves to the next apparatus.

(e) Step-Up. The apparatus is a platform or ledge 18 inches high and of such size. At accommodate 10 Marines. The Marine faces the platform and on the command **EXERCISE,** steps up onto the platform, bringing his trailing foot up beside the leading foot. He then steps back down to the original position, stepping down first with the same foot he initially used in stepping up. After 10 repetitions of the exercise, he changes the order of moving the feet to use the opposite leg for stepping up. Repeat this exercise until the command **STOP** is given. The Marine then moves to the next station.

(f) Leg Lift. The apparatus is a horizontal bar constructed as described in (a). To prevent the body from swaying, a horizontal back support is added 40 inches below the horizontal bar. The arms are kept fully extended. On the command **EXERCISE,** the Marine jumps up, grasps the bar with the palms forward and the back support behind him. The exercise is executed by raising the legs to a horizontal position then slowly lowering them to the vertical position. The Marine does not flex his knees. He does not swing the legs to the rear of a vertical position to gain momentum for raising them in the next repetition of the exercise. The movement is repeated until the command **STOP** is given. The Marine then moves to the next exercise. If unable to raise his legs to a horizontal position without flexing his knees, the Marine flexes his knees and draws them up to the chest, then lowers his legs to the vertical position.

(g) Rope Climb. The rope climb is 20 to 30 feet high with five ropes suspended from a horizontal bar which forms the uppermost part of the framework. To prevent the horizontal bar from sagging and to provide safety, only five ropes are attached to it. There are two frameworks per station. The ropes are 6 feet apart. Any method may be used to climb the rope, and the Marines climb as high as possible. Marines who are proficient should climb therope several times during the time allotted. Inexperienced Marines should be cautioned to take care during descent to avoid rope burns on their hands.

(2) Supplementary Stations. The following exercises are designed to expand the basic circuit by being inserted in specific places within the system. For each supplementary station used, there should be adequate room for 10 Marines to exercise.

(a) Bent Knee Situp or Bottoms-Up. These calisthenics are designed to strengthen the abdominal muscles. These exercises will be inserted between the pullup/chinup and twist grip stations. The primary stomach exercise is the situp. In case of inclement weather or other conditions that make ground contact undesirable, the bottoms up exercise is used. In situps on the command **EXERCISE,** the Marine lies on his back (supine position) with knees flexed and both feet flat on the ground and executes the exercise (minus an assistant) as described in MCO 6100.3_.The Marine then returns to the starting position, repeating the exercise until the command **STOP** is given. The Marine then moves to the next station. In bottoms up on the commandEXERCISE,the Marine assumes the front leaning rest and executes the bottoms-up exercise as described in Conditioning Drill 3. He continues this exercise at a moderate cadence until the command **STOP** is given. The Marine then moves to the next station.

(b) Pushup. This exercise is designed to strengthen the arm- and shoulder-girdle muscles. It should be included between the twist grip and pulley weight stations. Upon the command **EXERCISE,** the Marine executes the pushup as described in Conditioning Drill 1. He continues this exercise at a moderate cadence until the command **STOP** is given. The Marine then moves to the next station.

(c) Knee Bender. This exercise is designed to build leg muscles and is included between the pulley weight and barbell curl stations. On the command **EXERCISE,** the Marine executes the knee bender as described in Conditioning Drill 3.He continues this exercise at a moderate cadence until the command **STOP** is given. The Marine then moves to the next station.

(d) Trunk Twister. This exercise strengthens the major muscles of the trunk and is included between the step-up and pullup/chinup stations. On the command EXERCISE, the Marine executes the trunk twister as described in Conditioning Drill 1. He continues this exercise at a moderate cadence until the command **STOP** is given. The Marine then moves to the next station.

3703. MOVEABLE STRENGTH CIRCUIT

a. Description and Objective. The exercises in this circuit are progressive and the course is planned to gain and hold the interest of the participating groups. The circuit consists of a series of stations, with each station designed to develop a particular group of muscles. Along with muscular development, correct posture and deep rhythmic breathing should be stressed on this circuit at all times.(See figs. 3-17 and 3-18.)

Figure 3-17 is available in the illustration section at the end of the book.

b. Equipment. The equipment is set up in files. Six files of 8 stations will accommodate a platoon

c. Formation. The platoon marches to the area where the equipment is positioned and forms a file within each lane of stations, covering down on a piece of equipment. Movements are made on the double, the important factor being that no time is wasted in getting to work.

d. Progression. Initially 40 to 45 seconds per station is adequate. As individuals become stronger the time should be increased in 5-second increments until a minute to a minute and a half is reached.

e. Leadership

(1) The leader stands in front of the barbell station and controls the rotation from this position. The leader supervises the entire group, with the assistance of several instructors who move about in the platoon correcting and encouraging the Marines.

(2) The leader starts each group but does not count cadence nor lead them through the exercises. Each individual exercises rapidly but individually.

(3) As each Marine finishes his repetitions with the barbell, he places the barbells on the ground. The leader calls, READY, followed by the command, FALL OUT ONE. All will double-time to the station directly in front of them, while the individuals on the barbell stations will do an about face to the rear station in their lane.

f. Movable Circuit Stations. The Marine can obtain best results on the movable circuit if the exercises on the various stations are given in the following manner:

(1) Station 1—The Barbell. (See fig. 3-17, 1.)The exercises at this station stress proper posture; deep, rhythmic breathing; and development of the muscles of the arms, shoulders, and upper body. An instructor teaches the proper methods of lifting before the exercises begin. The methods are to lift with the legs, to keep the back straight, and to merely grip with the hands. Two recommended exercises are given below. The instructor will use only one exercise per period .He may use either exercise. At successive periods, the other exercise is used.

(a) Exercise 1--Two-Hands Military Press. (See fig. 3-18, 1.)Grasp the barbell with both hands, knuckles up at shoulder width, and lift to the chest. Steadily press to arm's length overhead; lower to the chest resisting weight all the way. Inhale as the weight is pressed up and exhale as the weight is brought down.

(b) Exercise 2--Two-Hands Regular Curl. (See fig. 3-18, 1.) Lift the weight to the waist, with the palms of the hands out, heels together, stomach in, chest lifted and arched, shoulders back, elbows in close to the sides; inhale deeply and curl the weight to the shoulders, using the arms only, at the same time keeping the elbows close to the sides; exhale Emphasize posture and the use of the arms only. This is a very valuable exercise for the development of the biceps and the grip and should be repeated from 8 to 16 times, depending on the ability of the Marine.

(2) Station 2--The Jump Rope. (See fig. 3-18, 2.) This exercise develops strength and agility in the legs and stamina of the whole body. It makes the Marine agile on his feet and increases his footwork efficiency and timing. Each Marine should progress until able to jump rope at least 3 minutes at top speed.

(3) Station 3--The Twist Grip. (See fig. 3-18, 3.) The twist grip is an excellent exercise for the hands and forearms, and adds greatly to the Marine's ability in hand-to-hand combat. The handle is gripped and twisted, winding the rope until the weighted can is level with the height of the hands, which are held horizontal. The weight is lowered in the same manner; the Marine resists the weight all the way, occasionally stopping the twisting motion and alternately removing first one hand, then the other, from the handle. A variation of the above exercise is to wind the handle with the palms up and the arms bent and with the elbows held close in to the sides. Each Marine should maintain a good posture and keep the stomach muscles taut throughout this exercise.

(4) Station 4--The Incline Plane. (See fig. 3-18, 4,)The use of the incline plane is a very strenuous exercise and well designed for the development of the abdomen. Although 6 to 10 repetitions are sufficient for the beginners, more can be added as ability increases. Marines with hernias or recent operations will be excused from participation at this station.

(5) Station 5--The War Club. (See fig. 3-18, 5.) The war club is a simple and effective means of exercising the principal muscle groups of the body, especially those of the trunk, back, and shoulders. To gain the maximum benefit from this exercise, the Marine must keep both feet flat on the ground at all times. Throughout the exercise period, the weight is swung from arm's length as follows:

(a) As in chopping wood, first on one side, then on the other.

(b) As a batter warming up with a number of bats.

(c) In large circles, first with one hand and then with the other.

(6) Station 6--The Bicycle Ride. (See fig. 3-18, 6.) The bicycle ride is well suited to exercising many of the muscle groups of the body, particularly those of the abdomen. Vary the speed of the exercise, but keep everyone "riding" the entire period. A variation exercise may be performed by placing the legs together, raising them slowly to a height about 2 feet from the ground, and then lowering them slowly to the ground.

(7) Station 7--The Step-up. (See fig. 3-18, 7.) The step-up exercises the legs. The step-up is performed by initially stepping up with the left foot, followed by the right, then stepping down with the left foot followed by the right. Continue for 20 seconds, then change to the right foot as the lead foot for 20 seconds.

(8) Station 8--The Isometric Pull. (See fig. 3-18, 8.) Two trainees work at this station with a cable pull and perform the following exercises:

(a) Initially start with one Marine in the supine position and one Marine sitting. The sitting Marine lowers the upper body to the ground and pulls the partner up to the sitting position. The partner then performs this same action and this is continued for 20 seconds at a rapid rate.

(b) During the last 20 seconds, the same action takes place but, in this case, the Marine in the supine position resists the pull of the partner for approximately 5 seconds before allowing to be pulled up into the sitting position.

3704. CIRCUIT-INTERVAL TABLE

a. Description and Objective. The circuit-interval table is designed to develop strength and endurance within a short period of time, with no equipment requirement, through a rapid and vigorous routine of exercise. Fifteen minutes is an adequate period to execute all exercises and to secure a vigorous workout with the circuit-interval principle.

Figure 3-18 is available in the illustration section at the end of the book.

b. Formation and Starting Positions

(1) A leader forms platoons or smaller groups in an oval or circular formation with 3- to 5-yard intervals between Marines. The Marines face to the right and move forward at quick time and then double time. (See fig. 3-19, A.)After running several platoon circle laps, the leader calls out the name of an arm and shoulder exercise from the list below, orders quick time and commands, for example, PUSHUPS. On this command, all Marines immediately hit the ground and individually and rapidly begin doing pushups. No cadence is counted. (See fig. 3-19, B.) After 30 seconds of exercise, the leader commands. **QN YOUR FEET, FORWARD, MARCH** The platoon resumes the quick time cadence and the leader, when ready, gives the necessary commands for double time. The double time is continued for one or more laps and the leader calls out the name of the next exercise and the process is repeated. This continues, with running between each exercise, until every body part has been exercised.

(2) The instructor controls the running and quick time to observe the effects of the exercise upon the Marines. Cadence, step, and precision are not important to the objective and the instructor should not use them. What is important is speed and the instructor should stress this. After the exercise period is started, the Marines do not stop. This circuit method emphasizes stress and recovery, the recovery occurring during the quick time periods.

c. Activities. The leader can use the following exercises. These exercises can be repeated if necessary, during a second round. The leader can use other calisthenic exercises in the circuit-interval table.

(1) Arms and shoulders-push-ups.

(2) Stomach situps.

(3) Back-squat thrusts.

(4) Legs-bicycle (on back).

d. Progression. Ihe progress is controlled by the leader, who must pace the running, quick-time movement, and exercise in such a way that Marines will receive a vigorous workout yet be able to participate throughout the 15-minute period. Marines who are in the initial stages of physical condition will not be able to double time or exercise as long as those who are better conditioned. The idea is to set a pace which can be increased during each workout, thus progressing gradually to a higher level of physical fitness.

e. Leadership. The platoon leader, platoon sergeant, or section leader can lead the group. The leader must execute the exercise with the unit to feel the effects and thereby adjust the pace.

f. Place in the Program. This activity may be scheduled whenever a short period of time is available. The only requirement is that enough space, indoors or out, be available to form the circle.

Figure 3-19 is available in the illustration section at the end of the book.

Section VIII. Basic Physical Skills and Obstacle Courses

3801. GENERAL

The purpose of this section is to list the basic military skills and the methods for their development. Many of these skills are best developed by obstacle courses but other drills are also discussed. The section explains types of obstacle courses, details of construction, and methods of negotiating the various obstacles.

3802. BASIC PHYSICAL SKILLS

a. Objective. The objective is to develop proficiency in the various military physical skills which are essential to personal safety and effective combat operations. In travel by foot over rugged terrain and in the execution of combat duties, Marines must be trained to perform basic skills, such as running, jumping, climbing, and carrying. During training, Marines will develop agility and coordination in these skills .Fast and skillful execution of these skills may mean the difference between success and failure on the battlefield.

b. Place in the Program. These skills are practiced throughout the entire program in many different activities. Many of these skills are best practiced on obstacle courses.(See par. 3803.)

c. Description. These basic skills are the minimum skills required by all Marines. The basic skills are as follows:

(1) Running. Running is used to strengthen the legs and develop cardiovascular endurance. Marines should be exposed to running in various situations: on roads, over rough ground, up and down hills, cross-country, and running over low obstacles.

(2) Jumping. In broad jumping, the take-off foot is planted firmly and the spring comes from the extension of this leg as the other leg reaches for the far side of a ditch or similar obstacle. (See fig. 3-20.) The arms are forcibly raised forward and upward to assist in propelling the body up and forward. Landing may be on one or both feet depending upon the length of the jump. In vertical jumping downward from a height, the jumper should aim his feet at the desired landing spot and he should also jump with the knees slightly bent and feet together, with the trunk inclined slightly forward. As the feet touch the ground, the shock is absorbed by bending the knees into a full squatting position. If the height is too great or the ground too hard to absorb the shock, then the jumper should forward roll or side roll thus eliminating some of the momentum.

Figure 3-20 is available in the illustration section at the end of the book.

(3) Dodging In combat situations, it is often times necessary to change directions quickly. To execute this movement while running, a lead foot is firmly planted, left foot if the direction is to the right and right foot if the direction is to the left. The opposite foot is moved toward the new direction. The knees are slightly flexed during the movement and the center of gravity is low and balanced. (See fig. 3-21.) At the time of the change of direction, the head and trunk are turned quickly in the new direction.

Figure 3-21 is available in the illustration section at the end of the book.

(4) Climbing and Surmounting. All Marines should know how to climb and surmount various types of obstacles. (See fig. 3-22.)

Figure 3-22 is available in the illustration section at the end of the book.

(a) Vertical Climbing, as in Climbing a Rope or Pole. The hands grasp the rope or pole overhead with the palms toward the face. Gripping the object, the body is pulled upward with the arms and shoulders, assisted by the feet which grip the object and assist by pushing downward. If shoulder girdle strength and body coordination are not adequate to permit alternating the hands, the arms act together in pulling upward.

(b) Climbing as in Surmounting a Wall. Ingoing over a wall, the body should be kept as close to the top as possible, since in combat operations it is important to offer as small a target as possible to the enemy. If an individual climbs a wall while carrying a rifle, both hands should be freed by slinging the rifle over the back. There are two methods commonly used for surmounting a wall of moderate height, but only one for dropping from it.The methods are as follows:

1 Run, Jump, and Vault. Approach the wall at a run, jump forward and upward at the wall and place one foot against it as high up as possible. Use the foot in contact with the wall to help push the body upward while grasping the top of the wall with the hands. Pull the body up with the arms, assisted by pressure of the foot against the wall and swing the legs over, propelling the body weight over the wall.

2 Hook and Swing. Approach the wall at a run and jump forward and upward. Hook one elbow over the wall, locking the arm in place by pulling up until the top of the wall is underneath the armpit. Grasp the top of the wall with the other hand. Draw the leg which is closer to the wall up as far toward the top as possible. Then swing the other leg over the top of the wall.The body is then carried over with a rolling motion. A variation of this leg action can be used by Marines who are unable to draw up the leg as described. While hanging with both legs fully extended, start a swinging motion with the legs together. When the legs have enough momentum, swing the outside leg over the top of the wall with a vigorous kick, then follow with the body.

3 **Dropping.** All drops from a height are executed in the same manner regardless of the method used to gain the top. One hand is placed against the far side of the wall while the other hand grasps the top. From this position, the body is rolled over the wall and vaulted away from it with the legs swinging clear. As the body passes over the wall and drops, it should at all times face the wall. This will keep the rifle and other equipment clear. Break the fall by retaining a grasp on the top of the wall as long as possible.

(c) Climbing Ladders and Cargo Nets. Rope ladders, stationary vertical ladders, and cargo nets employ the same general technique. The important element is to grasp the side supports firmly in the hands about shoulder height and place the feet on a rung which would cause the body to be fully extended. In movement upward, one hand is moved upward and a new grasp is secured and, at the same time, the opposite leg moves up a rung. As the knee straightens, the body is elevated. This process is repeated using the opposite arm and leg. Alternation continues in this manner until the climber reaches the objective.

(5) Traversing Horizontal Objects. The traversing of horizontal objects puts heavy stress on the arms and shoulder girdle area as the feet are usually suspended in the air with all of the body weight on the arms and shoulders.(See fig. 3-23.)

(a) Traversing Horizontal Ropes or Pipes. The hands grasp the horizontal support overhead with the palms facing. To propel the body forward, one hand is released and moved forward to secure a new grasp. At the same time, the opposite side of the body is swung forward (some people are able to "walk" in the air, keeping the body to the front and moving the legs in time with the arms as in walking on the ground). The other hand is then released and moved forward; this alternation is continued until the objective is reached.

Figure 3-23 is available in the illustration section at the end of the book.

(b) Traversing Horizontal Ladders. In this situation, the movement is the same as used in traversing a rope or pipe. The hands, however, are placed on the rung with the palms away from the face. Other than this difference the technique is the same.

(6) Crawling. Crawling in combat situations is an often used skill. Crawling may be high or low.(See fig. 3-24.)

Figure 3-24 is available in the illustration section at the end of the book.

(a) High Crawl.

In the high crawl, the Marine moves on hands and knees, moving one hand and the opposite knee and then continuing to move the hands in alternation with the opposite knee following the companion hand.

(b) Low Crawl. The Marine is in the prone position usually with the forearms and palms of the hand on the ground. He propels forward by bending the knee of one leg and pushing with the inside edge of the shoe. At the same time, the opposite arm moves forward and pulls to the rear. The body remains low and movement is continued by bending the opposite knee and pushing, and at the same time sliding the opposite arm forward and pulling. Alternation of hands and legs continues until the objective is reached.

(7) Throwing. Throwing may be executed from the kneeling or standing positions. The object to be thrown is held in the hand, and the throwing arm is bent at the elbow; the hand is then moved to the rear until the hand is behind the ear. The body is turned so that the lead foot and balance arm on the side toward the target are pointing at the target. (See fig. 3-25.) The balance arm is used to sight over and align the throwing hand and the target. When properly aligned, the elbow is moved rapidly forward until it is at a point just in front of the body where the arm is straightened and the wrist snapped. This whip motion propels the object to the target. Underhand throws secure momentum by the thrower bending his knees and swinging the throwing arm to the rear. As the knees are straightened, the arm is forcefully swung forward from the shoulder and the object released.

Figure 3-25 is available in the illustration section at the end of the book.

(8) Vaulting. Vaulting is employed to overcome low barriers or fences. (See fig. 3-26.)The object to be surmounted is approached at an angle. The hand on the side next to the obstacle is placed on the top of the obstacle and, with a straight arm, the body weight is pushed upward. At the same time, the leg on the side next to the obstacle is thrown upward and over the top followed by the other leg (side approach). In landing, the weight comes down on the leading leg first followed by regaining the balance on both legs. The free arm serves as a balance. A direct (front) approach can be used at which time both legs go over the object together.

(9) Man Carrying. There are three basic individual means of carrying personnel in combat situations and one of these methods may be used in carrying objects.

Figure 3-26 is available in the illustration section at the end of the book.

(a) Fireman's Carry."A" stands sideways in front of "B", "A" bends his knees and leans forward, placing one arm through "B's" crotch, grasps the wrist of "B's" arm, which is hanging over the shoulder, and then "A" runs forward.

(b) Saddle-Back Carry. With his back toward "B", "A" stands in front of "B". "B" mounts "A's" hips and clasps his arms in front of "A's" chest. "A" grasps "B's" thighs.

(c) Single-Shoulder Carry. "A" stands facing "B"."A" assumes a semi-squatting position. "B" leans forward until "B" lies across "A's" right shoulder. "A" clasps his arms around "B's" legs and straightens up, lifting "B" from the ground. "A" then runs forward. This method may also be used to carry heavy objects.

(10) Balancing. Balancing the body while walking or running on a narrow object when crossing obstacles is a skill which requires practice and confidence. Balance is required in negotiating a log placed across a stream, in crossing a narrow beam or rail, and in similar situations.(See fig. 3-27.) To perform this skill, place the feet on the object to be crossed, hold the arms to the sides at shoulder level, and fix the eyes on the object approximately 5 yards in front of the feet. Generally, it is not a good practice to look down at the feet. Walk the beam by placing first one foot and then the other in the center of the beam, thereby moving forward, using the arms to aid in maintaining balance.

Figure 3-27 is available in the illustration section at the end of the book.

(11) Falling. Injury can be avoided if Marines are taught to fall properly. They should know how to use the body momentum to their advantage during a fall rather than to try resisting that force. (See fig. 3-28.) If enough force is present, such as occurs during a fall while running or in jumping downward from a height, individuals can extend their hands to catch the weight. At the same time, duck the head and roll forward onto the feet. The key to falling without injury from the standing position is relaxation and rolling on the outside of the leg, hip, and buttocks to take the brunt of the fall.

Figure 3-28 is available in the illustration section at the end of the book.

3803. OBSTACLE COURSES

a. General. Obstacle courses are a valuable part of physical readiness training. The challenge presented by the obstacles assists in developing and testing the basic physical skills. In many combat situations, success will depend upon the Marine's ability to perform one or more of these skills, often while carrying field equipment and when fatigued. In this section, two types of obstacle courses will be discussed-Conditioning Obstacle Course and Confidence Obstacle Course.

b. Course Safety Precautions. Commanders and course instructors should take certain precautions to prevent injury on obstacle courses. A few of the precautions are:
(1) Inspect the course for faulty construction of obstacles, protruding nails, rotten logs, condition of the landing pits, and other hazards to safety.
(2) Conduct warm up exercises before the unit runs the course.
(3) Explain and demonstrate the correct techniques for negotiating all the obstacles before allowing the Marines to try them.
(4) Give Marines at least two weeks of conditioning exercises before scheduling the obstacle and/or confidence courses.
(5) Ensure that negotiation of the higher and more difficult obstacles are under the supervision of an instructor.
(6) Do not permit individuals who have neither practiced the basic skills nor run the conditioning obstacle course to participate in the confidence obstacle course.
(7) Weather conditions may cause footing or handhold surfaces to be slippery. If such is the case, postpone training on the course.

3804.CONDITIONING OBSTACLE COURSES

a. Description and Objective. The Conditioning Obstacle Course is commonly known as the Obstacle Course. This course consists of fairly low obstacles which are designed to be negotiated quickly. The obstacles serve to test various basic skills, and running the course is a test of the Marine's physical condition. After receiving instruction and an opportunity to practice the skills, Marines run the course against time.

b. Area and Equipment
(1) Complete standardization of obstacle courses should not be attempted since topographical conditions always vary. Commanders should use ingenuity in constructing a course, making good use of streams, hills, trees, rocks, and other natural obstacles. Since the course is eventually run at high speed, it should not be dangerous.

(2) The course should be wide enough for six or eight men to run simultaneously, encouraging competition. The lanes for the first several obstacles should be wider and the obstacles themselves easier than those that follow. This avoids congestion until the contestants scatter out over the course. The last two or three obstacles should not be too difficult and should not involve high climbing. This prevents injuries and falls resulting from fatigue.

(3) The total distance of the course should range from 300 to 450 yards and include from 15 to 25 obstacles. Normally the obstacles should be 20 to 30 yards apart and arranged so that those which exercise the same groups of muscles are separated.

(4) The obstacles should be substantially built. Peeled logs, 6 to 8 inches in diameter, are ideal for many of the obstacles. Sharp points and corners should be eliminated. Landing pits for jumps or vaults should be filled with sand or sawdust to prevent injuries.

(5) The course should be constructed and marked so that it is not possible to sidestep or detour obstacles. However, it is desirable to provide alternate obstacles of varying degrees of difficulty.

(6) The course should be in the shape of a horseshoe or figure eight so that the finish is close to the start and signs should be placed to indicate the course route.

c. Leadership. Before Marines run an obstacle course, they should be instructed in the proper technique of negotiating each obstacle. In each case this technique should be explained and demonstrated in detail, with emphasis on avoiding injury. Every individual should be given an opportunity to practice on each obstacle until he becomes reasonably proficient at negotiating it. Before the course is run against time, it is advisable for individuals to make several runs at a slower pace. During such practice or trial runs, the instructor should observe the performances and make appropriate corrections. Marines should never be permitted to run the course for time until they have practiced on all obstacles. The best method of timing the runners is to have the timer stand at the finish and call out the minutes and seconds as each individual finishes. If several watches are available, each wave may be timed separately. If only one watch is available, the different waves should be started at regular intervals, such as every 30 seconds. If an individual fails to negotiate an obstacle, a previously determined penalty should be exacted.

d.Types of Obstacles

(1) Jumping-Type Obstacles. These obstacles may be ditches which are cleared with one leap, trenches which the individuals can jump into, heights which require jumping downward, or hurdles which an individual can jump over. (See fig. 3-29.)

Figure 3-29 is available in the illustration section at the end of the book.

(2) Dodging-Type Obstacles. Obstacles of this type are usually mazes consisting of posts set in the ground at irregular intervals. (See fig. 3-30.)The intervals between posts should be rather narrow so that the Marines must pick their way carefully through and around them. Lane guides may be constructed to guide the Marines to dodge and change direction. Obstacles may be put into a maze pattern to cause the Marines to change direction.

Figure 3-30 is available in the illustration section at the end of the book.

(3) Vertical Climbing and Surmounting Type Obstacles. These obstacles may be climbing ropes, either plain or knotted and 1-5 inches in diameter; cargo nets or walls 7 or 8 feet high; or vertical poles 6 to 8 inches in diameter and 38 feet high.(Seefig. 3-31.)

Figure 3-31 is available in the illustration section at the end of the book.

(4) Horizontal Traversing-Type Obstacles. Horizontal obstacles may be pipes, beams, ladders, or ropes. (See fig. 3-32.)

(5) Crawling Type Obstacles. Obstacles which require crawling may be constructed of large pipe sections, low rails, and wire. (See fig. 3-33.)

Figure 3-32 is available in the illustration section at the end of the book.

(6) Vaulting-Type Obstacles. Obstacles of 3 to 3.5 feet in height such as low walls or fences may be used as a vaulting obstacle.(See fig. 3-34.)

(7)Balancing-Type Obstacles. Beams, logs, and planks may be used as balancing-type obstacles.(See fig. 3-35.)These items may be used to span water obstacles and dry ditches, or raised off the ground somewhat to simulate natural depressions.

Figure 3-33 is available in the illustration section at the end of the book.

3805.CONFIDENCE OBSTACLE COURSES

a. Description and Objective. The Confidence Obstacle Course is commonly known as the Confidence Course. This course is composed of higher and more difficult obstacles than those used in the conditioning course. The confidence obstacle course is designed to give Marines confidence in their mental and physical capacities and to cultivate their spirit of daring. They are encouraged but not compelled to negotiate this course and the course is not run against time. The negotiation of a confidence course, however, is strenuous enough to be an excellent physical conditioner. Marines should NEVER attempt to take the obstacles at high speed and should not compete for speed. The obstacles vary from fairly easy to extremely difficult. Some are of considerable height to accustom Marines to climbing such heights without fear. Considerable emphasis is placed on obstacles that train and test an individual's balance.

Figure 3-34 is available in the illustration section at the end of the book.

b. Area and Equipment

(1) The confidence course accommodates four platoons, one platoon at each group of six obstacles. The course should be made up of about 24 obstacles, numbered and marked as follows:! to 6, white numbers on red background; 7 to 12, black numbers on a white background;

13 to 18, white numbers on a blue background; and 19 to 24, white numbers on a black background.

(2)A few simple pieces of equipment should be provided for individuals who do not have the strength, courage, or ability to negotiate the obstacles.

Figure 3-35 is available in the illustration section at the end of the book.

c. Formation. The obstacles should be divided into groups of six, and each group is designated by a different color. Each platoon starts at a different color. Individuals are separated into groups of 8 to 12 at each obstacle. At the starting signal from the company commander, they proceed numerically through. Anyone may skip an obstacle who is afraid to try. Individuals proceed from obstacle to obstacle until time is called, then assemble as ordered.

d. Leadership. If the Marines are new to the confidence course, an instructor will demonstrate or will give a brief orientation at each obstacle, including an explanation and demonstration of a method of negotiating it. Marines are encouraged to try the various obstacles, but they are not compelled to do so. No compulsion is to be used. The manner of negotiating any obstacle is left to the discretion of the Marine. However, the instructor assists anyone who experiences difficulty. Instructors must supervise closely at all times to prevent injuries, as some of the obstacles are quite high. Also, some of the obstacles should not be used when slippery or wet. The example of instructors and especially selected demonstrators will serve to inspire the individuals to greater effort.

e. Negotiating the Obstacles. Although personnel need not conform to any one method of negotiating the obstacles, there should be some uniformity in the approach to them. A general method of negotiating the obstacles is indicated below.

(1) Red Group This group contains the first six obstacles. (See fig. 3-36.)

(a) The Belly Buster Individuals may vault, jump, or climb over. Warn them that the log is not stationary.

(b) Reverse Climb. Climb the reverse incline and go down the other side to the ground.

(c) The Weaver. Move from one end of the obstacle to the other by weaving the body under one bar and over the next.

(d) Hip-Hip. Step over each bar, either alternating legs or using same lead leg each time.

(e) Balancing Logs. Step up on log, retaining the balance, walk or run along log.

(f) Island Hopper. Jump from one log to another until the obstacle is negotiated.

(2)White Group. This group is composed of the second six obstacles. (See fig. 3-37.)

(a) Tough Nut. Step over each X in the lane.

(b) Slide for Life. Climb the tower, grasp the rope firmly, and swing the legs upward. Hold the rope with the legs to distribute the weight between them and the arms. Braking the slide with the feet and legs, proceed down the rope. Warn Marines that there is danger of getting rope burns on their hands. When the rope is slippery or wet, this can be a dangerous obstacle.

(c) Low Belly Over. .Mount the low log and jump onto the high log, both arms grasping over the top of the log, the stomach area in contact with it. Swing the legs over the log and lower the body to the ground.

(d) Belly Crawl Move forward under the wire, belly down, to the end of the obstacle.

(e) The Dirty Name Mount the low log and jump to or reach the higher logs in succession, then jump or drop to the ground. Warn the Marines about the height of the final log.

(f) The Tarzan Mount the lower log and walk the length of it and each successive, higher log until reaching the horizontal ladder. Grasp two rungs of the ladder and swing the body into the air. Negotiate the length of the ladder by releasing one hand at a time and swing forward, grasping a more distant rung.

(3) Blue Group. This group is formed by the third group of six obstacles. (See fig. 3-38.)

(a) High Step over. Step over each log, alternating the lead foot or using the same lead foot.

(b) Swinger. Climb onto the swinging log and over to the ground on the opposite side.

(c) Low Wire. Move under the wire on the back, using the hands to raise the wire to clear the body.

(d) Swing, Stop and Jump. Gain momentum with a short run, grasp the rope, and swing the body forward to the top of the wall. Release the rope while standing on the wall and jump to the ground.

(e) Six Vaults. Vault over the logs, using one or both hands.

(f) Easy Balancer. Walk up one inclined log and down the one on the other side to the ground.

(4) Black Group. The last group is formed by the final six obstacles. (See fig. 3- 39.)

(a) Incline Wall. Approach the underside of the wall, jump up and grasp the top and pull the body up and over. Slide or jump down the incline to the ground.

(b) Skyscraper. Jump or climb to the first floor, climb up the corner posts or assist each other to any desired floor. Descend to the ground in any desired manner.

(c) Jump and Land. Climb up the ladder to the platform and jump to the ground.

(d) Confidence Climb. Climb the inclined ladder to the vertical ladder. Go to the top of the vertical ladder, then down the other side to the ground.

(e) Belly Robber. Step on the lower log and assume the prone position on the horizontal logs. Crawl over the logs to the opposite end of the obstacle.

(f) The Tough One Climb the rope or pole on the higher end of the obstacle, then go down the ladder and across the log platform. Climb over or between the logs at the end and go down the rope or pole to the ground. Vault over the final log.

Figure 3-36 is available in the illustration section at the end of the book.

Figure 3-37 is available in the illustration section at the end of the book.

Figure 3-38 is available in the illustration section at the end of the book.

Figure 3-39 is available in the illustration section at the end of the book.

Section IX Individual Exercise Programs

3901.GENERAL

Marines at times will **be** stationed on independent duty and consequently will be responsible for their own physical fitness program. This section will assist them in understanding the need for exercise and will aid in the planning and execution of an individual exercise program. Exercise activities included are the Bench Conditioner, 6-12 Plan, Weight Training, and Isometric Contraction. Keeping physically fit is a problem that faces every Marine. Even though we are frequently engaged in training that requires some physical effort, in many cases, it is not enough to prepare us to meet the intense physical demands of combat. Attaining a satisfactory level of physical readiness is not an insurmountable objective. Available time appears to be the most difficult obstacle to the development of physical readiness. In most cases, regular physical training programs are centralized, requiring the individual to temporarily leave the work area. The problems involved in setting an hour aside two or three times each are numerous. However, most of us can devote a few minutes each day to physical fitness with little, if any, impact on our daily work schedule, especially if it does not require us to leave our work area.

a. Type of Program. There are many good physical fitness programs available to the individual or group. Regardless of the type or duration, to be effective, the program must contain exercises that are strenuous and are challenging to the individual. Space will not permit the inclusion of all available means of individual exercise. The programs selected for this chapter have met the requirement of minimum space and minimum time.

b. Need to Augment Program. These programs are quite strenuous and will develop a satisfactory level of physical readiness. However, if the Marine desires additional development of endurance, it is recommended that he supplement these programs with a 15- to 30-minute period of wind sprints and double timing on an alternating daily basis.

c. Progressive Training. If Marines are performing duties which require little or no physical activity, they must plan a physical conditioning program that assures a moderate beginning, moderate but steady progression, and sufficient warmup before starting the vigorous exercise. To avoid organic of bodily harm, a Marine activity without adequate warmup. He should conduct conditioning programs on a daily basis over an extended period of time, never on an unduly accelerated or crash basis.

3902.THE BENCH CONDITIONER

a. Description and Objective. The bench conditioning program uses a modified bench to employ both isotonic (moving) and isometric (stationary) exercises are as the nucleus of the program. The exercises are designed to develop strength and endurance in all the major muscle groups of the body. The principles of progression, overload, and balance are employed when the exercises are performed properly.

b. Area and Equipment. The conditioning apparatus can be constructed in any unit motor pool with welding equipment found in most salvage yards.(See figs. 3-40, 3-41.)Additionally, there are available any number of commercially produced apparatus available. The important thing is that in utilizing this equipment the following exercise routine be adhered to.

c. Starting Level and Progression.. The program consists of two tables, each with 10 exercises. The Marine controls progression by required repetitions or, in some cases, by application of maximum effort. Each table can be completed within 15 minutes.

d. Starting Position. To start the program, the Marine begins with the Table 1 and executes each exercise for the required number of repetitions as indicated. The Marine controls the starting level and progression. When the Marine executes the maximum repetitions for Table I within a 15-minute exercise period, he progresses to Table II. To maintain this level of development, the Marine should also execute the maximum repetitions for Table II within a 15-minute exercise period. The Marine should keep substitution of exercises to a minimum. However, it he completes a full 15 minutes of strenuous exercise and exercises all muscles, then there should be no appreciable difference in the overall development.

e. Bench Conditioning, Table I

(1) Exercise 1: Side-Straddle Hop. This is a two-count warmup exercise done at moderate cadence. The starting position is the position of attention. On count ONE jump slightly into the air, swinging the arms out to the sides and up to a vertical position, hands touching.(See fig. 3-40, A.)At the same time, spread the feet wider than shoulder-width apart. On count TWO, using a slight flexing of the knees and ankles, jump slightly into the air and return to the starting position by swinging the arms back down to the sides. Twenty repetitions of this exercise is the standard dosage throughout the program.

(2) Exercise 2: Hand Walk. Remove the lower horizontal bar. Adjust the upper horizontal bar so that it is high enough to permit a "dead hanging" position, with the feet off the ground.(See fig. 3-40, B.j From the "dead hanging" position, release one hand and drop the arm to the side of the body. Then raise that arm and regrasp the horizontal bar. Release the bar with the other hand and drop that arm to the side. Repeat this as many times as possible.

(3) Exercise 3: Situps. Lie down with the fingers interlocked and placed behind the head. Hook the toes under the foot braces. Raise the trunk and upper body to an upright sitting position, twisting it to the left and then forward and downward until the right elbow touches the left knee.(See fig. 3-40, C.)Lower the body to the starting position .Sit up again but twist the body to the opposite direction as before, touching the left elbow to the right knee. Again lower the body to the starting position. The starting dosage is 20 situps. The Marine should continue the progression until he has attained 40 situps.

Figure 3-40 is available in the illustration section at the end of the book.

(4) Exercise 4: Double Step-upStarting at one end of the bench, step up onto the bench, and walk across it. Step down from the other end; turn around and repeat the process to return to the starting point.(See fig. 3-40, D.)Each return to the starting point constitutes a repetition. The starting dosage is 20 repetitions. Maximum dosage is 35 repetitions. This exercise should be done at a rapid cadence.

(5) Exercise 5: Isometric Bar Lift. Adjust the lower bar so that it is slightly higher than the beltline. Placing the feet on the footplates at the base of the frame, grasp the lower bar so that the hands are spread shoulder-width apart. Assume a crouched position and lift with maximum effort using the arms, back and legs.(See fig. 3-40, E.)Starting dosage is 4 repetitions of a stress time of 5 seconds followed by a 5-second rest prior to the next repetition. The Marine obtains progression by lengthening stress periods to 6 and later 7 seconds. Do not increase the number of repetitions.

(6) Exercise 6: Knee Lift. Adjust the upper bar to the same height used in Exercise 2.Adjust the lower bar so that is stops rearward movement of the hips when the "dead hanging" position is assumed.(See fig. 3-40, F.) Keeping the arms extended, flex the legs and raise the knees as high as possible. Hold this position for 5 seconds, then return to the starting position. After 2 seconds in the starting position, raise the knees again. Each return to the starting position constitutes one repetition. The dosage is five repetitions. Progression is the same as in Exercise 5.

(7) Exercise 7: Isometric Pull. Adjust the lower horizontal bar to a position where it is slightly higher than the beltline. Grasp the handles and pull outward.(See fig. 3-40, G.)Apply maximum effort and hold for approximately 5 seconds. Relax for 5 seconds between repetitions; perform four repetitions. Moving the body closer to or farther away from the bar will change the stress from the upper arms to the forearms. Progression is the same as in Exercise 5.

(8) Exercise 8: Isometric Compression. Maintain the position as in exercise 7.(See fig. 3-40, H.)Grasping the handles in the same manner, press in **with** maximum effort and hold for approximately 5 seconds. Relax for 5 seconds between repetitions. Progression is the same as in Exercise 5.

(9) Exercise 9: Isometric Press. Remove the lower horizontal bar. Adjust the upper horizontal bar until it is about 6 inches lower than the extended arms can reach. Stepping on the footplates at the bottom of the frame, grasp the bar with both hands and push up.(See fig. 3-40,1.)Keep both the legs and arms slightly flexed and the back straight. Apply maximum effort for 5 seconds then relax for 5 seconds. Complete four repetitions. Progression is the same as in Exercise 5.

(10) Exercise 10: Pushups. Grasping the foot braces with both hands, assume the front leaning rest position.(See fig. 3-40, J.)Keeping the back and legs straight, lower the body until the chest is lower than the **hands,** then return to the starting position. The Marine should complete the maximum possible number of repetitions.

f. Bench Conditioning, Table II. There is no limit on the maximum number of repetitions attainable in Exercises 3, 4, and 6 of Table II. The only limit imposed is that the entire program of 10 exercises outlinted in either table should not exceed 15 minutes.

(1) Exercise 1: Side-Straddle Hop. This is a two-count warm up exercise done at a moderate cadence. The starting position is the position of attention. On count ONE, jump slightly into the air, swinging the arms out to the sides and up to a vertical position, hands touching.(See fig. 3-41, A.)At the same time, spread the feet wider than shoulder-width apart. On count TWO, using a slight flexing of the knees and ankles, jump slightly into the air and return to the starting position by swinging the arms back down to the sides. Twenty repetitions of this exercise is the standard dosage throughout the program.

(2) Exercise 2: Pull-up. Adjust the horizontal bar so that is is high enough to permit a "dead hanging" position with the feet off the ground. Grasp the bar with both hands, palms facing forward. By flexing the arms, raise the body to a position where the chin is higher than the bar.(See fig. 3-41, B.)Then lower the body to the "dead hanging" position. Repeat as many times as possible.

(3) Exercise 3: Bench situps. Sit on the bench and hook the feet under the foot braces. With the fingers interlocked behind the head, lean back until the head touches the floor.(See fig. 3-41, C.)Return to the starting position. The starting dosage is 15 situps.

(4) Exercise 4: Step-Up. Face the bench and step up on it with one foot, bringing the trailing foot up next to the leading foot. Step back down again, leading with the same foot used first in stepping up.(See fig. 3-41, D.)Perlorm half of the total repetitions, then change the sequence of moving the feet to use the other leg in stepping up, and repeat the same amount of exercise. The starting level is a total of 40 step-up. **This exercise should be done at a rapid cadence.**

(5) Exercise 5: Isometric Bar Lift\djust the lower bar so that it is slightly higher than the beltline. Placing the feet on the footplates at the base of the frame, grasp the lower bar so that the hands are spread shoulder-width apart. Assume a crouched position and lift with maximum effort using the arms, back, and legs.(See fig. 3-41, E.)Starting dosage is four repetitions of a stress time of 8 seconds followed by a 5-second rest prior to the next repetition. Progression is obtained by lengthening stress periods to 10 seconds. Do not increase the number of repetitions.

Figure 3-41 is available in the illustration section at the end of the book.

(6)Exercise 6: Leg Lift. Adjust the bars and assume the starting position as shown in figure 3-41, F. Keeping arms and legs extended, raise the legs to a horizontal position and hold in that position for 2 seconds. Then lower the legs slowly to the starting position. Five repetitions is the starting level.

(7) Exercise 7: Isometric Pull. Adjust the lower horizontal bar so that it is slightly higher than the beltline. Grasp the handles and pull outward.(See fig. 3-41, G.)Apply maximum effort and hold for approximately 8 seconds. Relax for 5 seconds between repetitions; perform four repetitions. Moving the body closer to or farther away from the bar will change the stress from the upper arms to the forearms. The Marine obtains progression by lengthening the stress period to 10 seconds.

(8) Exercise 8: Isometric Compression. (See fig. 3-41, H.)Maintain the position as in exercise 7.Grasping the handles in the same manner, press in with maximum effort and hold for approximately 8 seconds. Relax for 5 seconds between repetitions; perform four repetitions. The Marine obtains progression by lengthening the stress period to 10 seconds.

(9) Exercise 9: Isometric Pres. Remove the lower horizontal bar. Adjust the upper horizontal bar until it is about 6 inches lower than the extended arms can reach. Stepping on the footplates at the bottom of the frame, grasp the bar with both hands and push up.(See fig. 3-41,1.)Keep both the legs and arms slightly flexed and the back straight. Apply maximum effort for 8 seconds, then relax for 5 seconds. Complete four repetitions. The Marine obtains progression by lengthening the stress period to 10 seconds.

(10) Exercise 10: Inclined Pushup. Assume the front leaning rest position with the feet on the bench.(See fig. 3-41, J.)Keeping the back and legs straight, lower the body until the nose touches the ground. By extending the arms, raise the body to the starting position. Repeat as many times as possible.

3903. THE 6-12 PLAN

a. Description and objective. The 6-12 plan of physical fitness has been developed to assist in regulating quantity and progression and to provide a convenient set of exercises. This is a basic program and will take 18 weeks to complete if you follow the moderate progression as to the time prescribed for each level of achievement. The time can be shortened as explained below. This plan consists of six basic exercises a day which can be completed in 12 minutes. There are six tables of six exercises each, thus allowing you to progress from table to table. The plan is progressive, fits any age group, contains balance and variety, and applies the principle of overload in a safe and gradual manner. Begin at Table I, Progression Guide, with the number of repetitions as indicated by age.

b. Formation. If just starting an exercise program, do not rush through the first table. Remember, individuals should remain at each level for about a week before moving upward. The time allotment stated for each exercise at the bottom of the tables is a guide; some people may take more and some less time on the individual exercises. At the end of a one-week period, when the individual can comfortable perform the six exercises in 12 minutes, move on to the next level. To a certain degree, the individual must be the judge of his ability to progress form level to level and table to table. If starting with a certain degree of fitness, some of the beginning tables may present little challenge.

c. Starting Level **and Progression.** There are three levels of achievement for each age group, indicated as A, B, and C. Start at the C level for the appropriate age group. At the end of a one-week period, or when the individual can do all exercises at that level within 12 minutes, progress to the B level. At the end of the second week, or when the individual can accomplish that level within 12 minutes, progress to the A level. At the conclusion of the third week or when the individual can achieve the A level within the time limitation, move on to table II.(See Tables I through VI, Progression Guide.)

d. Maintenance Level. Attempt to work through all six tables. If this proves to be too difficult, then maintain exercise at the—

(1) A-level on Table IV, Progression Guide it in the 45 to 49, 50 to 59, or over 60 age group.

(2) A-level on Table V, Progression Guide if in the 17 to 29, 30 to 39, or 40 to 44 age group.

e. Precautions. To achieve the maximum benefit, perform each exercise exactly as specified. Read the descriptions and study the illustrations. Do not slight the movements. Use a sensible approach and follow these four points as they apply before starting or during your exercise program.

(1) If you have the slightest doubt about your ability to participate in this exercise program, consult a physician.

(2) Stop immediately if you notice unusual breathlessness or chest pain while taking part in these exercises. If these conditions persist, consult a physician.

The following Tables of Progression Guides and Exercises are the progression of the 6-12 Plan program.

Table I. Progression Guide is available in the illustration section at the end of the book.

Table I. Exercises.is available in the illustration section at the end of the book.

Table II. Progression Guide.is available in the illustration section at the end of the book.

Table II. Exercises.is available in the illustration section at the end of the book.

Table III. Progression Guide.is available in the illustration section at the end of the book.

Table III. Exercises.is available in the illustration section at the end of the book.

Table IV. Progression Guide.is available in the illustration section at the end of the book.

Table IV. Exercises.is available in the illustration section at the end of the book.

Table V. Progression Guide.is available in the illustration section at the end of the book.

Table V. Exercises.is available in the illustration section at the end of the book.

Table VI. Progression Guide.is available in the illustration section at the end of the book.

Table VI. Exercises.is available in the illustration section at the end of the book.

(3) Unless you have exercised regularly and know yourself to be in good physical condition, start at table I with the C level appropriate to your age.

(4) If you are out of shape, admit that fact to yourself. Hide your pride; after all, you are in the privacy of your own quarters. Set your goal for the longer, steadier pull toward fitness. Resist the urge to pass over the lower numbered tables to find a table that will test your fitness. You are not trying to test, but rather to develop.

3904.WEIGHT TRAINING

Weight or barbell training should not be confused with the more common types of weight lifting used as a competitive sport. Weight lifting is designed to develop specific muscle groups so that the individual is capable of lifting a large amount during a single lift. In contrast, weight training is the systematic development of all the major muscle groups by the use of calisthenics reinforced with weight to provide resistance.

3905.WEIGHT LIFTING

a. Description and Objective. The weight lifting program is progressive and applies the principle of overload in a safe, gradual manner. The exercises of the table can be completed in 15 minutes. The objective of the exercises is to develop strength and muscular endurance, and muscle tone of the five major muscle groups:legs, arms, back, trunk, and shoulder girdle.(See fig. 3-42.)

b. Warmup. A warm up exercise is important to prepare the body for the more vigorous exercises that are to follow. Ten repetitions of the high jumper exercise are excellent for a warm up period. Also, an exercise which will require fast body movement is needed.To provide such exercise, 3 to 5 minutes of rope skipping is recommended to increase the individual's development potential.

c. Formation. Marines should take care in the completion of these weight lifting exercises. They should ensure that their back is straight during the lifting phase of all exercises. When exercises require assuming the standing position with the weight, they should always grasp the weight while in a squatting position and then rise to a standing position.

d. Progression. Each exercise has a starting number of repetitions and specified pounds of weight. After each 4th or 5th day of exercise, the Marine should increase the repetitions by one until he has reached the maximum of 10 repetitions. At this time, the Marine should increase the weight by 5 pounds. Then he should repeat the process again with the initial number or repetitions.

e. Weight Training

(1) Exercise 1:**Squat**(See fig. 3-42, A.)Starting level—six repetitions, 50 pounds (commonly called the flatfoot deep knee bend). Place the bar upon the shoulders. Stand with feet about 18 inches apart. Keeping the feet flat, lower the body into the low squat position. Stand erect and repeat. Exhale while lowering into the squat position and inhale while standing. This constitutes one repetition.

(2) Exercise 2:Waist Bender(.See fig. 3-42, B.)Starting level-6 repetitions, 40 pounds. Assume the standing position with the bar across the shoulders and the feet a shoulder-width apart .Bend forward at the waist until the upper body is parallel to the ground; return to the starting position. Each return to the upright position constitutes one repetition.

(3) Exercise 3:Curl (See fig. 3-42, C.)Starting level—6 repetitions, 40 pounds. Grasp the barbell with the palms facing to the rear and assume the standing position with the feet a should-width apart. With the barbell held in front of the hips, flex the elbows and lift the weight until the bar touches the upper chest. Lower the barbell back to the hip level position. Inhale deeply with the upward movement and exhale on the downward movement. Each time that the bar touches the chest will constitute one repetition.

(4) Exercise 4:Side Bender(.See fig. 3-42, D.)Starting level—6 repetitions per side, 40 pounds. Assume the standing position, with the bar across the shoulders, with feet a shoulder-width apart. Bend to the left as far as possible and return to the starting position. Repeat six times and then execute the same procedure to the right for six repetitions.

(5) Exercise 5:Standing Press(See fig. 3-42, E.)Starting level—6 repetitions, 45 pounds. Grasp the bar with the palms facing forward and assume the starting position. Curl the weight to the upper chest position. Inhale deeply and press the bar upward to an overhead position. Exhale while lowering the bar to the chest position. Each time that the bar is pressed upward constitutes one repetition.

(6) Exercise 6:Upward Row. .see fig. 3-42, F.)Starting level—6 repetitions, 40 pounds. Grasp the bar, hands close together, palms to the rear, and assume the standing position. Starting with the bar held in front of the hips, flexing the elbows and the shoulder girdle muscles, lift the bar straight up to an overhead position. Inhale deeply while lifting the bar. Exhale while lowering the bar to the hip position. Each time that the bar returns to the hips constitutes one repetition.

(7) Exercise 7:Shoulder Curl. See fig. 3-42, G.)Starting level—6 repetitions, 25 pounds. Grasp the bar, palms down, and assume the standing position. Keeping the elbows locked, curl the bar, pivoting the arms at the shoulders until the bar is in an overhead position and as far to the rear as possible. Return the bar in the same manner to the hip position. Each time that the bar returns to the hip position constitutes one repetition.

Figure 3-42 is available in the illustration section at the end of the book.

3906.ISOMETRIC EXERCISING

a. Description and Objective. Isometric exercising is the application of maximum effort during an exercise period. The isometric principle is to apply force gradually over a 5- to 10- second period until the maximum effort is applied. Relaxation follows for approximately 5 seconds and then force is applied again. The Marine continues this process at the prescribed level for each exercise for a period of 15 minutes or less. The objective of isometric exercises is to create muscle growth. Isometric exercises are the fastest means of creating muscle growth. They are founded on the fact that a muscle will grow only so fast regardless of the type or duration of the activity. The principle of overload—that the muscles develop commensurate with demand—reinforces that fact. However, isometrics will not develop cardiovascular or muscular endurance. Consequently, Marines who choose to use isometrics in their exercise routine must also include running or some other type of aerobic activity.

b. **Area and Equipment** isometric exercises may be designed to be performed with or without equipment.

c. Use With Other Programs. The isometric exercises presented in this section will provide a variety from which to choose. In addition to the exercises contained in this section, the application of isometric force is used in some of the exercises in section II.

d.Door Frame Exercises. The following exercises are designed for use with a standard door frame found in all offices or barracks.(See fig. 3-43.)

(1) Exercise l:Arm Press. Stand in the doorway with the legs straight, knees locked. Using the arm muscles, press hard upward against the top of the door frame. Repeat for three repetitions applying gradual effort to maximum contraction.

(2) Exercise 2:Leg Press. Stand in the doorway with the hands on the top of the door frame, elbows locked .With knees bent, press hard with the leg muscles .Repeat for three repetitions beginning with a gradual effort and increasing to maximum contraction. A low platform may be necessary to reach the top of the door frame and still maintain a bent knee position.

(3) Exercise 3:Side Press. Extend both arms to the side of the doorway. Palms are shoulder high, facing outward. With both arms, press hard against the sides of the door frame. Repeat for three repetitions. Begin gradually and increase to maximum contraction.

(4) Exercise 4:Lateral Raise. Extend both arms to the sides of the doorway, arms down, palms facing inward. With the back of the hands, press hard against the sides of the door frame .Repeat for three repetitions. Begin with a gradual effort and increase to maximum contraction.

Figure 3-43 is available in the illustration section at the end of the book.

(5) Exercise 5:Neck Press?lace the forehead against the door frame, hands clasped behind the back Using the neck muscles, press hard against the door frame. Repeat for three repetitions, then reverse position so that the back of the head is resting on the door frame. Again do three repetitions .Begin gradually with both exercises and increase to maximum contraction.

(6) Exercise 6:Door Pull Stand facing the edge of the open door and grasp the doorknobs. Pull outward with both arms (if doorknobs are not available grasp the edge of the door).Applying outward pressure, move the body toward and away from the door .Repeat the three repetitions. Begin with gradual effort and increase to maximum contraction.

Chapter 4
COMBAT WATER SURVIVAL
Section I. Marine Corps Water Survival Program

4101.MILITARY SWIMMING

Military swimming emphasizes strokes that result in staying power rather than speed. Marines need to swim easily, quietly, and with adequate vision. Strokes should be used that allow the Marine to carry basic combat equipment and to tow or push a wounded buddy while keeping the face out of the water to allow breathing. The best strokes to use are the side stroke and the breast stroke. Marines must be proficient in floating, drown-proofing, and in using their basic combat equipment to make flotation devices. They must be drilled on using the Kapok and Mae West buoyancy compensators and on emergency egress from helicopters, landing craft, and amphibious assault vehicles.

4102.PROGRAM DEVELOPMENT

Commanders of the Landing Force Training Commands (Atlantic/Pacific) are tasked with developing and maintaining the individual water survival and swimming training program. They are responsible for training all Marine Corps water safety and survival instructors and integrating this training in to their combat readiness training program. All Marines are deployable and as soldiers of the sea must constantly practice the skills necessary to achieve mission accomplishment. Commanders must be imaginative in program development, integrating basic skills development which enhances the total mission accomplishment. An example is teaching waterproofing of the Marine's backpack so that it can be used as a flotation device in streams or rivers. Once learned, this skill would be practiced first in a pool and then in a stream or a river. Routine inspections during deployments would ensure that backpacks could be used as flotation devices.

4103.QUAL STDS AND TRAINING GUIDELINES

The inherent nature of Marine Corps operations and training requires that Marines achieve an ability to survive in water. Water survival and swimming training is designed to reduce fear of water, instill self-confidence, and develop a Marine's ability to survive in water. As a minimum, every Marine should be qualified as swimmer, third-class (S3).This qualification should be met during recruit training or Officer Candidate School. Once qualified, a Marine need not be requalified unless requalification at a specific level (S3 or higher) is required by unit mission, military occupational specialty or duty assignment, or for other reasons determined by the commander. The following classifications and standards apply to Marine Corps water survival and swimming training. Abbreviations will be used for service, medical, and training record entries citing MCO 1510.29A, Individual Water Survival and Swimming Training, as authority.

a. Qualification standards and test procedures are described in paragraph 4104.

b. Swimmer classification and abbreviations are:

(1) Unqualified - UQ.

(2) Swimmer, third-class (S3).Minimum aquatic skill level for all Marines.

(3) Swimmer, second-class (S2)-

(a) Minimum standard for all naval aviators, naval flight officers, and aircrewmen with written waiver from the commanding officer.

(b) Requirements for military occupational specialty 1803 (assault amphibian vehicle officer), 1833 (assault amphibian vehicle crewman), 1302 (engineer officer), 1371 (combat engineer), 1381 (shore party specialist), and 1379 (engineer operations chief).

(4) Swimmer, first-class (S1)-

(a) Naval aviator, naval flight officer, and aircrewman qualification requirement.

(b) Requirements for military occupational specialty 0321 (reconnaissance man), 8652 (reconnaissance man, parachute jump qualified), 8653 (reconnaissance man, self-contained underwater breathing apparatus qualified), 8654 (reconnaissance man, parachute/self-contained underwater breathing apparatus qualified), 9952 (self-contained underwater breathing apparatus Marine (officer/enlisted)), and 9953 (parachutist/self-contained underwater breathing apparatus Marine (officer/enlisted)).

(5) Water Survival Qualified (WSQ).Ultimate water survival/swimming goal for Marines.

c.Instructor and assistant instructor classifications are-

(1) **American Red Cross Advanced Lifesaver.** Individuals who are currently certified as American Red Cross advanced lifesavers may assist qualified instructors during Marine Corps water survival/swimming training and testing. Certification is valid for 3 years.

(2) American Red Cross Water Safety Instruction. Individuals who are currently certified as American Red Cross water safety instructors may conduct swimming training and qualify individuals as 3d, 2d, and 1st class swimmers. Certification is valid for 2 years.

(3) United States Marine Corps Water Safety and Survival Instruction. Those personnel certified as water safety and survival instructors by one of the landing force training commands may conduct Marine Corps water survival and swimming training and American Red Cross swimming and water safety training. These instructors may also participate as assistant instructors during the conduct of landing force training command-sponsored water safety and survival instructor courses. Certification is valid for2 years.

(4) American Red Cross Water Safety Instructor Traine. Individuals qualified who have been certified as instructor trainers by the American Red Cross may train American Red Cross water safety and basic swimming instructors. Certification is valid for 2 years.

(5) United States Marine Corps Water Safety and Survival Instructor Traine. Individuals qualified who are certified as instructor trainers through special training courses conducted jointly by the American Red Cross and the Marine Corps, under the direction of the Commandant of the Marine Corps, may train water safety and survival instructors and American Red Cross water safety and basic swimming instructors. Certification is valid for 2 years.

d. Training Guidelines Instructors engaged in training Marines as swimmers will use the current edition of the American Red Cross swimming and water safety courses instructors' manuals as the authoritative textbooks for swimming procedures, strokes, breaks, and holds.

4104.WATER SURV/SWIM QUAL STD AND TEST PROCS

a. The following qualification standards and test procedures are applicable to Marine Corps water survival and swimming training. The utility uniform without cover, boots/oxfords or socks will be worn for all swimming tests through swimmer, first class. The utility uniform with boots/oxfords will be worn as prescribed for the water survival qualification.

(1) Swimmer, Third-Class (S3) To qualify as a swimmer, third-class, a Marine must enter the water feet first from a minimum height of 5 feet and remain afloat for 5 minutes. During this time the Marine must swim 50 yards using any stroke or combination of strokes. This test should be taken by all Marines as early as possible in their initial training. Those unable to pass this test will be classified as UQ and should be given instruction in fundamental swimming skills. Those who barely meet requirements are swimmers who need help and should receive additional instruction in fundamental swimming skills, whenever possible.

(2) Swimmer, Second-Class (S2.) To qualify, a Marine must enter the water feet first from a minimum height of 10 feet and remain afloat for 10 minutes. The Marine must swim 100 yards and use the three basic survival swimming strokes (side stroke, backstroke, and breast stroke) for a minimum distance of 25 yards each.

(3) Swimmer, First-Class (S1)A prerequisite to qualification as swimmer, first-class, is successful completion of the test for swimmer, second-class. To qualify as a swimmer, first-class, a Marine must be able to do each of the following:

(a) Approach a person of approximately the same size while in the water, demonstrate a release, get that person in a carry position, and tow the "distressed swimmer" 25 yards.

(b) Enter water feet first and immediately swim underwater for 25 yards. Swimmer is to break the surface twice for breathing during this distance at intervals of approximately 25 feet.

(c) Remove trousers in water, inflate for support, and remain motionless for a minimum of 1 minute.

(d) Swim 220 yards using any survival swimming stroke or any combination of survival strokes.

(4) Water Survival Qualification (WSQ)A prerequisite to water survival qualification is successful completion of the test for swimmer, first-class. To be water survival-qualified, a Mannemust:

(a) Enter water by jumping from a height of 10 feet.

(b) Stay afloat in full utility uniform (less cover) for 1 hour. Boots/oxfords will be removed after 5 minutes in the water but will be retained by the swimmer.

(c) Traverse 75 yards in deep water in full utility uniform (less cover) with boots/oxfords.

b. All water survival and swimming training will include instruction in the following:

(1) Employment of safety lines or other comparable expedients to assist in swift stream/river crossings.

(2) Adverse physiological effects caused by cold water (hypothermia) and the precautionary measures to be taken prior to exposure to such an environment.

(3) Employment of a standard and expedient flotation devices in a water survival situation.

(4) Removal of individual field equipment after unscheduled water entry. Entry will be made from a minimum height of 10 feet with a complete set of field equipment, properly packed as an expedient flotation device and waterproofed. The ALICE pack will then be removed, the rifle laid across the top of the pack, the helmet remaining or put back on the individual's head, and the individual will traverse 25 yards to the side of the pool or water's edge.

c. The additional water survival qualification (flight physiology training) and requalification required of personnel assigned to flight status will be conducted in accordance with OPNAVINST 3710.7_,Promulgation of NATOPS General Flight and Operating Instructions. Swimming qualification (first-class, SI) is a prerequisite to water survival (flight physiology) training and is a one-time requirement.

Section II. Water Sports and Competitive Activities

4201.GENERAL

The commander, when developing the combat readiness program, needs to look carefully at the mission. If the mission requires Marines who must perform mission-essential water-related activities of long duration, he may want to look to water sports/competitive activities as a method of developing endurance. Endurance and confidence can only be developed by extended time in the water. While defined training can accomplish this goal, the goal can be enhanced through training which is integrated with the physical conditioning program. This approach will provide variety and enthusiasm. Two activities which will accomplish this goal are water relays and water polo. The important consideration is that the unit achieve total participation by all members of the unit.

4202.WATER RELAYS

Development of water relays will generally conform to the guidelines in chapter 5, section IV. When contemplating relays, the commander should review chapter 5 and adapt the principles to the size of the unit and the facility to be used.

4203.WATER POLO

The official rules provided in this paragraph will need to be adapted to the unit's unique situation, its size, and the size of the facility. Commanders should make every effort to have all members of the unit participate.

a. NCAA Rules This style of play is based on two popular sports: swimming and basketball. Playing is done on the surface of the water by teams of seven players each. The size of the playing area is 25 yards by 14 yards, which is the size of the typical high school or college indoor pool. The deeper the water, the better. If the entire playing area is deep-6 feet or more—this is ideal, although most indoor pools have a shallow end. At each end of the pool is a goal. In deep water, the goal is 10 feet across and 3 feet high. In shallow water, where the goalie can stand on and jump from the bottom, the goal is 10 feet across and 5 feet high. One of the seven players on each team is the goalie. He is the only player on the team who can stand on or jump from the bottom (if it is shallow enough to permit this), catch and pass the ball with both hands at the same time, or hit the ball with a clenched fist. However, the goalie may not swim across or pass the ball across the middle of the pool.

b.Equipment. The goalie on the visiting team must wear a white cap with the number 1 clearly marked on it; his teammates, who are designated as guards and forwards, must wear white caps numbered 2 through 7, with the substitutes wearing white caps numbered 8 through 21.The goalie on the home team must wear a dark-colored cap with the number 22 clearly marked on it; his teammates designated as guards and forwards, must wear similar dark-colored caps numbered 29 through 42.The guards and forwards on each team may swim freely up and down the pool as they see fit, interchanging positions as often as they wish. They may not stand on or jump from the bottom, touch the ball with more than one hand at a time when catching or passing or shooting it, or enter inside the opposing team's 2-yard line unless preceded by the ball. The ball is similar to a soccer ball, except that it is yellow and covered with a waterproofing substance that makes it easier to handle with one hand.

c. Playing Time. In intercollegiate competition the game consists of four 7-minule quarters, the teams changing ends after every period of play. In high school competition the game consists of four-,five-, or six-minute quarters, depending on the ages of the participants. Here again the teams change ends after every period. There is a very brief interval between each quarter. At the start of each period the teams line up at their respective ends, and with ablast of the whistle, the head referee throws the ball into the middle of the pool. The fastest swimmers on each team then race to gain possession.

d. Officials. The head referee walks along one side of the pool watching for infractions. He is aided by the assistant referee, who patrols the opposite side of the pool. The referees are both equipped with whistles and two official's flags. These flags should be 12 inches square and should correspond in color to the caps worn by the respective teams. Whenever a referee sees an infraction, he blows his whistle and signals with his flag indicating that the player nearest to the ball with the appropriate colored flag will be given possession of the ball. Whenever the whistle is blown, all players should react by looking immediately to see which color flag is being waived which will indicate which team should be moving into the defense. Other officials are a timer and a scorer. As in all sports, it is important that the officials be skilled and competent.

e. Technical Fouls. The following are some of the common technical fouls: starting before the referee blows his whistle to open play; holding onto or pushing off from the side of the pool during play; taking hold of the ball underwater when tackled by an opposing player; swimming inside an opponent's 2-yard line unless preceded by the ball; touching the ball with both hands at the same time (goalie excepted); standing, walking on or jumping from the bottom when taking an active part in the game. When a technical foul occurs, thereferee blows the whistle and with his flag awards possession. The player on the team awarded possession nearest the point of infraction then has 5 seconds to put the ball into play by passing to a teammate or by dropping the ball into the water and swimming after it.

f. Personal Fouls. The following are some of the more common personal fouls: committing any of the technical fouls mentioned above for the purpose of scoring or preventing a goal; holding, ducking, pulling, pushing off from, swimming over, or impeding the arm or leg movement of any opponent who is not touching the ball; and splashing water in the face of an opponent. When a personal foul occurs, the referee will follow the same procedure as with a technical foul. While doing this, the referee will also call out clearly and loudly the number of the player who was guilty of the infraction and a personal foul will be marked against that player by the official scorer. When a player accumulates five personal fouls, he fouls out of the game and must be replaced by a substitute.

g. Penalty Shot. A penalty shot can be awarded by either of the officials or the scorer when:

o An offensive player inside the opponents' 4-yard line not touching the ball is held, ducked, pulled-back, kicked, or struck.
o A team has accumulated a total of ten personal fouls. When the former occurs, the head or assistant referee should immediately blow the whistle and by holding a two-flagged stick in a vertical position above his head, signal that a penalty shot has been awarded. When the latter occurs, the scorer should use a buzzer located at the scorer's tabic to signal that one team has accumulated ten personal fouls, thereby entitling the other team to a penalty shot. A penalty shot is taken from the 4-yard line in front of the goal. All players except the defending goalie must leave the 4-yard line until the shot is taken. No player can be within 1 yard of the shooter. After ascertaining that the shooting player is on the 4-yard line and the goalie is onThegoal line, the referee will ask the shooter to lift up the ball. When he docs, the referee will give a sharp blast from the whistle. At the whistle, the shooter must shoot without delay and without any faking at the goal. The goalie may try to block the shot. If he does, or the shot is otherwise missed, it is immediately in play and action continues.

h. When a Goal is Scored. When a goal is scored, either from a shot taken by a player out in the field or by penalty shot, the defending goalie must then pass the ball to the nearest referee. The referee will then pass the ball back to the goalie, who puts it into play by a pass to one of his teammates as soon as possible.

Section III. The Battle Swimming Test

4301.GENERAL

The battle fitness test is the evaluation contained in the Guidance for Basic Warrior Training (BWT) as part of the Basic Warrior Training Concept Plan.

4302.THE BATTLE SWIMMING TEST

The battle swim test consist of the following:
a. Be able to climb a three meter high board, while wearing a field uniform, carrying a M-16, web gear, and four canteens. Jump off blindfolded, recover and swim 100 meters with the M-16.
b. Repeat the same process, without the blindfold, and upon entering the water a second time, recover and retrieve all submerged gear.
c. Remain afloat and tread water for five minutes.
d. Demonstrate a knowledge of how to cross a stream using field expedient techniques and a single strand rope bridge.
e. Successfully complete helicopter egress training.

Chapter 5
COMPETITIVE CONDITIONING ACTIVITIES
Section I. Organization of Competitive Activities

5101.LEADERSHIP OF COMPETITIVE ACTIVITIES

a. Place in the Program. Competitive conditioning activities consist of dual combatives, relays, team contests, and team sports in which individuals or teams compete against an opponent to win. Competition is one of the best ways of maintaining interest in the physical training program during the sustaining stage. Organized competition provides enjoyable, vigorous physical activity that has proved to be one of the best supplements to conditioning drill activities. The benefits of competition are the development of aggressiveness, unit pride and identity, teamwork, and the will to win. These activities help develop cardiovascular and muscular endurance, strength, and coordination. Scheduling competitive activities in an orderly and progressive manner is desirable. The progression should be from relays to dual combatives, to team contests, and finally to team athletics. These activities should be conducted as part of the program after a basic level of conditioning has been developed, usually during the slow-improvement stage of conditioning. Muscles and joints should be strengthened by preconditioning to withstand the strain placed upon them by sudden stops and turns, body contact, bearing of weight, and falls. Competitive activities, however, should not be allowed to dominate the physical training program.

b. Area and Equipment. Some of the competitive activities included in this chapter require specific types of areas and equipment; others do not. The area requirement can usually be satisfied on available training fields. When items of equipment are required or specific courts or field layouts are to be marked off, such information is included.

c. Leadership. The principal factor for success as a leader of competitive activity is an energetic, dynamic, enthusiastic approach. The leader's attitude is reflected by the group, so the commander must carry on the activity in a snappy and vigorous manner. Confidence on the part of the commander will create an impression of decisiveness and certainty. Confidence grows out of experience and a thorough knowledge of the activity. Mastery of subject is the first step in developing confidence, assurance, and poise.

(1) The following suggestions are offered for leaders of competitive combatives, contests, and sports:

(a) Get the activity underway quickly by selecting and teaching only the essentials.

(b) Use rules to add to the enjoyment of the activity and do not allow them to interfere with the spirit of competition.

(c) Stop the activity before interest begins to lag.

(d) In team contests, clearly distinguish sides by attempting to maintain unit identity.

(e) Always insist on fair play, enforcing the rules impartially.

(2) The following procedure is recommended for presenting a competitive activity.

(a) Name the activity.

(b) Briefly explain the objective of the activity and give only the pertinent rules.
(c) Have a demonstration at slow speed and answer questions.
(d) Organize groups into teams and appoint captains.
(e) Arrange teams in the proper starting positions.
(f) Conduct the activity.

d. Competitive Units. Units for competition should be the same organizations in which Marines train: battalions, companies, platoons. In most situations, the unit is the squad.

e. Provisions for Instruction. One of the most effective methods of maintaining interest and participation in competitive activities is to provide instruction in activities with which most Marines are unacquainted. Such instruction can be conducted during regular physical training periods. Careful planning is required to keep all individuals continuously engaged in vigorous activity. The materials in the following sections should be used as a guide for instructional purposes.

f. Officiating and Control. Every effort should be made to provide good officials for all competitive activities. Poor officials quickly cause dissatisfaction among participants in team activities and create a situation which can turn an organized activity into a brawl. Each company should have or develop several competent and qualified officials available for games on company and platoon levels.

5102.MILITARY FIELD MEETS

When units reach the latter part of the slow-improvement and sustaining stage of training, interest in the program may lag. A change in course content can arouse the desire to participate. An event which does not require a high degree of skill yet demands strenuous activity is ideal for the military field meet. This is a series of team contests conducted on a station-to-station basis during a given period of time. Team contests carried on simultaneously provide essential training, vigorous exercise, and stimulating competition in an atmosphere that is enjoyable for all. Because of the healthy rivalry that a field meet arouses, it is an excellent form of inter-unit competition.

a. Objective and Advantages. The objective of the military field meet is to provide activity for everyone in the participating units. Activities are chosen that will develop aggressiveness, teamwork, a will to win, and competitive spirit, and that will stimulate interest and build esprit de corps. The military field meet can **be** included as part of the physical training program or as part of the off-duty recreational program. It is a form of contest that can be conducted in nearly all circumstances because it can be easily modified, requires a minimum amount of equipment, and can be readily organized.

b. Level of Competition. The military field meet is flexible. It can be adjusted to large or small groups. A company-size unit is the most desirable, but it may also be administered within a larger unit. If it is conducted within a company, the participating units will be squads. If it is held within a larger unit, platoons compose the teams. A larger-size unit requires more extensive organization and administration.

c. Selection of Events. In organizing a military field meet, select events that are simple and easy to administer. All rules and regulations should be clearly understood by everyone. No event should require previous practice. In selecting the events, consider the interest and capabilities of the Marines and available equipment and facilities. Select events which will require that all members of the units participate. Events should not be dominated by the athletes; rather, MOS-related events which have significance in producing skills and combat readiness should be used.

d.Equipment and Facilities. The site must be large enough to permit events to be grouped about a central control point. Use available facilities such as parade decks, Softball fields, or tracks. A public address system is desirable at the control point for the initial orientation of teams and for subsequent announcements of time lapses, cumulative scores, and final standings. A tally board is necessary so that all units can constantly monitor how the events are progressing.

e.Administrators. Efficient administration of the military field meet depends on the referees, judges, and scorers. For these positions, choose individuals within the units who have had athletic or officiating experience. Prior to the day of the event, all administrators should be briefed and assigned a specific task in order that they may become familiar with the rules and organization of the contests they will conduct. The following officials should be available:

(1) A primary instructor or supervisor who is in charge of the control center and who is responsible for the successful operation of the athletic carnival. The supervisor must have an assistant to act as a timer and scorer.

(2) One assistant instructor in charge of each event. This person should be-

(a) Familiar with the rules of the event.

(b) Effective as a leader to ensure proper supervision and control over the participating teams.

(c) Enthusiastic to provide proper motivation.

(d) Self-confident of the ability to judge infractions of the rules. The assistant instructor must be fair in judgment and penalize without hesitation when infractions occur.

(3) Runners between event stations and the control point. These individuals collect and deliver scores.

f. Team Organization for Competition. The size of the teams is determined by the level on which the field meet is organized. Maintaining the integrity of the unit promotes esprit de corps, but this does not preclude grouping two squads into one team.

g.Conduct of Events

(1) The assistant instructor at each station takes charge of the group and gives a brief explanation of the major rules of the event. The assistant instructor speaks clearly and distinctly from a position to be seen and heard by all. A short demonstration is desirable if it will help clarify the event.

(2) The assistant instructor should make certain that teams can be clearly distinguished by the use of, for example, T-shirts and fatigue jackets, or caps and no cap, or colored jerseys. The assistant instructor should teach a whistle response (teams stop play immediately upon hearing whistle), get the event started as quickly as possible, and make any necessary corrections as the contest progresses. The rale of good officiating is to use a minimum of calls, yet maintain control of the contest. Penalize when necessary, but refrain from disqualifying contestants or teams.

(3) Keep the activity moving as rapidly as possible. When the central control point sounds the whistle to stop the play, all competition ends immediately. The assistant instructor then assembles the group, forwards the team scores to the central point, and, upon a signal from the central control point, rotates the teams to their next station. It is essential that rotation and orientation of teams be quick and orderly to reduce time spent between contests.

(4) Upon completion of the final event, assistant instructors move their teams to the central control point for the announcement of winning teams, presentation of awards (if any), and final critique.

h. Scoring System. The system for determining the winner of the military field meet should be simple and efficient. At the completion of competition, the scorer totals the points that each team has scored in all contests. The scorer then subtracts the number of points scored against a team (penalty points for nonparticipation,etc.) from the total. The resulting scores are placed in a column with the highest score at the top and the lowest at the bottom. The team with the highest total is the winner. This type of scoring system encourages a team to ensure total unit participation.

Section II.Combatives

5201.DESCRIPTTON AND OBJECTIVE

Combatives are strenuous, short competitive contests in which two individuals attempt to overcome each other in a bout of skill and strength. These contests help to develop Marines' resourcefulness, confidence, strength, agility, coordination, and will to win. Any level ground area can be used. However, extremely hard ground should be avoided as some of the combatives require ground contact. A whistle is needed to control the bouts during the competition since voice commands may go unheeded. The extended rectangular formation is used for dual combatives. Combatives are conducted on an informal basis. Marines are allowed to remain at ease between activities and are allowed to brush themselves off after being on the ground. To get the most out of combatives, individuals must be urged to overcome their opponents as quickly as possible as would be required in combat.

a. Benefits for Marine Training. The possibility of close contact with an enemy in combat faces Marines at all times. They must be trained to react aggressively and violently. Combatives may be used as an introduction to such hand-to-hand contact and should be followed by hand-to-hand combat training. Marines enjoy competition, and this type of activity is a welcome change from formal conditioning activities.

b. Instructor Responsibility

(1) The instructor tells the individuals that all combatives begin and end on the whistle signal. The instructor demonstrates each activity before having it performed, explaining it in simple terms.

(2) After stopping one activity, the instructor gets everyone in place for the next.

(3) The instructor must closely supervise combatives to insure that contestants do not use unfair or unsportsmanlike tactics. To avoid unnecessary injury, instructors must see that the bouts are closely controlled and opponents equally paired. Adjustments should be made in apparent cases of mismatched abilities.

5202.COMBATIVES TABLES

There are three tables of combatives, each of which can be completed in 15 minutes. These tables become progressively more difficult from lower to higher numbers.

a.Combatives Table I

(1) Open-Hand Slap Boxing(See fig. 5-1, A. individuals assume a boxer's stance, palms open, fingers extended and joined. Each contestant tries to slap the opponent about the head and upper body with the open hand. This is a good warm- up activity.

(2) Wrist Tug-of-War(See fig. 5-1, B.)Two individuals sit on the ground with the soles of their feet in contact. Each grasps the opponent's wrists so that the hands are directly over the feet. At the whistle, each individual tries to pull the opponent from sitting to standing position.

(3) Arm-Lock Wrestling.(See fig. 5-1, C.)Two individuals sit back-to-back with legs spread and arms locked at the elbows. Each right arm is inside the opponent's left. Each tries to force the opponent's left arm or shoulder to the ground. The individual who first wins three bouts is the winner.

(4) Bulling(See fig. 5-1, D.)Two individuals assume the Westmoreland wrestling hold, each grasping the opponent's neck with the right hand and the opponent's right elbow with the left hand. Each tries to force the other to move one foot by pushing, pulling, or otherwise manipulating. The individual who first wins two boutsis the winner.

(5) Indian Wrestling(See fig. 5-1, E.)Two individuals lie on the ground, side by side, with their heads in opposite directions. They link right elbows. On the instructor's signal or by mutual agreement, each individual raises the right leg approximately straight and far enough to engage the opponent's heel. To start the contest, each person usually raises the leg three times rhythmically and on the third time engages the opponent's heel. Each tries to roll the other over backward. The right leg is used for three bouts, then the left leg for three bouts.

Figure 5-1 is available in the illustration section at the end of the book.

b. Combatives Table II

(1) Wrist Bending.(See fig. 5-2, A.) Opponents pair off, face each other, raise their arms forward, and, with palms forward, interlock their fingers. At the starting signal, each individual attempts to bend the opponent's wrist. The arms are kept up and forward and are not allowed to swing around and down to the sides. The individual who first wins two bouts is the winner.

(2) Back-to-Back Push.(See fig. 5-2, B.)Two individuals stand back-to-back with elbows locked. Each right arm is inside the opponent's left arm. At the starting signal, each pushes backward, trying to move the opponent. Opponents are not allowed to lift and carry each other; only pushing is permitted. The one who pushes the opponent the farthest wins the bout. The individual who first wins two bouts is the winner.

(3) Hop and Pull Hand.(See fig. 5-2, C.) The individuals are matched in pairs. Each grasps the opponent's right hand and, hopping on the right foot, attempts to pull the opponent off balance. Contestants automatically lose if they touch their free hand or their lifted foot to the ground. For successive bouts, they alternate hands and feet.

(4) Westmoreland Wrestling. (See fig. 5-2, D.)Each contestant grasps the back of the opponent's neck with the right hand and the opponent's right elbow with the left hand. In this position, each attempts to pull, push, or force the opponent to touch the ground with any part of the body other than the feet. The individual who first wins two bouts is the winner.

(5) Crab Fight.(See fig. 5-2, E.)Two individuals sit on the ground facing in opposite directions with their hands on the ground behind them. At the whistle, they raise their hips and push with their shoulders and bodies, each trying to make the other's hips touch the ground. The individual who first wins two bouts is the winner.

Figure 5-2 is available in the illustration section at the end of the book.

c. Combatives Table III

(1) Hand Wrestling.(See fig. 5-3, A.) Two individuals stand facing each other. Their right feet are forward and braced side by side. They grasp right hands for the first bout, left hands for the second bout. Each pulls, pushes, makes sideward movements, and otherwise maneuvers to force the opponent to move one or both feet from the original position. The contestant who first wins two bouts is the winner.

(2) Back-to-Back Tug.(See fig, 5-3, B.) Two individuals stand back-to-back with both arms linked at the elbows. Each has the right arm inside the opponent's left arm. At the starting signal, each attempts to pull the opponent. Lifting and carrying are permitted. The contestants must maintain their original direction and keep their arms linked. After a predetermined time, the player pulled or carried the farthest is the loser.

(3) Wrestling to Lift off Feet.(See fig. 5-3, C.) Contestants face each other. Each places the right arm under the left arm of the opponent and around the body. The left arm is over the opponent's right shoulder. Each tries to lift the other off the ground. The individual who first wins two bouts is the winner.

(4) Arm Pull Between Legs.(See fig. 5-3, D.) Two individuals are paired off, back-to-back. Each bends forward and, extending the right arm between the legs, grasps the opponent's right wrist. At the starting signal, each person attempts to pull the opponent After a predetermined time, the player who has pulled the opponent the farthest is the winner of the bout. The one who first wins two bouts is the winner. Repeat with the left hand and then both hands.

(5) Rooster Fight.(See fig.5-3, E.) Each contestant grasps the left foot with the right hand from behind, and the right arm with left hand. Each hops on the right foot, and, by shoulder-butting the opponent or by feinting and sudden evasions, forces the opponent to let go of the foot or arm. The contestant who first wins two bouts is winner.

Figure 5-3 is available in the illustration section at the end of the book.

Section III .Relays

5301. DESCRIPTION AND OBJECTIVE

Relays are races in which each member of a team runs one leg of the race. The team effort decides the winner. Relays provide stimulating competition and contribute to the conditioning of personnel. They also develop aggressiveness, team spirit, and the will to win. Relays should be dispersed throughout the program for short periods of time to provide a change of activity.

a. Team Organization. Relays are conducted most efficiently in platoon-size groups. Teams of equal size must be organized. Competitive spirit is encouraged and teams are organized better by basing teams on units such as squads, crews, or sections. Team captains should be designated. Extra Marines may be used as officials. The number of individuals on a relay team should be limited to squad size. If larger teams are used, the runners will spend too much time awaiting their turns and too little time actually participating. Two to six teams are ideal for relay competition. It is difficult to keep track of winners when more teams compete.

b. Administration of Relays

(1) The time spent on any one relay should be relatively short. If one team achieves a substantial lead in a long relay, the competitive spirit and enthusiastic participation of the other teams may decrease. Several short relays are generally better than one long relay.

(2) To maintain competitive spirit throughout a number of relays, determine the teams that win, place, and show in each relay and their total points for all relays. This can be done by awarding points to all teams on the basis of position at the finish of each relay. The team with the greatest number of points is the winner of the entire set of relays.

(3) Difficulties commonly encountered in conducting relays may be avoided by the following procedures:

(a) The last player in a relay race should be conspicuously identified. For example, the last runner can tie a handkerchief around the head or arm, take off the shirt, put on a hat or take it off, or use some other means.

(b) Another way to keep track of the progress of the race is to have each player sit or squat as soon as each is finished.

(c) Judges at the starting line can keep the runners from starting too soon.

(d) To prevent contestants from turning before they run the full distance, they should be required to run around a peg, pole, or assistant instructor.

(e) Batons, handkerchiefs, tent pegs, or other objects should be passed from one runner to the next when relays are run on a circular track.

(4) Before a unit's first participation in a relay, inform participants of the rules and scoring system. Violation of the rules should not result in disqualification. Instead, impose point penalties. A point penalty is imposed by subtracting a point from the team total at the conclusion of the relay.

(5) Careful administration will prevent most violations. For efficient conduct of relays, follow this procedure:

(a) Announce the name of the relay.

(b) Form the individuals in relay position.

(c) Briefly explain the relay and the rules for running it.
(d) Demonstrate.
(e) Have a definite finish line, and ensure that all know where it is.
(f) Answer questions.
(g) Conduct the relay.
(h) Determine winner and award points.

5302. RELAY TABLES

a. Events. The following relays are grouped into a table of activities. Each relay table can be completed in 15 minutes. Thus, relays can be used as a sole activity or as a part of a longer period. Each table is planned for a platoon-size group (30-60Marines). Adequate warmup for participants is provided by conducting several repetitions of exercises 1 and 2 of a conditioning drill. The recommended relay tables require an area 40 by 60 meters in size. Each table provides a variety of activity. The tables are progressive in the overload applied and should be scheduled in numerical order although not necessarily on successive days.

b. Relay Table I

(1) **60-Meter Lane Relay** (See fig. 5-4, A.) Each team is assembled in single file behind the starting line. On signal, the first individual of each team runs to the turn-around line 30 meters away, then runs back and touches the next teammate waiting at the starting line. The winning team is the first team to get its last member across the finish line. If an individual starts before being touched by the preceding runner, the team may be penalized.

(2) Wheelbarrow Race.(See fig. 5-4, B.) The players of each team pair off and line up in single file behind the starting line. The first individual walks on the hands while the partner grasps the ankles. They advance to the distance line (25 meters) behind which they exchange positions and return to the starting line. The rear individual must always hold the partner's ankles. After the first pair returns across the starting line, the next pair starts.

Figure 5-4 is available in the illustration section at the end of the book.

(3) Squad Front Relay .(See fig. 5-4, C.) The teams form in a line along the starting line with a 10-foot interval between teams. The members of each team lock elbows so that they are linked together. At the starting signal, the teams run to the distance line (20meters) where the left flank individual acts as a pivot. The team swings around on the pivot and returns to the base line. If a team breaks its links, it must reform before continuing. The first team to completely cross the base line intact is the winner.

(4) Crab-Walk Race.(See fig. 5-4, D.) The players of each team line up in single file. The first person of each team assumes the crab-walk position with feet forward on the starting line, hands on the ground behind it. At the starting signal, contestants move forward to the distance line (10 meters). They touch the line with their feet and then return to the starting line in the reverse position with the head and hands leading. The second person may not start until the first individual touches the finish line.

c.Relay Table II

(1) 100-Meter Lane Relay.(See fig. 5-5, A.)This relay is conducted exactly as the first relay of table 1 except that the start and turn-around line are 50 meters apart. This relay provides progression in sprinting.

(2) Frog-Jump Relay. (See fig. 5-5, B.) Each team lines up in single file. The first individual assumes a squatting position on the starting line. At the starting signal, the contestant progresses to the distance line (15 meters) and back by leaping forward, catching the weight on the hands, and bringing up the legs to the squat position for the next leap forward.

(3) Simple Relay. (See fig. 5-5, C.)Each team lines up in single file. Place a marker on the distance line (20 meters) in front of each team. Each team member, behind the first one, grasps the belt of the individual ahead. At the starting signal, each team runs as a unit to the marker, circles it, and returns to the starting line. The first team to completely cross the starting line intact is thewinner.

(4) Fireman's Carry Relay.(See fig. 5-5, D.)The players of each team line up in pairs, one behind the other. One individual in each pair carries the partner to the distance line (30 meters), using the fireman's carry. At the distance line, individuals exchange places and return to the starting line. As a variation, the person to be carried lies on the ground. The one carrying lifts the partner to the proper position. This relay may be performed with the other carries described in guerrilla exercises in chapter 3.

d.Relay Table III

(1) 200-Meter Circle Relay .(See fig. 5-6, A.)A course is laid out in a circular, rectangular,or oval pattern that is 200 meters around. Each team provides one runner on the starting line. On signal, the runner races around the 200-meter track and touches the next teammate, waiting at the starting line, who runs the same course. Each team member runs one lap of the course.

Figure 5-5 is available in the illustration section at the end of the book.

(2) Bear-and-Crab Race.(See fig. 5-6, B.) Each team lines up in single file. At the signal to start, the first contestant in each column assumes the bear-walk position (hands on ground in front), walks to the distance line (15 meters), and then runs back to the starting line. The returning contestant touches off the second individual and goes to the rear of the line. The second contestant assumes the crab-walk position (hands on ground behind) and crab-walks with feet leading to the distance line. This contestant also runs back to the starting line and touches off the next individual, who walks bear fashion. The rest of the members of each team alternate in this manner. The relay ends when the first person is back at the head of the line.

(3) Pilot Relay.(See fig. 5-6, C.) The players are grouped in threes, arms interlocked at the elbows, and outside players facing backwards. The middle person runs forward; the two outside individuals run backward. They run to the turning point (15meters), where they start back, this time with the middle person running backward and the two outside individuals running forward. The next set of three players starts when the first set crosses the starting line.

(4) Saddle-Back Relay.(See fig. 5-6, D.) Mark two parallel lines 15 meters apart. Each team selects a rider. The remaining members of each team count off. The even-numbered players from each team form in single files behind one line and the odd-numbered players from each team form in single files behind the other line directly across from their teammates. At the starting signal, the rider mounts the back of the first player of the team who carries the rider across the other line. There the rider changes mounts to the first player in the second line without touching the ground. This person carries the rider to the next player waiting in the first line. The relay continues until all of the mounts have carried the rider. A rider who falls off must mount again at the point of the fall. A rider who falls in changing mounts must get back on the original mount before making the change.

e. Relay Table IV

(1) 100-Meter Circle Relay (See fig. 5-7, A.)A course is laid out in either a circular, rectangular, or oval pattern that is 200 meters around. Each team is divided in half with each half positioned at starting lines on opposite sides of the track. Each runner races halfway around the track and touches a teammate who completes the lap. Each runner then waits in file at the halfway point until touched by the runner and then completes the second half of the lap. The first team to return all runners to their original starting line is the winner.

Figure 5-6 is available in the illustration section at the end of the book.

(2) In-and-Out Relay.(See fig. 5-7, B.) Each team lines up in a file with players 2 meters apart. At the starting signal, the first player runs back through the column in a zigzag fashion. He alternates, going to the right of one teammate and to the left of the next. Upon completing the run, the player lines up 2 meters behind the last one in the column. As soon as the first runner has passed the second person, the latter starts to run. This continues until all the players have realigned their original order. The team that finishes first is the winner. It may be desirable to have this relay continue until all individuals have run through their entire team two or three times in succession.

(3) Circle Race.(See fig. 5-7, **C.**) Each team forms a circle and holds hands with all individuals facing out except one who faces in and is the "driver."At the starting signal, the teams race to the distance line (20 meters) and back, keeping the circle intact. All the individuals in the circle must completely cross the distance line. The "driver" gives directions and orders. When the circle breaks, it must be reformed before it can continue. The first team completely over the starting line is the winner.

(4) Horse-and-Rider Relay.(See fig. 5-7, D.)Each team lines up in a single file. At the signal to start, the second player in each column leaps upon the back of the first who carries the rider across the distance line (30 meters). At the distance line, the rider dismounts and runs back to the starting line. There the rider picks up the third individual in the column, and carries this player to the distance line where the first player has remained. This continues until the last person is carried across the distance line.

Figure 5-7 is available in the illustration section at the end of the book.

Section IV. Team Contests and Athletics

5401. DESCRIPTION AND OBJECTIVE

Team contests are competitive activities in which Marines as a team compete with another team to win. They are guided by simple rules and organization. The function of team contests is to provide competition and an opportunity for body contact and to contribute to the development of physical readiness. In competing and working together as a team, individuals develop aggressiveness, the will to win, and teamwork. Team athletics deserve a prominent place in the physical training program because they contribute to increased combat efficiency. Because of the competitive nature of athletics and their natural appeal, individuals take part with enthusiasm. Athletic teams formed at intramural and higher levels are a strong unifying influence and provide one of the best means of developing esprit de corps.

a. Area and Equipment A level training field is sometimes the only area required. Many contests need no equipment. In contests requiring equipment, the need is for standard items such as logs, balls, nets, goals, and similar types of equipment. Specific requirements for area and equipment are listed with each contest description.

b. Progression. Team contests of a strenuous nature should be introduced after a basic period of conditioning has been completed and individuals are in the slow improvement stage of conditioning. Progression can take place from the less active to the more vigorous contests, and then from the non-contact to the contact or combative-type contests.

c. Necessity for Preconditioning Individuals must undergo conditioning prior to participation in athletics. Muscles, organs, joints, and ligaments not accustomed to stress and strain from sudden stops and starts, falls, body contact, rapid turns, prolonged running, and other rigors of athletic competition are subject to injury. Although athletics should not be introduced until players are physically prepared, there is still opportunity to engage in competition through lead-up contests. Individuals learn many of the skills required for athletics while participating in team contests.

d. Benefits. Athletics are beneficial primarily to sustain interest in the program and to maintain an achieved level of physical fitness. Athletics are a supplement, not a substitute for other types of conditioning activities which should continue. All of the desirable traits of physical fitness cannot be developed through athletics, yet their contribution is significant. For athletics to make a proper contribution to physical conditioning, the selected sports must bevigorous. The team contests recommended here are carefully selected for their simplicity, aggressiveness and applicability to squad- teaching teamwork. They are generally strenuous and involve the whole team throughout the entire duration of the contest.

5402. TEAM CONTESTS

a. Pushball

(1) **Players.** 10 to 50 players on a side. (See fig. 5-8.)

(2) Equipment A large pushball 5 to 6 feet in diameter.

(3) Area. A field 240 to 300 feet long, 120 to 150 feet wide. Mark a center line in the middle of the field parallel to the end lines. Mark a line 45 feet on either side of this center line and parallel to it, extending it across the width of the field. Mark another parallel line 15 feet from each end line, extending it across the width of the field.

(4) Game. Four 10-minute quarters are played. Give 2-minute rests between quarters and 5-minute rests between halves. The object of the game is to propel the ball over the opponent's goal line by pushing, rolling, passing, carrying, or any other way except kicking.

(a) The ball is placed on the center line with the opposing captains 3 feet from the ball. The rest of the players are 45 feet from the ball, on their half of the field. On the referee's starting whistle, the captains immediately play the ball with their respective teams coming to their assistance.

(b) At quarter time, the ball remains dead for 2 minutes at the spot where it was when the quarter ended. At half time, the teams exchange goals. The play is then started as it was in the beginning.

Figure 5-8 is available in the illustration section at the end of the book.

(c) Players may use any means of interfering with an opponent's progress except striking and clipping. Clipping is throwing the body across the back of an opponent's leg, while the opponent is running or standing. Legal use of force may be applied to all opponents whether or not they are playing the ball. For striking an opponent, the offender is removed from the game. The team penalty is half the distance to their goal. The penalty for clipping is the same.

(d) When any part of the ball goes out of bounds, it is a dead ball. The teams line up at right angles to the side lines and 3 feet apart at the point where the ball went out. The referee then tosses the ball between the teams.

(e) When, for any reason, the ball remains in one spot for more than 10 seconds, the referee declares the ball dead. The ball is then put into play as it is for an out-of-bounds situation.

(5) Scoring. A goal is scored when the ball, or any part of it, is propelled across the opponent's end line. A goal counts 5 points. The team scoring a goal has the privilege of trying for a point after the goal. To try for this extra point, the ball is placed on the opponent's 5-yard (or 15-foot) line. The teams line up on either side of this line separated by the width of the ball. Before the whistle blows, one player on each team may place hands on the ball. On the referee's signal, the ball is put into play for 1 minute. If any part of the ball is driven across the goal line in this 1-minute period, the offensive team scores 1 point. The defending team may not score during the opponent's try for the extra point.

b. Line Rush

(1) Players. Any number up to 50 on each side. (See fig. 5-9.)

(2) Equipment. None.

(3) Area. A field, 75 by 100 feet.

(4) Game. One team lines up behind one goal line and the other in midfield. On the starting signal, the team standing behind the goal line seeks to cross the goal line at the opposite end of the field within 30 seconds. The team in the center seeks to prevent it by catching and holding the runners. At the end of 30 seconds, the teams change.

(5) Scoring. Count the number of individuals who have crossed the far goal at the end of 30 seconds. After each team has had from three to five tries, the scores are added and the winner declared. A player scores 1 point when any part of the body is across the goal line.

Figure 5-9 is available in the illustration section at the end of the book.

c. Human Tug-of-War

(1) Players. 10 to 20 on a team.(See fig. 5-10.)

(2) Equipment. None.

(3) Area.40 to 60 feet.

(4) Game. Draw a line in the center of the area. Divide the players into two equal teams. Place them in single file on opposite sides of the center line facing each other. Each individual places the arms around the waist of the teammate in front. The two leaders of the opposing teams grasp each other around the waist. On signal, the teams try to pull each other over the center line within 30 seconds.

(5) Scoring. The team pulled across the center line loses. If neither team is pulled over the center line, but one team breaks its file, that team loses the match.

(6) Variation. Use a 3/4- or 1-inch rope. Space the leading players on each team 10 feet apart, each holding one end of the rope. The team pulled across the center line loses.

Figure 5-10 is available in the illustration section at the end of the book.

d. Master of the Ring

(1) Players. Any number. (See fig. 5-11.)

(2) Equipment. None.

(3) Area. A clearly marked circle large enough to contain all the players.

(4) Game. All the players stand inside the circle. At the signal, all players attempt to throw each other out of the circle. All tactics are fair except unnecessary roughness. When any part of the body touches across the line, the player is out and leaves the circle at once. Several officials are needed to spot the players who cross the line.

(5) Scoring The player who remains in the circle when all the others are out is the master of the ring.

(6) Variation 1. The players are divided into two equal teams. Each team is clearly marked. On signal, each team tries to throw the opponents out of the circle. The winning team is the team that eliminates all opponents from the circle.

(7) Variation 2.The players are divided into two equal teams. Each team sends only one individual into the circle. When one person has been forced out of the circle, the losing side sends in only one player. The team which eliminates all opponents is the winner.

(8) Variation 3.A pit, approximately 4-feet deep, is used rather than a circle on level ground. The contest may then be an individual or team activity.

Figure 5-11 is available in the illustration section at the end of the book.

e. Log Pivot Circle. Space teams far enough apart so each team can pivot in a circle without colliding with any other team. Each team holds a log in the bend of the arms in front of the chest. (See fig.5-12.) At the command **CIRCLE RIGHT, MOVE**,the left-flank individual holds the pivot while the log is carried around 360 degrees in a counter- clockwise motion, back to the original position. This movement may also be performed to the left by facing in the other direction and pivoting clockwise(**CIRCLE RIGHT, MOVE**).Other commands may be used such as: **CIRCLE HALF-RIGHT, CIRCLE HALF- LEFT** and so on. The first team to complete the prescribed movement is the winner.

Figure 5-12 is available in the illustration section at the end of the book.

f. Log-Rolling Race. Each team tries to roll its log a measured distance by pushing it with the hands and driving the body forward with the legs. (See fig. 5-13.)The first team to get the entire length of the log across the finish line wins.

Figure 5-13 is available in the illustration section at the end of the book.

g. Prone Pushing Contest.. Two teams lie prone, facing each other with a log between them. (See fig. 5-14.)Both teams place their hands against the log, keeping their arms straight. Then, by driving with the legs, each team attempts to push the other a measured distance to the rear.

Figure 5-14 is available in the illustration section at the end of the book.

h. Shuttle Relay Race. Each team in this race divides into two parts, A and B. Members of part A run 50 yards with a log held under their right arms. (See fig 5-15.)At the distance line, they give the log to their teammates in part B who bring it back to the starting line. The team pair finishing first is the winner.

Figure 5-15 is available in the illustration section at the end of the book.

5403.CROSS-COUNTRY AND DISTANCE RUNNING

a. Value of Running. Long-distance running gives some benefits that cannot be obtained in the same degree from any other sport. It builds powerful leg muscles, increases lung capacity, and develops endurance. For these reasons, cross-country and distance running should be included in the physical training program. These sports require only a few miles of open space which is generally available. Short cross-country runs and middle-distance runs can be used to supplement other activities, particularly team sports or the sports that develop precision or agility rather than endurance. Short cross-country runs can be scheduled once a week, gradually increasing the distance as the physical condition of the runners improves.

b. Cross-Country Runs. A cross-country run is a distance run held on a course laid out along roads, across fields, over hills, through woods, and on any irregular ground. A flat cinder or dirt track is not a suitable surface for cross-country running. Some runs are as long as 6 miles while others are as short as 3 or 4 miles. If cross-country running is being used to supplement other activities, the 3-mile course is long enough for most. Cross-country runs should be used only after individuals reach the sustaining stage of conditioning. This type of running should then be scheduled occasionally to provide variety. Cross-country running has the advantage of allowing mass participation. Interest can be stimulated by putting the runs on a competitive basis.

c. Practice Methods. Conditioning is essential to distance and cross-country running. Championship distance running depends on stamina, which can be developed only through constant training. An individual of only average ability can become an outstanding distance runner by steady and careful training. Hiking is an excellent method for getting into condition before the season opens. Long walks build up leg muscles. During the first month of the season, training should be gradual, starting with short distances, and increasing day by day. At first the legs will become stiff, but the stiffness gradually disappears if running is practiced for a while every day. To prevent strain, it is essential to limber up thoroughly each day before running.

(1) In training a large group, leaders should be stationed at the head and the rear of the column. They should make every effort to keep the runners together. After determining the abilities of the unit in cross-country running, it is advisable to divide them into three groups. The poorest-conditioned group is started first, the best-conditioned group, last. The starting time of the groups should be staggered so that all of them come in at about the same time.

(2) In preliminary training, running is similar to ordinary road work in that it begins with rather slow jogging, alternating with walking. The speed and distance of the run is gradually increased. As conditioning improves, occasional sprints may be introduced. Atfirst, the distance run is from one-half to 1 mile. It gradually increases to 2 or 3 miles. On completing runs, require runners to walk for 3 or 4 minutes before stopping to permit a gradual cooling off.

d. Facilities and Equipment. There should be at least one timer with a stopwatch (preferably three) for timing the runners. A course 3 or 5 miles long should be measured and marked by one of the three following methods:

(1) Fasten directional arrows to the top of tall posts and place them at every point where the course turns. Such signs should also be placed at every other point where there may be doubt as to the direction of travel.

(2) Place a lime line on the ground over the entire course.

(3) Place flags, clearly visible to the runners.

- **(a)** A red flag indicates a left turn.
- **(b)** A white flag indicates a right turn.
- **(c)** A blue flag indicates the course is straight ahead.

e. Rules/Scoring. A cross-country team consists of seven individuals, unless otherwise agreed. In dual meets, a maximum of 12 on one team may be entered, but 7 or fewer are scored. First place scores 1 point, second place 2, third place 3, and so on. All who finish the course are ranked and tallied in this manner. The team score is then determined by totaling the points scored by the first five of each team to finish. The team scoring the least number of points is the winner. If fewer than five (or the number determined prior to the race) finish, the places of all members of that team are disregarded. If two or more teams score the same number of points, the event is called a tie.

5404.TEAM ATHLETICS

Team athletics add variety to the combat readiness program. At no time should they be allowed to replace intramural activities or to provide practice time for intramural or base teams. If team athletics are integrated into the program, the goal should be total participation and competition. Therefore, the games selected can be played by the skilled or the unskilled. They require limited equipment, yet will teach competitiveness, aggressiveness, and teamwork while they improve conditioning.

a.Soccer. Soccer is one of the best athletic activities for developing endurance, agility, leg strength, and great skill fulness in using the legs. The most popular sport in the world, soccer is the national game of many European, Asian, and Central and South American countries. In recent years, it has become popular in the United States. A soccer ball is the only equipment needed for the game, and Marines can learn to play it easily. Players do not need much skill to participate, but the amount of skill they can develop is almost unlimited.

(1) Place in the Program. Soccer should be introduced during the latter part of the slow improvement stage of physical training. It can be used as a competitive activity in the sustaining stage. It is primarily a spring or fall sport. Any level field is suitable for competition. The boundaries for the soccer field are similar to the dimensions for a football field. (See fig. 5-16.) Goal posts are essential, but they are easily constructed and usually temporary so that they may be removed when not in use.

Figure 5-16 is available in the illustration section at the end of the book.

(2) Basic Skills

(a) Passing. Passing with the feet is the primary means of moving the ball. Short passes are easier to control. Emphasis should be continually placed on the skill of passing.

(b) Dribbling. The ball is dribbled by a series of kicks with the inside or outside of the foot. Do not kick with the toe. Keep the head over the ball when kicking and propel it only a short distance at a time. Keep it close to the feet. If a player lets a ball move far from the feet, an opposing player can easily take it away.

(c) Instep Kicking. The instep kick is the basic soccer kick. The toe does not come in contact with the ball. The toe is pointed downward, and the instep (the area under the shoe laces) is applied to the ball with a vigorous snap from the knee.

(d) Inside-of-the-Foot Kicking. The ball is kicked with the inside of the foot, and the leg is swung from the hip. The toe is turned outward, and the sole of the foot is parallel with the ground as the foot strikes the ball. This kick is used for short passes and for dribbling.

(e) Foot Trapping. The foot trap is the method of stopping the ball by trapping it between the ground and the foot. Place the sole of the foot on top of the ball at the instant it touches the ground, keeping the foot relaxed. This is an effective way to stop a fast-moving ball.

(f) Shin Trapping. .The shin trap is a method of stopping the ball with the shins. Stand just forward of the spot where the ball should strike the ground and allow it to strike the shins in flight or on the bounce. Use either one or both legs from the knee down so as to absorb the bounce.

(g) Body Trapping. .Intercept the ball with any part of the upper body except the arms and hands. Keep the body relaxed and inclined toward the ball. To keep the ball from bouncing, move backwards from it as it strikes the body. This drops the ball at the feet in position for dribbling or passing.

(h) Heading. Heading is a technique for changing the direction of the flight of a ball by butting it with the head. Tense the neck muscles and jump up to meet the ball. Butt the ball with the forehead at about the hairline to reverse its direction. Use the side of the head to deflect it to the side.

(3) Offensive and Defensive Positions. Forwards usually play on the offensive half of the field. Fullbacks usually play on the defensive half of the field. Halfbacks are the backbone of the team. They move forward on the offense and back on defense. The goalkeeper almost always remains within a few feet of the goal.

(4) Abridged Rules

(a) A soccer team is composed of 11 players. (See fig. 5-16.)

(b) The player propels the ball by kicking it with the feet or any part of the legs, by butting it with the head, and by hitting it with any portion of the body except the arms or hands.

(c) The goalkeeper is the only individual allowed to place hands on the ball, but only in the goalkeeper's area. The term "hands" includes the whole arm from the point of the shoulder down.

(d) A goal is made by causing the ball to cross completely the section of the goal line lying between the uprights and under the cross bar.

(e) Each goal scores 1 point for the team scoring the goal.

(f) The penalty for a foul committed anywhere on the playing field (except by the defensive team in its penalty area) is a free kick awarded to the opposing team.

(g) All opponents must be at least 10 yards from the ball when a free kick is taken.

(h) The penalty for a foul committed by the defensive team in its penalty area is a penalty kick.

(i) A penalty kick is a free kick at the goal from the spot 12 yards directly in front of the goal. The only players allowed within the penalty area at the time of the kick are the kicker and the defending goalkeeper.

(j) An official game consists of two 30-minute halves,

(k) Teams change goals at the end of every quarter.

(l) In the event of a tie, an extra quarter may played. This may be followed by a sudden death period. If the tie still exists, the game can be decided by penalty kick.

(m) After a team has propelled the ball across a side line and out of play, the ball is put back into play by a throw-in from the side line by a member of the opposing team. The ball is thrown in from the point where it crossed the side line as it went out of bounds. The ball must be thrown in with both hands using an over-the-head motion, keeping both feet on the ground.

(n) When the offensive team propels the ball across the defensive team's goal line but not in the goal, the defensive team is awarded a goal kick-a free kick taken from within its own goal area. The ball must come out of the penalty area to be in play.

(o) When the defensive team causes the ball to go behind its own goal line, excluding the portion between the goal posts, the offensive team is awarded a corner kick—a free kick taken by a member of the offensive team at the quarter circle, at the cornerflag post nearest to where the ball went behind the goal line. The flag post must not be removed.

(p) The game is started and, after a goal has been scored, is resumed by placing the ball in the center of the midfield line. Players must be on their side of the line until the ball is kicked. The ball must be kicked forward and must move at least 2 feet to be legal. The first kicker may not touch the ball twice in succession at the kick off. The opposing team must be 10 yards from the ball until it moves.

b.Speedball. Speedball offers vigorous and varied action with plenty of scoring opportunities. It is easy to learn and provides spontaneous fun. Little equipment is needed-a ball is all that is absolutely necessary. Speedball combines the kicking, trapping, and intercepting elements of soccer; the passing game of basketball; and the punting, drop kicking, and scoring pass of football.

(1) Place in the Program.Speedball, like soccer, should be introduced into the physical training program during the latter part of the toughening stage and used as a competitive activity in the sustaining stage. Speedball may be played any time weather permits but is primarily a spring or fall activity.

(2) Method of Play.The game is played as follows:

(a) Two teams of 11 each play the game under official rules, but any number of players may successfully constitute a team.An inflated leather ball, usually a soccer ball, is used. The playing field is a football field with a football goal post at each end. (See fig. 5-17.)

(b) The game starts with a soccer-type kickoff. The kicking team tries to retain possession of the ball and advance it toward the opposite goal by passing or kicking it. Running with the ball is not allowed, with the result that there is no tackling or interference. After the ball touches the ground, it cannot be picked up with the hands or caught on the bounce but must be played as in soccer until it is raised into the air directly from a kick. Then the hands may be used.

(c) When the ball goes out of bounds over the sidelines, it is given to a player of the opposing team and is put into play with a basketball throw- in. When it goes over the end line without a score, it is given to a player of the opposing team who may either pass or kick it onto the field.

(d) When two opposing players are contesting possession of a held ball, the official tosses the ball up between them as in basketball.

(e) Points are scored by kicking the ball under the crossbar of the goal post, drop kicking the ball over the crossbar, or completing a forward pass into the end zone for a touchdown.

Figure 5-17 is available in the illustration section at the end of the book.

(3) Basic Skills

(a) Shared Skills. Skills include kicking, passing, heading, and trapping (from soccer); punting, dropkicking and forward passing (from football); and passing, receiving, and pivoting (from basketball).

(b) Kickups. The kickup is a play in which a player lifts the ball into the air with the feet so as to legally play the ball with the hands. The kickup generally makes the transition from ground play to aerial play. The technique of making the play depends upon whether the ball is rolling or stationary. If the ball is rolling or bouncing toward the player, the foot is held on the ground with the toe down until the ball rolls onto the foot. Then the foot is raised, projecting the ball upward. If the ball is stationary, the player rolls it backward with one foot. Then the player places a foot where the ball will roll onto it. The player then raises the foot, projecting the ball as before. If a ball is rolling away, the player should stop it with a foot and play it as a stationary ball. The player can also raise the ball by standing over it with a foot on either side. The player presses the feet against the ball and jumps into the air, propelling the ball into the hands.

(4) Offensive and Defensive Play. The positions of players in speedball are much the same as in soccer. However, some of the positions have different names. Each team has 11 players. The forward line is composed of five players, the right end, right forward, center, left forward, and left end. The second line consists of right halfback, fullback, and left halfback. In the next line is the right guard and left guard. The player who defends the goal is the goal guard. The strategy employed in speedball during offensive play is very similar to that of soccer. There are two types of defensive formations in speedball: man-for-man and position defense. Man-for-man defense is recommended for beginning players.

(5) Abridged Rules

(a) Field. A speedball field is 100 yards long with additional 10-yard end zones on each end and 55 yards wide, the same as a regulation football field.

(b) Players. Eleven players are on a team. The goal guard has no special privileges.

(c) Time. Ten-minute quarters. Two minutes between quarters. Ten minutes between halves. Five minutes for extra overtime periods.(Begin the first overtime by a jump ball at center, same goals; change goals in the event of a second overtime period.)

(d) Winner of Toss. The winner of the toss has the choice of kicking, receiving, or defending a specific goal.

(e) Kickoff. A kick off starts the first and second half of a game and starts play after a score. The kick off is made from any place on the 50-yard line. The kicking team must be behind the ball when it is kicked. The receiving team must stay back of its restraining line (10 yards distance) until the ball is kicked. The ball must go toward the opposing team's goal before it can be played. Kick off out of bounds goes to the opponent at that spot. A kick off touched by the receiving team before going out of bounds, with no impetus added, still belongs to the receiving team. A kick off, in possession and control of the receiving team and then fumbled out of bounds, belongs to the kicking team at that spot.

(f) Starting Second and Fourth Quarters. The ball is given to the team that had possession at the end of the previous quarter, from out of bounds, as in basketball.

(g) Second Half. The first-half receiving team kicks off at the beginning of the second half.

(h) Fly and Ground Ball. The most characteristic playing rule of speedball is the difference between a fly ball (or aerial ball) and a ground ball. A player is not permitted to touch a ground ball with the hands and must play it as in soccer. A fly ball is one that has risen into the air directly from the foot of a player (for example, a punt, drop kick, place kick, or kickup).Such a ball may be caught with the hands provided the catch is made before the ball strikes the ground again.

(i) Kickup and Overhead Dribble. A kickup is a ball that is so kicked by a player that he can catch it himself. A bounce from the ground may not be touched with the hand because it has touched the ground since being kicked. This rule prohibits the ordinary basketball dribble, but one overhead dribble (throwing the ball into the air and advancing to catch it before it hits the ground) is permitted.

(j) Out of Bounds. If a team causes the ball to go out of bounds over the sidelines, a free throw-in (any style) is given to the opposing team. When the ball goes over the end line without scoring, it is given to the opponents who may pass or kick from out of bounds at that point.

(k) Tie Ball. In case two players are contesting the possession of a held ball, even in the end zone, a tie ball is declared. The ball is tossed up between them.

(6) Scoring

(a) Field Goal(3 points).A soccer-type kick, in which a ground ball is kicked under the crossbar and between the goal post from the field of play or end zone, is a field goal. (A punt going straight through is not a field goal for it is not a ground ball. The ball must hit the ground first.) A drop kick from the field of play that goes under a crossbar does not count as a field goal. A drop kick from the end zone that goes under the crossbar counts as a field goal;if it goes over the crossbar, it is ruled as a touch back, and the other team takes possession of the ball.

(b) Dropkick(2 points). A scoring drop kick must be made from the field of play and go over the crossbar and between the uprights. The ball must hit the ground before it is kicked (usually with the instep).

(c) End goal(1 point). This is a ground ball which receives its impetus (kicked or legally propelled by the body) from any player, offensive or defensive, in the end zone and which passes over the end line but not between the goal posts.

(d) Penalty Kick(1 point).A ball kicked from the penalty mark that goes between the goal posts and under the crossbar is a penalty kick. The penalty mark is placed directly in front of the goal at the center of the goal line.

(e) Touchdown(1 point). A touchdown is a forward pass from the field of play completed in the end zone. The player must be entirely in the end zone. If the player is on the goal line or has one foot in the field of play and the other in the end zone, the ball is declared out of bounds. If a forward pass is missed, the ball continues in play, but it must be returned to the field of play before another forward pass or drop kick may be made.

(7) Substitutions. Substitutions may be made any time when the ball is not in play. A player who is withdrawn may not return during that same period.

(8) Timeout. Three legal timeouts of 2 minutes each are permitted each team during the game.

(9) Fouls

(a) Personal Foul. Kicking, tripping, charging, pushing, holding, blocking, and unnecessary roughness of any kind, such as running into an opponent from behind, are personal fouls. Kicking at a fly ball and thereby kicking an opponent is a personal foul. Four personal fouls disqualify a player from the game.

(b) Technical Foul. Illegal substitution, more than three timeouts in a game, unsportsmanlike conduct, and unnecessarily delaying the game are technical fouls.

(c) Violation. Traveling with the ball, touching a ground ball with the hands or arms, double overhead dribble, violating tie ball, and kicking or kneeing a fly ball before catching it are violations.

(d) Penalties. The offended player shall attempt the kick.

(e) Officiation of Fouls

1 Fouls in the field of play allow no follow-up while fouls in the end zone always allow follow-up.

2 On penalty kicks, with no follow-up, only the kicker and goalie are involved.

3 On penalty kicks, with a follow-up, the kicking side is behind the ball and the defending side behind the end line or in the field of play. No one is allowed in the end zone or between the goal posts except the goal guard. The kicker must make an actual attempt at goal and cannot play the ball again until after another player plays it.

c. Volleyball. Volleyball is a popular sport. The game entails much physical activity, yet it is not strenuous. It is, therefore, a game for young and old alike, for beginners and for skilled players. It may be played indoors or outdoors on any type of terrain. As an informal activity, volleyball can be played by any number; as an organized activity, it provides, as few other sports do, a game for 12 to play in a limited area. While volleyball requires no great skill to play, it permits a high degree of proficiency. Players naturally get more enjoyment when they know the game and play it well. For this reason, instruction in the basic skills should be provided.

(1) Place in the Program Volleyball may be used occasionally as a competitive activity during the sustaining stage. Though it is a year- round sport, it should be included in the physical training program only when a more strenuous activity is impractical. Volleyball is an excellent off-duty activity.

(2) Instruction. Usually during the first three or four classes, a 10- to 15- minute period of instruction, followed by scrimmage, is enough to teach the basic skills, rules, and techniques of volleyball. More time, if available, can be given to teaching basic skills, but emphasize competitive play rather than formal instruction. Divide the unit into 6-person teams, organized on the basis of ability .Teams should be as nearly equal as possible.

(3) Basic Skills

(a) Passing the High Ball. The chest pass is the most effective method of playing the ball. To receive the ball, the knees are fixed, and the body is tilted forward. The elbows are raised sideward to a point in line with the shoulders. The wrists are extended in line with the forearm. The arms, wrists, and hands are rotated inward. To pass the ball, the hands are chest-high, thumbs pointing inward. The fingers are flexed, forming a cup, allowing them to contact the ball. On contact with the ball, the wrists are snapped while the fingers and elbows are pushed upward, sending the ball upward. A high ball is much easier to handle than a low one.

(b) Passing the Low Ball. A ball that is lower than the waist is one of the easiest to hit, but it is also a frequent cause of fouls (holding or carrying the ball). The best position for handling a low ball is to have the knees flexed and arms flexed at the elbows and rotated so the thumbs are pointing outward, the palms up. When the fingers contact the ball, the entire body reacts in a lifting motion. The arms and hands swing upward in a scooping action. It is important that the fingers, not the palms, contact the ball, and that the ball is batted, not thrown.

(c) Underhand Serve. Take a position behind the back line facing the net, left foot forward, holding the ball in the palm of the left hand. The left knee is flexed; the right knee is straight. Swing the right arm back and at the same time move the left hand (holding the ball) across the body in line with the right hip. Then swing the right arm forward hitting the ball off of the left hand with the palm of the right hand, raising the hips and arching the back in the same motion. Be certain to swing the right arm in a straight line, or the ball will be difficult to control. When the opposition is in formation, the server should try to place the ball in the right or left back area, and not near the net.

(d) Setup. A setup is a ball hit into the air near the net by one player, so that a teammate may hit or "spike" it sharply downward into the opponent's court. The chest pass is the best pass to use. The ball is sent approximately 10 feet into the air toward the spiker so it will descend from 4 to 20 inches from the net.

(e) Spiking. The spike is a leap into the air and a sharp downward hitting of the ball into the opponent's court. A spiker must be able to spring easily from the floor, judge the movement of the ball, and strike it with a downward movement of the arm. To jump from the floor, step off with one foot and jump with the other. Stand with the right or left side to the net, facing the setup player. Much depends upon the setup player to place the ball in the proper position. The spiker jumps into the air and strikes the ball above its center so as to drive it downward. A snapping movement of the arm and wrist will drive the ball forward and downward with power and control. Aim for a weak spot in the opponent's defense.

(f) Blocking. The block is a defense technique used to prevent a spiker from driving the ball across the net. It is an attempt by one or more defensive players at the net to block a hard-hit shot by using the force of the ball to send it immediately back into the opponent's court. In an effective block, forwards on the defensive team spring into the air at the time of the spike, placing both hands and arms in the expected path of the ball. An effective block tends to upset the offense and presents another element for the spiker to worry about. To be effective, the blocker must anticipate the path of the ball and time the block with the spike.

(4) Abridged Rules

(a) The volleyball court is 30 feet wide by 60 feet long. (See fig. 5-18.)

(b) The top of the net is 8 feet high.

(c) A volleyball team consists of 6 players.

(d) A match consists of the best two out of three games.

(e) The first team scoring 15 points wins the game, provided that they have 2 points more than their opponents.

(f) A deuce game is a game in which both teams score 14 points. The game is continued until one team obtains a 2-point advantage over the other.

(g) Only the serving team can score. If the serving team commits a fault, it loses the serve to the opposing team.

(h) The team members who receive the served ball rotate one position in a clockwise direction.

(i) The ball is put into play by serving from behind the back line.

(j) A served ball touching the net results in the loss of the serve. At any other time during play, a ball touching the net is still in play.

Figure 5-18 is available in the illustration section at the end of the book.

(k) The ball is out of play when it touches the ground or goes outside one of the boundary lines.

(l) All line balls are good.

(m) The players must hit or bat the ball; they may not throw, lift, or grasp it.

(n) A player may not touch the ball with any part of the body below the knees.

(o) A player may not play the ball twice in succession. In receiving a hard-driven spike, a defensive player may make several contacts with the ball even if they are not simultaneous. All such contacts, however, must constitute one continuous play, and all must be above the knees.

(p) The ball may be touched no more than three times on one side of the net before being returned across the net to the opposing team.

(q) A player must not touch or reach across the net.

(r) A player may touch the line under the net, but may not cross it.

Chapter 6
EVALUATION OF PERFORMANCE DURING TRAINING

6001.OBJECTIVES OF COMBAT READINESS

The objectives of the Marine Corps physical fitness program are-

a. To contribute to the health and well-being of every Marine through regular exercise and health education.

b. To develop Marines who are physically capable of performing their duties in garrison and in combat.

c. To develop Marines with a reserve level of physical fitness that will help them win in combat.

d. To provide a medium for developing the self-confidence of the individual Marine and thereby enhancing overall discipline, morale, esprit de corps, unit efficiency, and the desire to excel within the Marine Corps.

6002. THE PHYSICAL FITNESS TEST

The physical fitness test (PFT) has distracted attention from the proper goal of the physical fitness program. Simply stated, the goal of the physical fitness program is the success of Marines in combat. Unit commanders must not allow the PFT to become an end in itself. The PFT should be used within the commander's program of combat readiness training as a standard, an index of the physical fitness of individual Marines at a given time and place. The PFT is not difficult, and it must be kept in perspective. All athletes should be first class, and first class is attainable for all Marines. The conduct of the PFT is described in Marine Corps Order 6100.3_.

6003. OTHER METHODS OF EVALUATION

The commander should use other ways of evaluation to provide variety and imagination. Intangible objectives, such as confidence and aggressiveness, are hard to measure. The use of either inspection or observation is necessary to evaluate objectives of this nature. The commander has several other methods available when evaluating the physical condition of the command. The additional methods of inspection, observation, medical examination, and testing are summarized as follows:

a. Formal Inspections. Formal inspections, using inspection officers and standardized rating criteria, may help evaluate unit physical fitness.

b. Daily Observation. Routine observation of physical performance and appearance can serve as an indicator of an individual's or unit's physical readiness. However, mere observation is not a totally reliable or accurate means of evaluation.

c. **Physical Examination.** Medical examination may detect any individual disability or detrimental physical condition. It may guide in application of remedial, therapeutic, or limited exercise programs.

d. Other Physical Fitness Tests. Some other tests are described in detail in this manual for information purposes. Commanders are cautioned, however, not to place too great an emphasis on testing.

6004. RESPONSIBILITY

The commander is responsible for the physical fitness of the command and for the measurement and evaluation of its physical readiness. The goal of combat readiness training remains to prepare Marines and units for the physical demands of combat. The bulk of the commander's combat readiness training program should be such activities as obstacle and confidence courses, endurance courses, and progressive load-bearing marches which will prepare Marines for combat. Group activities which are competitive and combative in nature are advantageous and add imagination and enthusiasm to the program. Activities such as orienteering, which is not only physically demanding but also competitive and which teaches essential subject skills, are excellent and motivate all concerned. A commander's goal should be to provide an integrated training program that is so aggressive and demanding of Marines and leaders alike that a PFT administered at any time will reflect that each and every Marine is both physically and mentally prepared for the demands of combat.

6005. COMMANDER'S PHYSICAL FITNESS PROGRAM

The Commandant of the Marine Corps has directed that the physical fitness test shall be the universal measure of adequate individual physical fitness. The focus of a commander's physical fitness training program, however, should not be to prepare Marines to pass the physical fitness test, but to prepare them for the physical rigors of combat.

a. Administration. The physical fitness program requires each Marine to spend a minimum of 3 hours per week in physical fitness training, to be tested semiannually, and to obtain a minimum standard of third class. The attainment of a higher score is a laudable individual goal which should be encouraged but should not become a unit objective. Overemphasis could be detrimental to training required to develop the complete Marine. All commanders should exercise sound judgment as well as positive, aggressive leadership in striving toward the training objective.

b. Variety of Activities. The program must include the following elements:

(1) Physical Readiness Training. physical readiness training is a complete physical training program which develops and maintains the strength, endurance, and physical skills needed to sustain the individual during combat.

(2) Remedial Physical Conditioning. Remedial physical conditioning is a process by which physically substandard Marines are conditioned to meet prescribed standards.

(3) Competitive Conditioning Activities. Competitive conditioning activities consist of teams or individuals competing against an opponent to win. This involves a combination of sports and military skills designed to foster competitive attitudes and develop unit pride and esprit de corps.

(4) Occupational Conditioning. Marching long distances, dry net training, etc., should not be ignored as a means to fulfill the time requirement.

c.Time. In order to develop the desired level of physical fitness, unit commanders are advised that a minimum program of 3 hours per week is mandatory. The program must encompass at least three exercise periods weekly. Each period should include calisthenics, running, or other forms of vigorous activity .The minimum desirable length for each exercise period is 60 minutes. Shorter periods may be authorized by the unit commander if dictated by the local training situation. Unit commanders are encouraged to use the normal working day to satisfy this training requirement. However, commanders are authorized to conduct required physical fitness training during off-duty hours when the mission, workload, personnel status or other significant factors preclude fulfilling it during duty hours.

d.Action. Marine Corps Order 6100.3_ requires that commanders-

o Establish and maintain an effective physical fitness program which ensures that all Marines maintain an acceptable level of physical fitness.

o Ensure that all individuals are medically qualified prior to participating in the physical fitness program.

o Establish a minimum physical fitness program of at least 3 hours per week. Physical training may be authorized on an individual basis at the discretion of the unit commander.

o Ensure that all Marines participate in physical fitness conditioning activities commensurate with their medical qualifications and limitations.

o Conduct physical fitness testing for all Marines.

o Place all Marines who fail to pass the PFT on a daily command-supervised remedial physical conditioning program until they pass the test.

o Ensure that results of physical fitness testing are entered on section A of the fitness report. Comments are placed in section C in connection with attainment of superior physical performance or a medical excuse from the PFT.

6006. OTHER FORMAL TESTS

Commanders must realize that the best index of combat readiness is personal observation of performance on conditioning marches of 15 miles while under load. This manual presents three additional indicators of unit fitness. These indices are the physical readiness test, which was formerly the Marine Corps standard and is now routinely administered at Officer Candidate School; the airborne trainee physical fitness test which many Marines who attend U.S. Army formal schools are required to take; and the battle fitness test, the ultimate test of the combat readiness of Marines. Commanders may wish to include these tests or portions thereof within their combat readiness programs.

6007. PHYSICAL READINESS TEST

a. Purpose. The purpose of this test is to measure the Marine's ability to meet the minimum standard of physical readiness for duty in the field. This test is currently administered to all male officer candidates at Officer Candidate School, MCCDC, Quantico, VA. Female candidates do not take this test.

b. Test Area. Administration of this test requires an area suitable for crawling, jumping, and running 160 yards to include a trench that is 8 feet wide.

c. Equipment. The equipment required for this test includes—

o A platform, rail on a ledge, or a box 18 inches high,

o A 20-foot climbing rope,

o A stopwatch.

d. Conduct of Test. Adequate timers and other supervisory personnel must be provided by the company staff. All events should be conducted in a single session of one morning or afternoon. The events may be run in any sequence.

e. Uniform. The uniform for the physical readiness test is as follows: full utilities, cartridge belt, belt suspender straps, two canteens (full of water), rifle, and helmet. Depending on the heat condition, the PT shirt may be worn in place of the utility jacket. All equipment will remain secure throughout the test. Except for emergencies, canteens are not to be emptied.

f. Events

(1) Event 1: Climbing Uphill (Step-ups)This event simulates marching uphill at a rapid and steady pace. The Marine stands in front of the step. On the command GO, the Marine places either foot on the step and steps up. Hand pressure on the knee may be used if desired. The Marine then stands erect and steps backward and down, on one foot at a time. The same leg may be used for each repetition or the legs may be alternated. For 100 points, Marines will complete 60 repetitions in the maximum allotted time of 1 minute and 20 seconds (80 seconds).

*(2) Event 2: Rope Climb. T*his event resembles entering and leaving a hovering helicopter, using ropes in house-to-house fighting, and ascending and descending landing nets. The Marine assumes a sitting position on the ground at the bottom of the rope with hands at the highest point that can be reached. On the command GO, the Marine jumps up and climbs to the top. The Marine must touch the 20-foot mark with one hand, then descend without sliding or dropping. Marines must reach the 20-foot mark in the maximum allotted time of 30 seconds or less.

(3) Event 3: Evacuation. This event simulates reaching and evacuating a wounded Marine under fire. The Marine assumes the prone position at a distance of 50 yards from a casualty who is of approximately equal weight. On the command GO, the Marine springs up and covers the distance in a sprint. The Marine lifts the casualty to a fireman's carry and returns to the starting point. Assistance by the casualty is permitted. The tested individual must carry all equipment; the casualty has none. Marines must negotiate the course in a maximum time of 47 seconds or less.

(4) Event 4.'Advance by Fire and Maneuven. This event simulates advancing and assaulting during an attack. The Marine assumes the prone position at the starting point. On the command GO, the Marine then alternately crawls, runs in a zigzag manner, rolls, jumps, and assumes different firing positions for the next 130yards. Before reaching the objective, the Marine must "hit the deck," roll over, and assume the firing position three times. While assuming the firing position, the Marine must pull back the charging handle, take aim, and squeeze the trigger before rising from the ground. After rising from the ground for the third time,the Marine continues running and jumps to clear the 8-foot trench in one leap. The Marine then finishes the course in

the maximum allotted time of 2 minutes. The Marine is allowed multiple attempts only within the specified time limit.

(5) Event 5: Forced March. This event is a combat run of 3 miles without halts and against time. The Marine must double-time. Marines must complete the 3-mile course in a maximum allotted time of 30 minutes or less.

g. Event Standards

(1) To successfully pass the physical readiness test, the Marine must pass each event with a minimum of 80 points for each. Points are then assessed for each event in accordance with the scoring tables. (See fig. 6-1.)

(2) If a Marine fails one event, the highest possible score that the Marine may receive for the entire test is 79 percent. (See example 1 in fig. 6-2.) If two events are failed, the highest possible score for the entire test is 69 percent, regardless of the total points accumulated. (See example 2 in fig. 6-2.) If three events are failed, the highest possible score is 59 percent. Failure in four or more events will result in a score of zero for the entire test.

(3) To convert the total points for all five satisfactorily passed events to a final percentage score, total the accumulated points for all five events and divide by five. (See example 3 in fig. 6-2.)

Figure 6-1 is available in the illustration section at the end of the book.

Figure 6-2 is available in the illustration section at the end of the book.

6008. AIRBORNE TRAINEE PHYSICAL FITNESS TEST

a. Use and Composition of the Test.

(1) Use. The airborne trainee physical fitness test is a means of determining the physical ability of the applicant for acceptance to and retention in the airborne course of instruction.

(2) Test Events The test battery consists of five events as follows: chinups, knee bender, pushups, situps, and an endurance run.

b. Method of Scoring and Standards

(1) Scoring. The Marine will be scored by a trained scorer who is thoroughly familiar with the minimum standards for the test events. The Marine will be scored on a pass or fail basis. The performance on each event may be recorded on the reverse side of DA Form 705 (Physical Fitness Testing Record).The examinee records the personal information on the face of the card as specified on the first, second, and third lines. The face of the card is identified by writing diagonally across the lower half of the card "Airborne Trainee PPT—See reverse side."Using line 7 on the reverse side of the scorecard, the scorer enters the test title in the block entitled "(Other (Specify))" and completes the required information on the remainder of that line. The "Remarks" section can be used to record the test event titles and the applicant's performance on each test event.

(2) Standards. To successfully pass the test, the examinee must reach the standard in each test event. The standards follow:

(a) Chinups

(b) Knee bender 80(2-minute period)

(c) Pushups 45(2-minute period)

(d) Situps45(2-minute period)

(e) Endurance runs:2 miles in 15:54 minutes or less in athletic gear and 4 miles in 32 minutes in utilities and running shoes.

c. Uniform for Testing

(1) Examinees. The prescribed uniform for test participation is boots and the work uniform of the season. No headdress is worn. When climatic conditions permit, jackets or outer shirts may be removed.

(2) Officials. Scorers and other test officials should be uniformly and distinctively dressed for contrast with Marines being tested.

d. Test Administration Procedure

(1) Preparation for Test. The administration of the test to a large group makes it mandatory that the test effort be organized and efficiently operated. All testing is not completed with large test groups; on certain occasions, individuals and small groups are tested. Care must be exercised to administer the test uniformly and to standardize the conduct of all elements of the test. Regardless of the size of the test group, the following elements of sound test administration should be included:

(a) An orientation to include the purpose, method of administration, scoring of the test, preparation of the score card, and required standards.

(b) A correct demonstration of each event to ensure that there is no misunderstanding of the proper form and required standards.

(c) Completion of all five test events in one test period with all Marines taking the events in the same sequence.

(d) Adequate rest periods between the test events to allow for recovery before the next event.

(2) Method of Administration with a Large Group. With a minimum of 12 lanes per test event, 14 officials can administer the test battery to 150 or 200 Marines in 2 hours. The officials are designated as follows: one officer in charge, one demonstrator,and 12 scorers. If more or fewer Marines are to be tested, a greater or lesser number of officials will be required. The following procedure is recommended:

(a) Conduct an orientation and ensure the examinees have properly completed their score cards.

(b) Assign Marines to lanes and caution them to remain in the same lane order throughout the test.

(c) Explain and demonstrate the chinup event, administer it, and score it. Then proceed to the knee bender and pushup events, and administer them in the same manner.

(d) Grant a 5- to 10-minute rest period after the pushup event. Advise against excessive consumption of water during the break period.

(e) Explain and demonstrate the situp event, administer it, and score it as prescribed. Then move to the run area, explain the running event, and complete it.

(f) Retain the scorecards at the completion of the running event.

(3) Method of Administration with a Small Group. A similar procedure is followed for the testing of individuals and small groups. The informality usually associated with small groups must not conflict with sound test administration. With fewer examinees, fewer officials are required.

e. Description and Explanation of Test Events

(1) Test Event l:Chinups

(a) Purpose. This event is devised to test arm and shoulder flexor strength.

(b)Equipment There is one horizontal bar per lane, made of plumber's pipe or a gymnasium horizontal bar 1 1/2 inches in outside diameter. The bar must be rigidly supported at a height of 8 feet above the ground. The upright supports must be 5 feet apart. There must be a movable stand at each bar for short Marines to stand on to reach the bar.

(c) Officials There is one scorer per lane.

(d) Organization. The examinees, holding their scorecards, stand in order behind the restraining line in their respective lanes. The scorers take each scorecard when the Marine is called forward for the test.

(e) Starting Position The bar is grasped with the palms turned toward the face, the thumbs underneath the bar. The body is fully extended in a "dead" hanging position with the arms straight and the feet above the ground.

(f) Movement. Pull the body directly upward until the chin is placed over the bar. Lower the body until the elbows are completely straight and the body is again in the "dead" hanging position. Repeat as many times as required.

(g) Instructions. Explain and demonstrate the fully extended "dead" hanging position with the proper grasp. Show that the chin is placed over the bar at the top of the movement and that the arms are fully extended, the elbows completely straight, at the bottom of the movement (the hanging position). Explain that the body must be kept from swinging and that it is permissible to raise the legs and flex the hips when pulling up, but that any kicking, bicycling, or jerking motion with the trunk or legs is not acceptable. Inform the examinees that no penalty is exacted for hanging on the bar to rest in the bottom position but that this is not to their advantage. Tell them that half-completed chinups are not counted, and that the scorer will repeat the number of the last correct chinup when incorrect execution is detected.

(h) Administration and Scoring. Caution the examinees to assume the "dead" hanging position and wait for the scorer's command to begin. The scorer is at the examinee's left with a clear view of the bar. If the examinee begins to swing widely, the scorer should stop the swinging by extending the left arm across the front of the examinee's body, being sure not to hinder the execution of the chinup. The scorer counts aloud the number of chinups correctly executed. When a chinup is not correctly executed, the scorer repeats the number of the last correct one. The scorer records the number of correct chinups on the scorecard and returns the card to the examinee.

(2) Test Event 2:Knee Bender

(a) Purpose. This event measures the strength and endurance of the leg muscles.

(b) Equipment. None.

(c) Officials There is one scorer per lane.

(d) Organization. Marines stand in numerical order behind the restraining line in their respective lanes. The scorer takes each scorecard when the Marine is called forward for the test.

(e) Starting Position. The feet are spread less than shoulder width apart, hands on hips, thumbs in the small of the back, elbows back.

(f) Movement. Do a knee bend and at the same time bend slightly forward at the waist and thrust the arms between the legs until the extended fingers touch the ground. The hands are about 6 inches apart. The bend is approximately a three-quarters bend. From this knee bend position, recover to the starting position by moving the body upward, straightening the knees, and returning the hands to the hips. Repeat as many times as required.

(g) Instructions. Explain and demonstrate the correct starting position. Be certain examinees understand the correct knee bend and that only the tips of the fingers touch the ground. Tell them the scorer will repeat the number of the last correct knee bender when incorrect execution is detected. Some of the common errors are failure to correctly bend the knees, failure to touch the ground, and failure to assume the completely erect position after the bend has been executed.

(h) Administration and Scoring. The scorer stands to one side so as to see that the knees are properly bent and the fingers touch the ground as prescribed. From this position, the scorer can view the examinee to see that a properly erect position is assumed after each knee bend. The scorer counts aloud the number of correctly executed knee bends. When a knee bend is done incorrectly, the scorer repeats the number of the last correct one. The scorer records the number of correct knee bends on the scorecard and returns it to the examinee.

(3) Test Event 3:Pushups

(a) Purpose. Pushups measure arm and shoulder extensor strength.

(b) Equipment. None.

(c) Official. There is one scorer per lane.

(d) Organization. Marines stand behind the restraining line in their respective lanes until the scorer calls on them to perform. The scorer takes each scorecard when the Marine comes forward.

(e) Starting Position. The front-leaning rest position is the starting position. The body is straight from head to heels, palms are flat on the ground directly underneath the shoulders, and elbows are straight and locked. The body weight is supported on the hands and toes throughout the event.

(f) Movement. Bending only the elbows, lower the body in one straight plane until the chest touches the scorer's hand. Straightening and locking the elbows, raise the body in one straight plane, returning to the original front-leaning rest position. Repeat as many times as required, keeping the body in a straight line from head to heels.

(g) Instructions. Explain and demonstrate that the arms are straight at the beginning and completion of the movement and that the chest must touch the scorer's hand, but the stomach and thighs must not touch the ground. Also explain that the whole body must be maintained in a straight line as it is lowered and raised; that is, there is to be no breaking at the hips or shoulders so that any body part is lowered or raised in advance of the other or as a separate segment .Dipping or rolling through the shoulders is illegal, as is lowering or raising the body with one arm or shoulder at any time. Resting is not permitted during repetitions. Instruct

Marines that the scorer will repeat the number of the last correct pushup when incorrect execution is detected.

(h) Administration and Scoring. It is recommended that Marines assume a prone position while placing their feet and hands in the proper positions. This permits them to rest while the scorer gets into position and, at the same time, provides a feel of the body in a straight plane from head to heels. The scorer lies on the right hip and side to the right of the examinee. This gives the scorer a clear view of the examinee's body to see any errors. The palm of the scorer's right hand rests flat on the ground underneath the lowest part of the examinee's chest. By keeping the right forearm flat on the ground at an angle in front of the examinee's right arm, the scorer's position will not prevent the examinee from lowering the body completely. The scorer's left hand is free to test the straightening of the elbow at the completion of the movements and to point out body segments being lowered or raised separately. When in position and ready, the scorer has the examinee assume the starting position and begin doing pushups. The scorer counts aloud the repetitions done correctly and repeats the number of the last correct pushup if one is done incorrectly. There is no penalty if the contour of the examinee's body causes the hips to protrude slightly out of line, provided that the whole body is raised and lowered simultaneously. The scorer enters the number of repetitions on the scorecard and returns it to the examinee.

(4) Test Event 4: Situps

(a) Purpose. This event primarily measures abdominal strength.

(b) Equipment. None.

(c) Official. There is one scorer per lane.

(d) Organization Marines stand behind the restraining line in their respective lanes until the scorer calls on them to perform. The scorer takes each scorecard when the Marine comes forward.

(e) Starting Position. The examinee lies flat on the back with knees flexed, both feet flat on the ground. The correct angle of the thighs to the ground is 45 degrees. If the heels are too near the buttocks, the applicant will not be able to sit up. Legs are spread shoulder width apart. The examinee interlaces fingers and places them behind the head in contact with the ground. The feet are not held by another person.

(f) Movement. Bend forward at the waist and raise the upper body until the head is directly over the knees. Heels are not to leave the ground. Elbows remain in the same plane to the head and body throughout the event. The upper body is slowly lowered to the starting position until the head touches the ground. Repetitions are done at a slow cadence with no rest periods.

(g) Instructions. Explain and demonstrate the correct starting position and the proper execution of the situps to be sure that Marines understand the movement. Warn them that their knees must remain flexed during each situp, the heels cannot leave the ground at any time, and they may not roll up on one side and push up with one elbow. Tell them they must do the repetitions at a slow cadence with no rest periods. Instruct Marines that the scorer will repeat the number of the last correct situp when incorrect execution is detected.

(h) Administration and Scoring. When the performer is in position and ready, the scorer has the performer assume the starting position and begin doing situps. The scorer counts aloud the correct executions. When a situp is improperly done, the scorer repeats the number of the last correct one. No situp is credited if the hands are unclasped from behind the head, if the back is used to bounce up from the ground (which means the shoulders would not touch the ground), or if one shoulder or elbow is used to push up. The examinee is not penalized if the heels slide forward slightly so long as the knees remain flexed and the heels maintain contact with the ground. The scorer enters the number of repetitions on the scorecard and returns it to the examinee.

(5) Test Event 5:Endurance Run

(a) Purpose. This event measures cardiovascular endurance.

(b) Equipment One stopwatch or watch with a sweep second hand.

(c) Area. A large training field on which a quarter-mile track has been staked out or a level road over flat terrain may be used as a running surface. A 1-mile route is designated with wooden stakes marking the start point, finish point, and one-quarter mile intervals.

(d) Officials For large groups, a scorer times the event and controls the conduct of the run, and a guide runs with the group and sets the pace.

(e) Organization The run is conducted with groups of Marines in a column formation. Company-sized units may run at the same time with platoons serving as running groups. The scorer issues the command to assume double-time.

(f) Starting Position. Marines are assembled in proper column formation (column of twos, threes, or fours, as appropriate to the size of the group) with short men to the front. When all is ready, the column is moved forward a short distance before the running period is started.

(g) Movement. At the command DOUBLE-TIME, MARCH, the examinees retain their places in the column formation and execute the command. Length of steps is about 40 inches. The scorer has the group execute the run. The formation is maintained during therun.

(h) Instructions. Marines are instructed to maintain formation while running and are informed that the guide will set the proper pace. They will be instructed in the command used to control the column in the execution of the test. The scorer should announce the 4-minute, 2-minute, 1-minute, and 30-second remaining time intervals.

(i) Administration and Scoring. The event may be administered as previously prescribed to **a** large group, to several Marines, or to an individual. An individual examinee usually does not require a guide or pacer. If the event is administered on a training field, the scorer may stand in the center as the runner(s) circles about the field. This method of administration relieves the scorer of running with each group to be tested. Scoring is based on successful completion of the run as prescribed. The scorer should announce the remaining times as prescribed in (h).

6009. BATTLE FITNESS TEST

The battle fitness test is a physical fitness evaluation which is currently under development as part of the basic warrior training concept plan.

Chapter 7
THE HUMAN BODY

7001.GENERAL

To implement an effective physical conditioning program, a leader must understand how the human body functions and how exercise affects the body .This chapter provides only an introductory discussion of anatomy and functioning.

a. Body Functioning During the Stages of Conditioning. For more on the stages of conditioning, see chapter 1.
(1) During the toughening stage, the waste products of muscle activity (lactic acids) collect more rapidly than the blood can remove them. This acid waste builds up in the muscle tissue and irritates the nerves in the muscle fiber, causing pain and stiffness. As the exercise program continues, more blood is carried through the muscle, removing the waste materials more rapidly and eventually causing the soreness to disappear.
(2) During the slow improvement stage, the blood circulation in the muscles increases, and the body as a whole becomes a more efficient machine. The improvement is rapid in the first few weeks, but as a higher level of skill and conditioning is reached, the improvement becomes less noticeable. The body reaches its maximum level of performance after 6 to 10 weeks and should then be maintained at this peak.
(3) During the sustaining stage, it is possible to maintain this state of conditioning through 15 to 20 minutes of exercise a day, but the exercise must be quite strenuous.

b. Diet. Regular exercise increases the appetite. If the desire for greater amounts of food is satisfied by a balanced diet, the body benefits.
(1) There are two main types of foods: body-building and energy- producing.
(a) Body-building foods consist of proteins, which build up tissue and repair wear and tear.
(b) Energy-producing foods include carbohydrates and fats. Carbohydrates provide a quick source of energy, while fats act as a reserve store of energy.
(2) In addition, food contains vitamins, mineral salts, and water. During hot weather and strenuous training periods, the body requires greater fluid intake.
(3) Diet should be supplemented with proper rest to allow the digestive system to digest the lood.
(4) Occasionally, especially during early stages of conditioning, strenuous exercise may cause vomiting. Although vomiting is not a frequent occurrence, it is not usually a cause for concern.

7002. SYSTEMS OF THE BODY

The systems of the body include the skeletal, muscular, circulatory, respiratory, endocrine, digestive, genitourinary, and nervous systems. Each has a different function, but all must work in cooperation with one another to insure a sound body. Of these systems, the first four are the most affected by exercise and are discussed in succeeding paragraphs.

7003. THE SKELETON

a. Bones. The skeleton (see fig. 7-1) is composed of about 206 bones, which-
o Provide structure for the body and support for the attachment of muscles,
o Protect vital organs, such as the brain, lungs, and heart,
o Manufacture red blood cells, which carry oxygen through the body.

Figure 7-1 is available in the illustration section at the end of the book.

b. Joints. A joint is a place of union between two or more bones. Joints can be-
o Immovable, such as in the face and head.
o Slightly movable, such as in the vertebrae or sternum.
o Freely movable, such as in the shoulder, hip, knee, ankle, and elbow. The bones in a freely movable joint are held in place by ligaments. Freely movable joints are of the greatest importance in physical training because exercise improves their mobility and stability.

c. Cartilage and Ligaments. The joints in the body, where bones connect, are supported by cartilage and ligaments.
o Cartilage is a tough, elastic, translucent tissue that acts as a shock absorber or buffer between bones. Examples are the discs between the vertebrae, the tissue attaching the ribs to the sternum, and buffers in the knee joints. Damaged cartilage does not heal.
o Ligaments are connective tissue that bind bones together. They contract and extend but are not elastic. Consequently, ligaments can be torn or strained. Damaged ligaments will heal to some extent.

7004. THE MUSCLES

a. General. Muscles are tissue, or an organ composed of tissue, which contract or extend to produce movement. Muscles are of three classifications: involuntary, voluntary, and cardiac. The voluntary and cardiac muscles are the most affected by physical training; physical training has little effect on involuntary muscles.(Seefigs. 7-2, 7-3, 7-4.)
o People have no control over the involuntary muscles, an example of which is the diaphragm,
o Voluntary muscles are the larger skeletal muscles which cause movement and which the individual can control.
o Cardiac muscle is found only in the heart and for all practical purposes is an involuntary muscle. Physical training has a significant effect on the fitness of cardiac muscles.

b. Muscle Structure. Muscle is composed of cells which are grouped into bundles called fibers. Groups of fibers form muscles. These bundles of fibers are held in place by sarcolemma, a thin, sheath-like material that surrounds the muscle bundles and secretes a fluid that lubricates the muscle tissue. The fused ends of the sarcolemma form the tendons which attach the muscles to the bones.

c. Attachment of Muscles. The arrangement of muscles on the skeleton provides the proper angle of pull to make movement possible. Voluntary muscles are usually attached to the skeleton in two places.

d.Action of Muscles. To produce motion and to do work a muscle usually shortens its fibers. The movement may be flexion such as bending the arm at the elbow, or extension, such as the straightening the arm. Some muscles raise the arms or legs, others depress the raised limb. Some muscles have the primary function of rotating the trunk from side to side. In addition, muscles act as stabilizers as well as prime movers.

Figure 7-2 is available in the illustration section at the end of the book.

Figure 7-3 is available in the illustration section at the end of the book.

Figure 7-4 is available in the illustration section at the end of the book.

7005. FUNCTIONING OF THE SKELETON AND MUSCLES

a. General. The skeleton and muscles function in close coordination. Muscles move the body, and bones support the muscles.

b. Effect of Exercise on Bones. Continuous exercise, particularly among younger people, usually strengthens the bones, allowing them to withstand greater stress and strain. Bones which are not used lose much of their strength, a factor which should be considered when individuals return to physical training after a prolonged period of inactivity. People in this category should refrain from activities which might result in bone breakage before the bone is exercised back to normal condition. The condition known as "march fracture" is usually a result of this condition.

c. Muscular Strength. When exercised vigorously to improve strength, muscle grows in size. In general, the larger the muscle, the stronger the muscle. Furthermore, trained muscle functions more efficiently than untrained muscle. Trained muscles contract more vigorously and with less effort. Regular and strenuous exercise also toughens muscle, making it firmer and able to stand more strain.

d. **Muscular Endurance.** Muscles naturally become fatigued during continuous, repeated work. Through specific exercises, such as chinups or situps, local muscle groups become exhausted long before an individual fatigues. Training to develop muscular endurance enables people to continue a relatively heavy load of exercise over a long period of time. Lengthened exercise periods improve muscular endurance.

e. Muscular Coordination. Improved speed and strength result in part from improved muscular coordination. An unskilled individual may use irrelevant muscles to perform a particular activity, increasing the amount of work without increasing the mechanical output. This increase in skill is highly desirable, but it should be offset by greater effort in duration or intensity to compensate for the loss in overload due to increased skill.

f. Muscular Fatigue. When the rate of work is increased, the energy required is proportionately much greater than the increase in rate. For example, if an individual doubles running speed, the amount of power demanded to do this is increased eight times.

g. Circulation in Muscles Regular, strenuous exercise causes the creation of new capillaries and the opening of inactive, latent capillaries, increasing blood circulation within the muscle by as much as 400 percent. This increases the supply of food materials and oxygen to the muscle, improving its endurance. This process takes about 8 to 12 weeks of regular conditioning in young adults and longer as age increases.

7006. THE CIRCULATORY AND RESPIRATORY SYSTEMS

a. The Circulatory System. The circulatory system transports blood to all parts of the body, removes waste products for disposal, and delivers protecting and repairing substances where needed. The circulatory system includes the heart and blood vessels.

(1) The Heart. The heart is a pump which forces blood through the blood vessels to the parts of the body. The heart is a little larger than a fist and is located in the left center of the chest between the lungs.

(2) Blood Vessels. The vessels carrying blood away from the heart are the arteries. They eventually divide into capillaries, the very small vessels through which diffusion and osmosis take place. The capillaries gradually increase in size, forming the veins which carry blood back to the heart.

b. The Respiratory System. The respiratory system performs the function of breathing, whereby oxygen is inhaled and carbon dioxide is exhaled. The respiratory system consists of the mouth, nose, trachea, lungs, and diaphragm.

o Trachea. The trachea, or windpipe, is a hollow, tube-like structure that carries air from the mouth to the lungs.

o Lungs. The lungs are elastic bags in the chest where the exchange of oxygen and carbon dioxide takes place.

o Diaphragm. The diaphragm is a thin, sheet-like muscle just below the lungs. During inhalation, the diaphragm flattens out and lowers, allowing the lungs to expand and fill with air. During exhalation, the diaphragm raises into a dome shape, helping to expel air from the lungs.

7007. CARDIOVASCULAR FUNCTIONING

a. General. Cardiovascular functioning is the combined functioning of the circulatory and respiratory systems. The chief organs of these systems, the heart and lungs, function together to provide oxygen to the body.(See fig. 7-5.)

Figure 7-5 is available in the illustration section at the end of the book.

b. Heart Action. The heart, the organ which propels blood through the blood vessels, is the chief organ of cardiovascular endurance. A person tires quickly if the heart lacks the capacity to circulate the blood through the body. The heart is a muscular organ and is developed by exercise just like any other muscle. The heart cannot be exercised alone; any exercise that puts a sufficient load on the heart, such as running or cycling, will also exercise other body parts. A conditioned heart tends to beat more slowly but pump more blood when at rest. This is known as an increase in "stroke volume,"a desirable condition because it enables the heart to pump more blood with a slower contraction rate.

c. Functions of the Lungs. Strenuous and regular exercise can improve the efficiency with which the lungs transmit oxygen to the blood by as much as 25 percent In a poorly conditioned person, some of the alveoli (air sacs) within the lungs are closed or collapsed. The forced breathing created by exercise over a period of weeks causes the air sacs to slowly expand, increasing the ability to absorb oxygen.

d. Relationship of Heart and Lungs. The heart pumps carbon dioxide-laden blood through the pulmonary artery to the lungs. In the lungs the carbon dioxide is exchanged for oxygen and the purified blood is returned to the heart by way of the pulmonary vein. The heart then pumps the blood through the aorta for circulation throughout the body. Blood moving into the muscles exchanges oxygen for carbon dioxide, after which time it is ready for the return trip to the heart.

e. Cardiovascular Functioning in High Altitudes. Marines to be employed in high altitudes should be acclimated in a similar area for 10 to 14 days prior to employment. Persons not accustomed to the rarefied air of higher altitudes tire more quickly and may collapse after rapid physical exertion. Air is much less dense at high altitudes than at sea level. Persons accustomed to sea level inhale only about 80 percent of the oxygen they are used to no matter how hard they breathe. Further, people accustomed to sea level or moderate altitude simply do not have enough red corpuscles in the blood to fulfill their needs at high altitudes. The red blood cells will increase over time.

f. Symptoms of Cardiovascular Malfunctioning. The brain is the first organ to react to a lack of oxygen. Unconsciousness results when the brain is denied sufficient oxygen. "Blacking out" is actually a defense mechanism to keep the body alive.

APPENDIX A
INSTRUCTOR HINTS AND INSTRUCTOR TRAINING

1. General. This appendix informs instructors on administration of exercise periods, commands, the extended rectangular and circle formations, methods of instruction and conduct of exercise activities, explanation of basic positions associated with the various activities, and instructor training.

2. Aim of the Physical Fitness Program. The aim of the physical fitness program is to prepare Marines physically for war. This can be conveniently split into two phases.

a. Preparation Phase

(1) Development of all-around physical fitness.

(2) Development of purposeful physical skills.

(3) Development of mental alertness.

(4) Development of character and leadership.

b. Maintenance Phase. Once the preparation is completed, the program must progress to a maintenance program. This phase revolves around the need to maintain the trained Marine's physical fitness for a particular role. This will include—

(1) Maintenance of a high standard of all-around physical fitness.

(2) Application of purposeful physical skills to combat skills.

(3) Development of leadership and self-confidence.

(4) Development of mental toughness and endurance.

3. Physical and Mental Development

a. Marines need to be developed physically and mentally. The physical training (PT) instructor must ensure that the program accomplishes this.

b. The system of the body functions best in response to progressive exercise. This means that Marines should master the easier exercises before they attempt the more difficult ones.

c. The body and mind are linked inseparably and must be considered as one in physical fitness training. Physical exercises assist in developing the mind.

d. The confidence in oneself and in one's comrades attained by physical achievement helps to promote the indomitable will to win. The Marine must be taught to realize this. It will help during arduous training, which at all times must be imaginative, realistic, and filled with enthusiasm. All physical fitness training must be associated with developing the qualities required of a Marine.

4. Planning and Preparation of ProgramsThe following factors must be considered when compiling a program (see also ch. 2)~

a. Aim or desired skill level.

b. Duration of the complete training program and number of physical training lessons.

c. Purpose of included exercises,drills, or practical applications.

d. Beginning/inventory fitness levels.

e. Medical facilities, medical personnel, and Marine Corps or local command regulations governing their proximity to the training.
f. Number of Marines to be trained.
g. Availability of instructors, assistants, and facilities.
h. Apparatus and equipment available.
i. Alternative training areas for varying weather conditions,
j. Stages of progression and tests of achievement.

5. Five Basic Principles of Program Design

a. Regularity. Regularity of exercise is far more important than the amount of exercise performed on a given day. A Marine should exercise a minimum of three times per week. Human muscle begins to atrophy after 3 days without exercise.

b. Balance. The program must be balanced in terms of total physical fitness. Imbalance is created when, for example, an individual merely runs. While the cardiovascular system is improved, general upper body strength may be lacking. Conversely, a person who just does weight training may develop strength, but not cardiovascular endurance.

c. Overload. If total body fitness is to be developed, then the muscular and cardiovascular systems must be made to perform more work than previously. This is done by increasing the frequency, intensity, and duration of the exercise. The body adapts to the amount of stress placed upon it and becomes stronger and more efficient.

d. Progression. To be successful, any type of training program must be progressive. This is an area where untrained persons conducting PT can cause injury to Marines resulting in failure of the PT program. Close supervision must be made of untrained instructors. The result of progressing too fast is that the class begins to reject PT because it is no longer enjoyable or motivating. The program should start at a slow pace, and the pace should increase as proficiency improves. Average persons will encounter three stages of progressiom the toughening stage, the slow improvement stage, and the sustaining stage. (See ch. 1.)

e. Variety. Some programs fail because the training becomes boring. Variety is essential and perhaps presents the greatest challenge to the commander and staff.

6. Grouping. Providing for different levels of physical fitness is particularly recommended in the early stages of conditioning. Older Marines and those in poorer physical condition should be expected to attain a group level of fitness, but they should be given more time to do it.

a. Homogeneous Grouping One simple method of providing for the difference in levels is to group them according to their condition. A two-group classification would divide Marines into highly conditioned and average groups. A finer classification could be obtained by dividing them into three groups—a highly conditioned, a moderately conditioned, and an unconditioned group.

b. Performance or Age Grouping. The segregation of Marines into different exercise groups should be based on physical fitness test scores or on the level of fitness they demonstrate. They may also be grouped at first according to age. A common classification by ages is under 30, 30 to 34, and 35 and above.

7. Preparatory Commands and Commands of Execution. The prepreparatory command describes and specifies what is required. The command of execution calls into action what has been prescribed. All preparatory commands are given with a rising inflection. The interval between commands is long enough to permit the average Marine to understand the first one before the second one is given.

8. Extended Rectangular Formation. The traditional formation for carrying on many physical training activities is the extended rectangular formation (fig. A-1). This formation is more compact than any other. It is the best type to use for large numbers of Marines because it is simple and easy to assume.

a. For the formation of oneplatoon, the base man paces off five paces from the stand, faces left and moves five paces, halts, and again faces left. With the base man positioned facing the stand, the platoon leader then commands. **FALL OUT AND FALL IN ON THE BASE MAN**. At this command, all Marines run to the designated area and reform. This procedure is preferred to marching the unit into position. If more control is desired, the unit may march at double time to the vicinity of the base man and then be directed to fall out and fall in on him. Much valuable time is wasted in the field by needless maneuvering of Marines at quick time in an effort to position the platoon or unit on the exact spot for the exercises.

b. A company-size unit assumes the extended rectangular formation from a column of threes or fours at normal intervals between squads. This extension can also be executed from a company mass without interval between platoons. In extending either a platoon or company-size unit, take your place at the head of the column and command:

(1) TAKE INTERVAL TO THE LEFT, MARCH. At this command, the Marines in the right flank file stand fast with arms extended sideward. All others turn to the left and run forward at double time. After taking a sufficient number of steps, all Marines face the front with both arms extended sideward. The distance between finger-tips is about 12 inches, and dress is right.

Figure A-1 is available in the illustration section at the end of the book.

(2) ARMS DOWNWARD, MOVE. At this command, the arms are lowered smartly to the sides.

(3) LEFT, FACE

(4) EXTEND TO THE LEFT, MARCH. At this command, those in the right flank file stand fast with arms extended sideward. All others turn to the left and run forward at double time. Spacing is the same as in (1) above, and dress is right.

(5) ARMS DOWNWARD, MOVE. Same as in (2) above.

(6) RIGHT, FACE.

(7) FROM FRONT TO REAR, COUNT OFF. At this command, the leading Marine in each column turns the head to the right rear, calls of **ONE** and faces the front. Successive Marines in each column call off in turn **TWO, THREE, FOUR, FIVE** in the same manner.

(8) EVEN NUMBERS TO THE LEFT, UNCOVER. At this command, each even-numbered Marine stride jumps sideward to the left, squarely in the center of the interval. In doing this, each one swings the left leg sideward, jumps from the right foot to the left foot, and smartly brings the right into position against the left.

c. To assemble the unit, you commandASSEMBLE TO THE RIGHT, MARCHAt this command, all return to their original position in the column at double time and reform on the base man.

d.It is recommended that the area for grounding equipment and arms be at the edge of, or well away from, the area to be used for exercising. To conserve time and ensure proper position of the unit, the base man or, if the unit is composed of several platoon-sizegroups, the various base men may precede the unit and establish their positions in relation to the instructor's stand.

9. Circle Formation.The circle formation is effective for conducting various exercise activities (fig. A-2). This formation has an advantage in that supervision of all Marines is facilitated and a moving formation is available which provides control. Guerrilla exercises, grass drills, and some forms of running are examples of activities which are more easily conducted in the circle formation than in the extended rectangular formation.

a. It is not advisable to have more than 60 Marines in a circle. When more must be accommodated, separate circles should be used.Concentric circles may be employed to reduce the size of the circle or to accommodate more persons. If concentric circles are formed, the different circles are made by designating squads for each circle.Each additional circle requires more Marines than the one inside it.For example, one squad of a platoon may form the inner circle and theremaimng three squads the outer circle.

b. When a platoon is to form a circle, the commands ai€IRCLEFORMATION, MARCH, FOLLOW MEUpon this command, the right flank squad of the column moves forward at double time with the leader of the platoon group gradually forming a circle in a counterclockwise direction. Each succeeding file falls in behind that on the right. After the rough outline of the circle is formed, the leader commandsPICK UP A **5**-YARD INTERVAL.

Figure A-2 is available in the illustration section at the end of the book.

c.The group may be halted and faced toward the center, or, if instruction is not necessary, the exercise activity may be executed without stopping the platoon.

lO.Leadership Techniques
a. Unless you experience all the exercises, you cannot appreciate how arduous they are, what movements are most strenuous and difficult, where the errors in performance are likely to occur, and what the proper cadence should be.

b. You must give everyone careful supervision and participate in the exercises to show that you can do them. When you participate, your assistant instructors should supervise because it is difficult for you to supervise and exercise at the same time.

c. Marines should never be kept too long in one position, especially a constrained one. They should never have to perform an exercise more times than they can do it without losing the proper form. You should insist upon proper form in the execution of all exercise activities.Even slight deviations from the proper form reduce the value of the exercise.

d. Avoid long explanations.As a rule, it should be necessary to give a full explanation of new exercises only. Explain the most essential features of an exercise first; add details later. Too many details at one time are more likely to confuse than to assist them.Minor corrections should be made to the entire class while the exercise is in progress (for exampHEADS **UP,** KNEES STRAIGHT). If necessary, follow this direction by the name of the person who is particularly at fault. If a Marine requires special attention, givethat person separate instruction by an assistant instructor to avoidwasting the time of the group.

e. The heavy demand on your voice can be lightened by training assistant instructors to assume some of the instruction. Using mass cadence is also an effective method of lessening the demand on your voice.

“Illustrations”

DAY	TIME	ACTIVITY
MONDAY	60 MIN	RUNNING ACTIVITY
TUESDAY	60 MIN	COMBATIVES
WEDNESDAY	60 MIN	CONDITIONING & LOG DRILLS
THURSDAY	60 MIN	CONDITIONING & RELAYS
FRIDAY	60 MIN	RUNNING & CONDITIONING
SATURDAY	60 MIN	RUNNING ACTIVITY

Figure 2-1. Sample A- -Single Activity.

DAY	TIME	ACTIVITY	REMARKS
MONDAY	20 MIN 20 MIN 20 MIN	RIFLE DRILL COMBATIVES RUNNING	PROGRESSIVE ACTIVITY SYSTEM
TUESDAY	30 MIN	GUERRILLA EXERCISES TABLE I	SINGLE ACTIVITY
WEDNESDAY	30 MIN	RIFLE DRILL	SINGLE ACTIVITY
THURSDAY	30 MIN	CIRCUIT INTERVAL TRAINING	SINGLE ACTIVITY
FRIDAY	20 MIN 20 MIN 20 MIN	STRENGTH CIRCUIT CONDITIONING DRILL CROSS-COUNTRY RUN	ROTATION ACTIVITY
SATURDAY	60 MIN	WARM UP RUN AND GRASS DRILL	SINGLE ACTIVITY

Figure 2-2. Sample B- -Various Time Periods, Activities, and Systems.

DAY	TIME	ACTIVITY
MONDAY	45 MIN	RUNNING – CROSS-COUNTRY
TUESDAY	75 MIN	TEAM ATHLETICS – TOUCH FOOTBALL
WEDNESDAY	45 MIN	RUNNING – CROSS-COUNTRY
THURSDAY	20 MIN 40 MIN	LOG DRILL DUAL COMBATIVES
FRIDAY	90 MIN	SPEED MARCH
SATURDAY	90 MIN	TEAM CONTESTS

Figure 2-3. Sample C--Sustaining Activities.

Route	Kilometers from SP	Recommended rate of march (kmph)	Remarks
SP: RJ 8th Div Rd—Superhighway	......	4	Bridge; hard surface; two guides.
RJ Jamestown—Hersey Rd	2,25	4	Traffic heavy; two guides; Jamestown Rd bears to right.
RJ Jamestown—Yankee Rd	7.10	4	Two guides
RJ Jamestown—Lightning Rd	10.50	4	Two guides
RJ Jamestown—Sunshine Rd	12.40	4	Light traffic; one guide Sunshine Rd; gravel, poor traction when wet.
RP; RJ Sunshine—Sedan Rd	14.80	4	Two guides

Figure 3-1.Example of Route Reconnaissance Report.

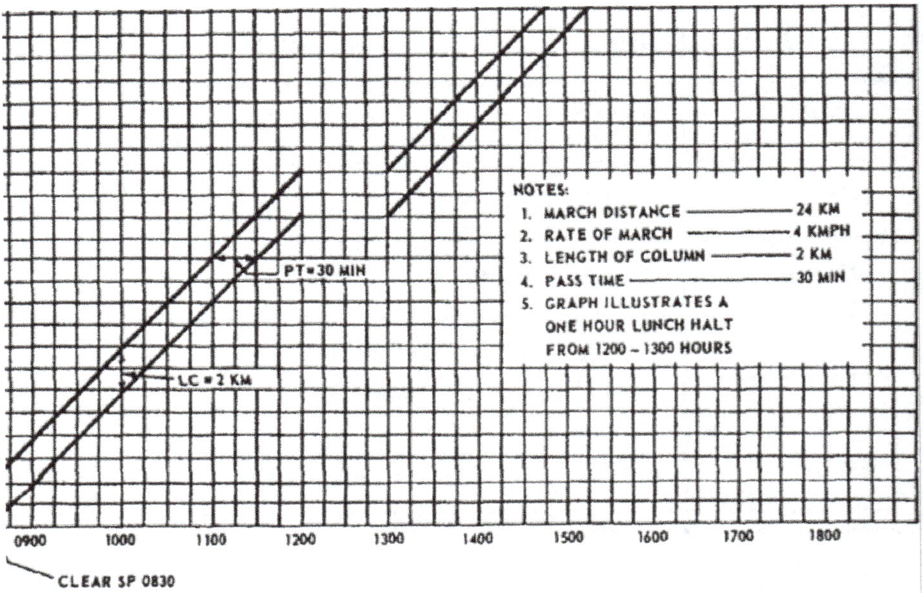

Figure 3-2.Road Movement Graph.

Figure 3-3.The Pace-Setter.

	SINGLE FILE	COLUMN OF TWO
2 m/MAN APART	24	12
5 m/MAN APART	54	27

GAP DISTANCES			
SERIAL:	(3-1) × 100 m	=	200 m
MARCH UNIT:	(12-1-2-) × 50 m	=	450 m
	TOTAL COLUMN GAP		650 m

Figure 3-4.Conditioning Drill One.

Figure 3-5.Conditioning Drill Two.

Figure 3-6.Conditioning Drill Three.

Figure 3-7. Rifle Drill.

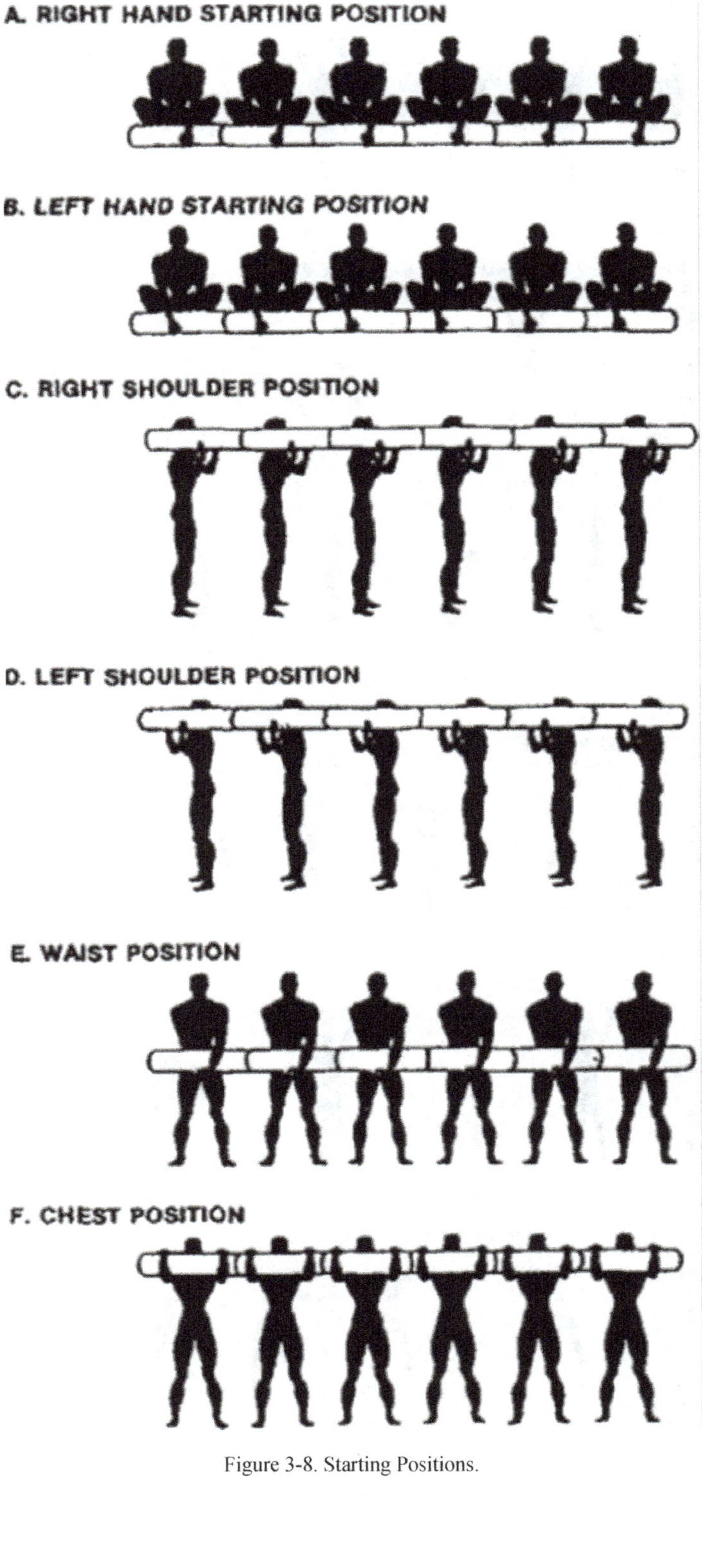

Figure 3-8. Starting Positions.

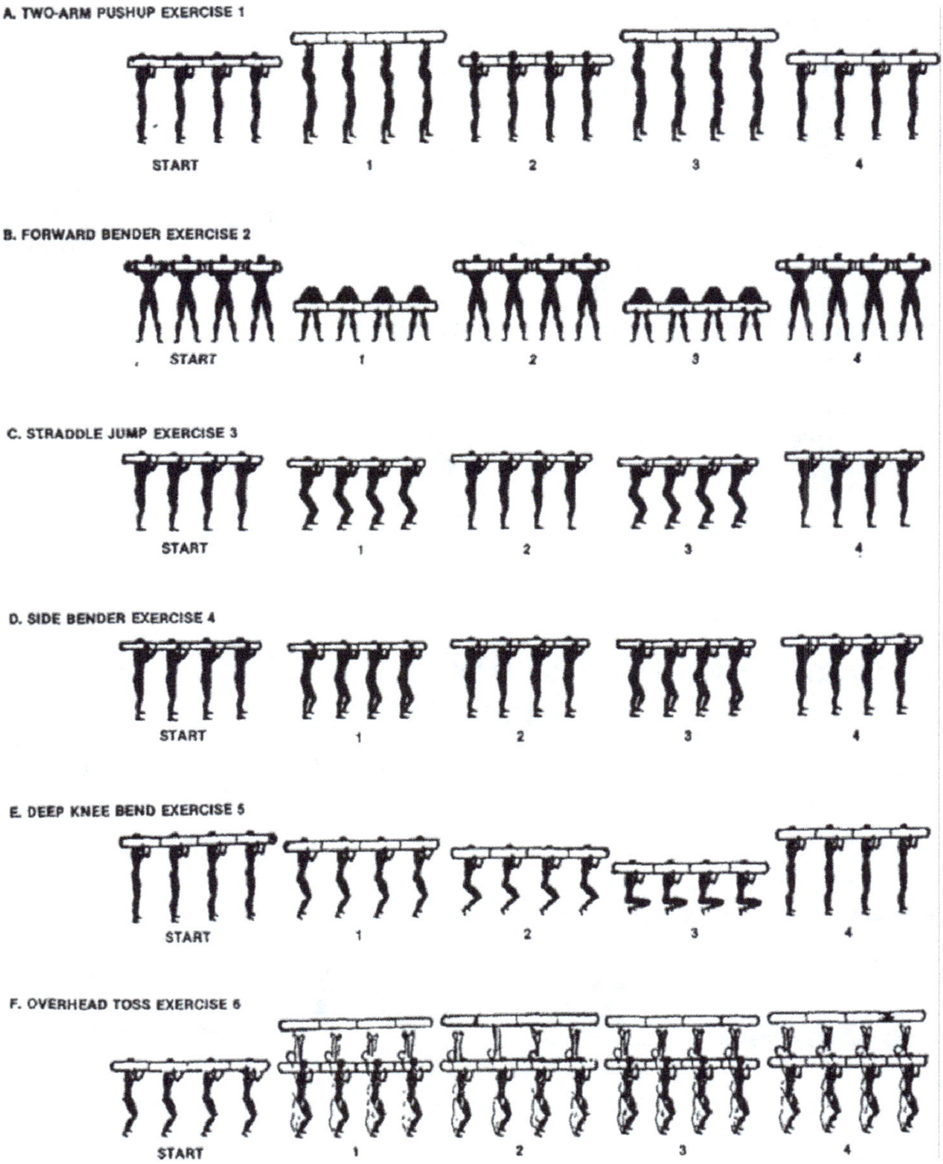

Figure 3-9. Log Drill

Figure 3-10.Basic Positions for Grass Drill.

Figure 3-11.Grass Drill One.

Figure 3-12.Grass Drill Two.

Figure 3-13.Guerrilla Table I.

Figure 3-14.Guerrilla Table II.

Figure 3-15.Proper Running Form.

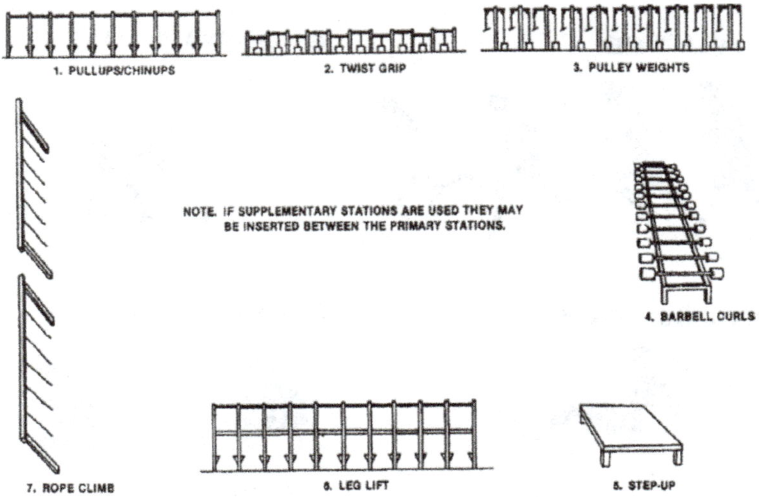

Figure 3-16. Fixed Strength Circuit.

STATION	ITEM	NO.	SPECIFICATION
1	BARBELL		1 ¼-INCH PIPE 5 FEET LONG WITH CONCRETE FILLED NO. 10 CANS
2	JUMP ROPE		¼- OR ⅜-INCH ROPE, 10 FEET LONG
3	TWIST GRIP		HANDLE 12 INCHES LONG, ROPE 4 FEET LONG, NO. 10 CAN CONCRETE FILLED.
4	INCLINE PLANE		¾-INCH PLYWOOD PLATFORM 2 FEET WIDE AND 6 FEET, 6 INCHES LONG, ELEVATED 10 INCHES AT ONE END. STRAP TO HOLD FEET DOWN.
5	WAR CLUB		HEAD IS 6 BY 12 INCHES, HANDLE IS 14 INCHES LONG BY 1 ¼ INCHES IN DIAMETER, ABOUT 20 POUNDS.
6	BICYCLE RIDE		PLYWOOD BOARD OR PLATFORM 2 BY 3 FEET WITH 2 BY 2 RUNNERS.
7	STEP – UP		A BOX OR STURDY PLATFORM 18 INCHES HIGH, 18 INCHES WIDE, 24 INCHES LONG.
8	ISOMETRIC PULL		TWO HANDLES 12 INCHES LONG WITH 4 FEET (BETWEEN HANDLES) OR LIGHT WIRE CABLE OR ¼-INCH ROPE

Figure 3-17.Movable Strength Circuit.

Figure. 3-18.Stations of Movable Strength Circuit.

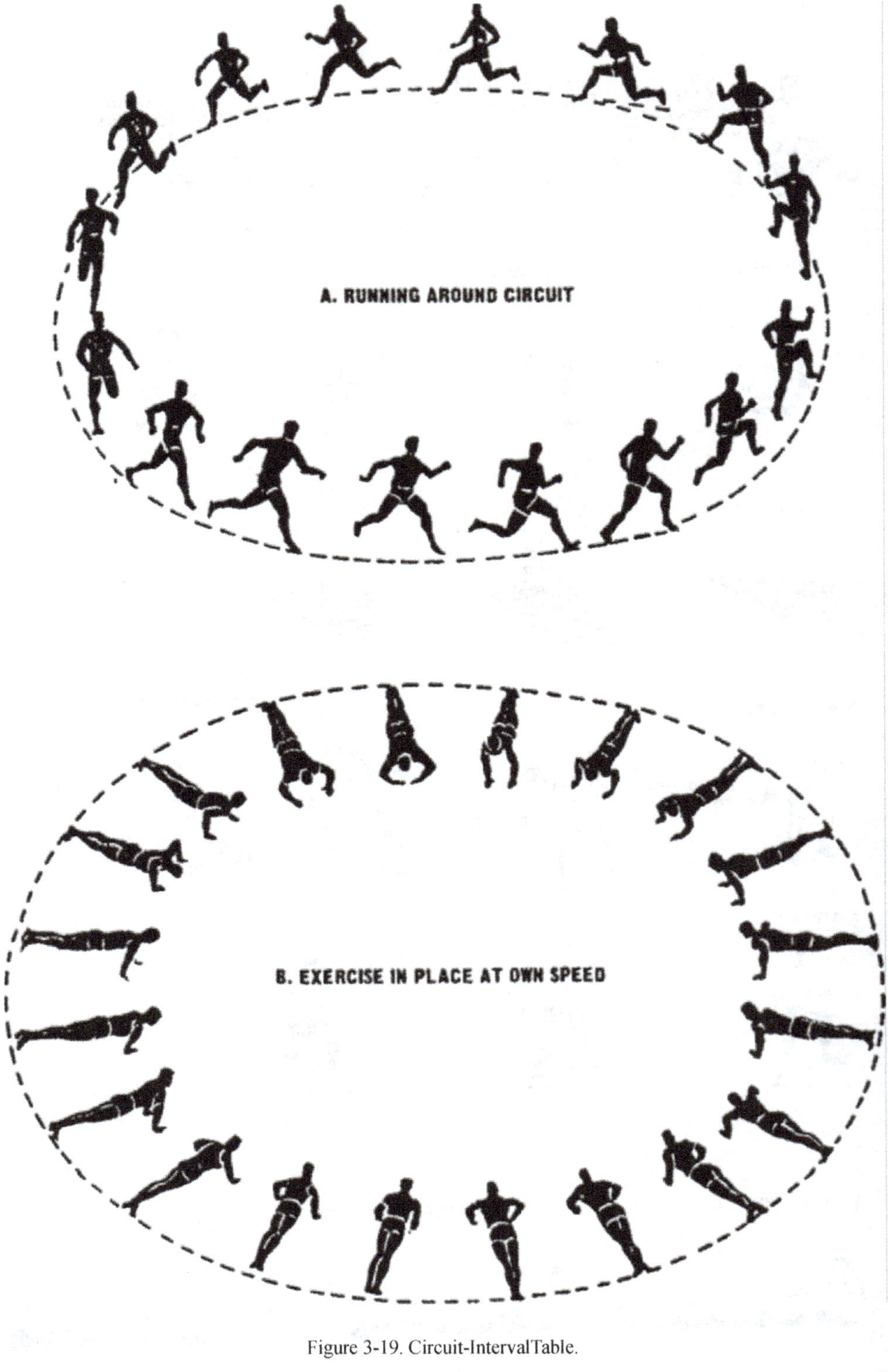

Figure 3-19. Circuit-IntervalTable.

Figure 3-20. Jumping.

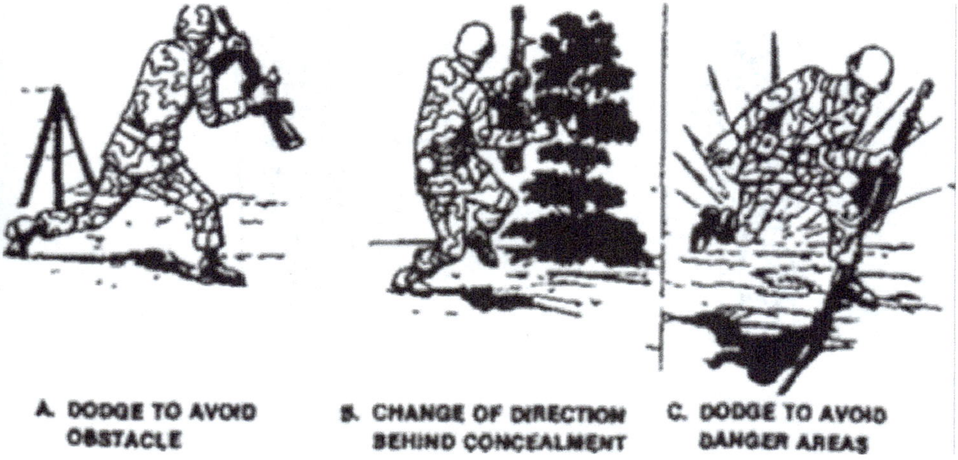

Figure 3-21.Dodging.

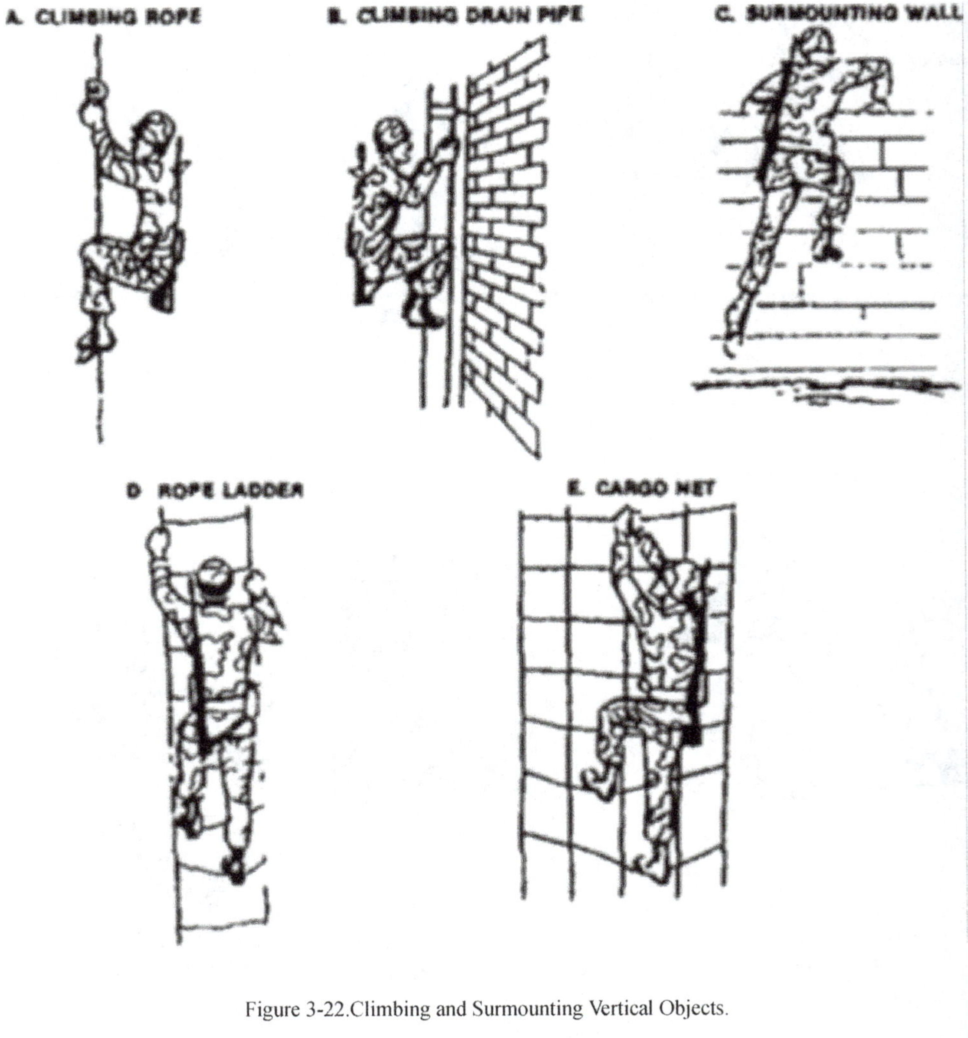

Figure 3-22.Climbing and Surmounting Vertical Objects.

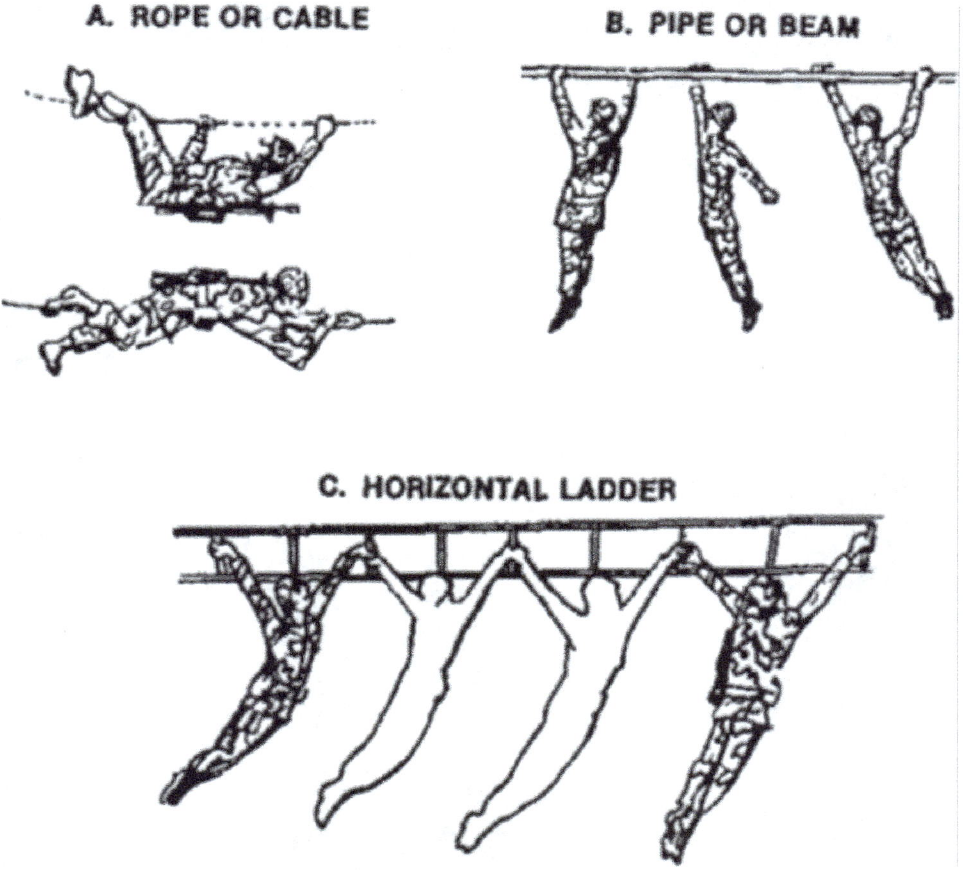

Figure 3-23. Traversing Horizontal Objects.

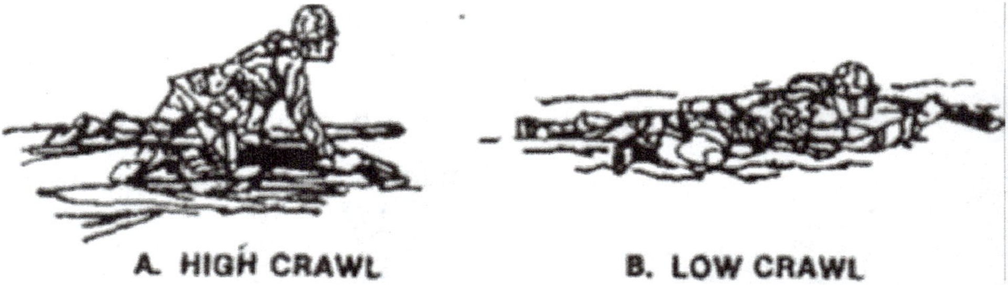

Figure 3-24.Crawling.

Figure 3-25.Throwing.

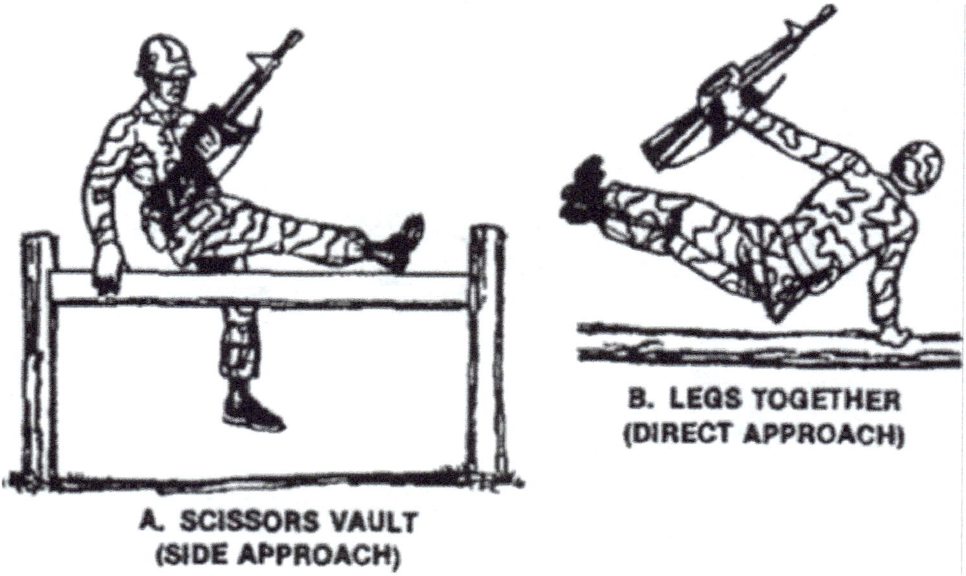

Figure 3-26. Vaulting

Figure 3-27.Balancing.

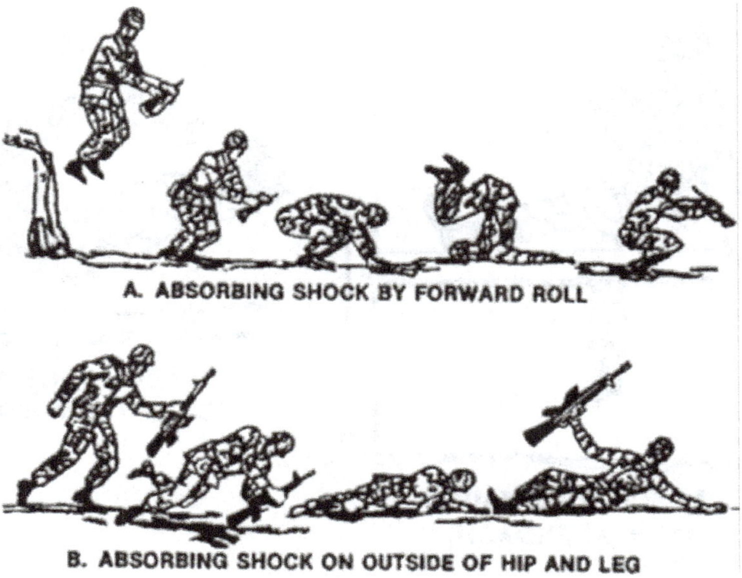

Figure 3-28.Falling.

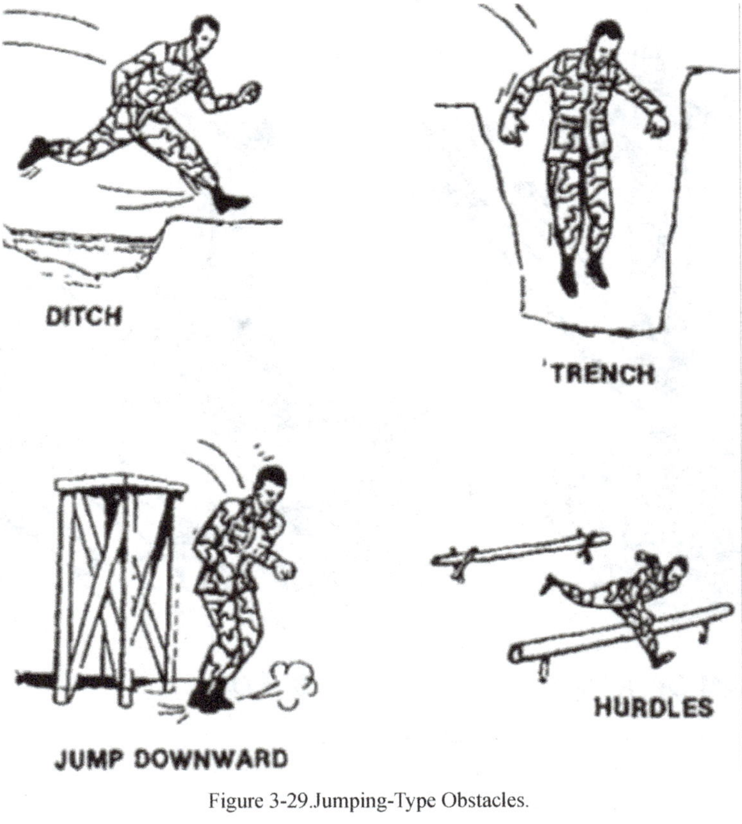

Figure 3-29.Jumping-Type Obstacles.

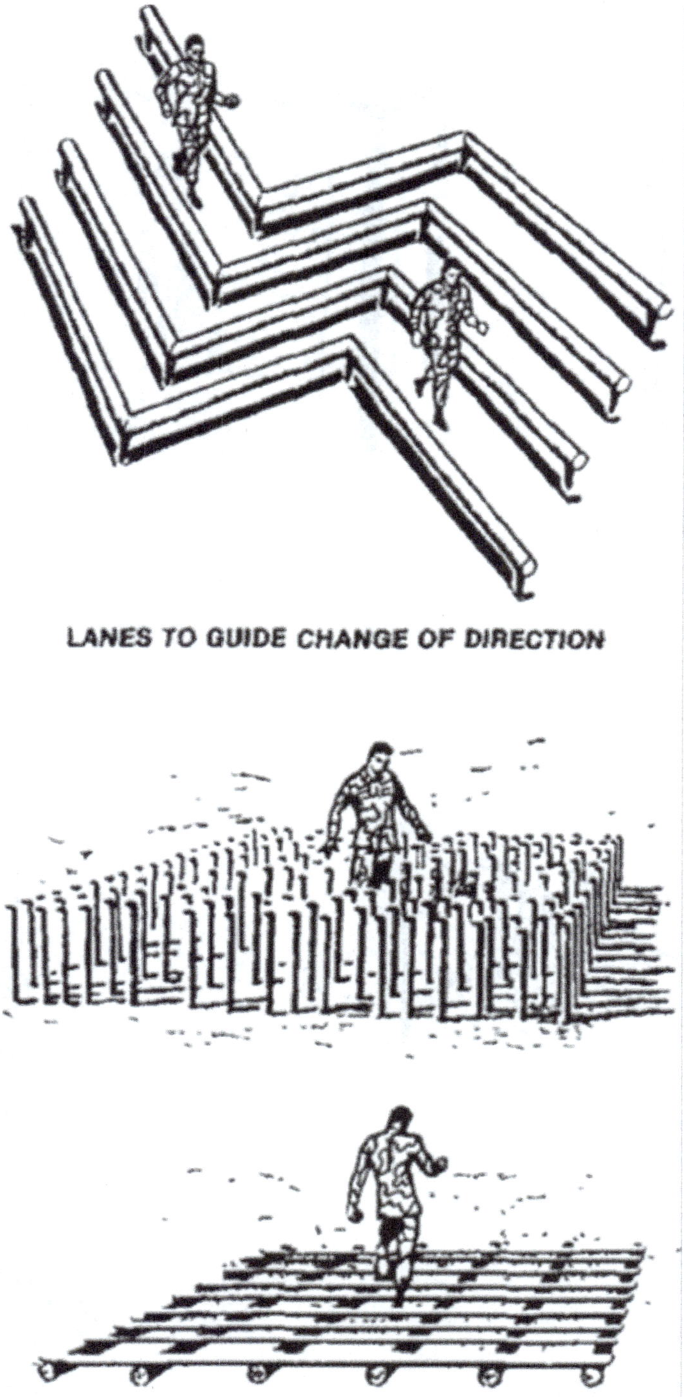

Figure 3-30.Dodging-Type Obstacles.

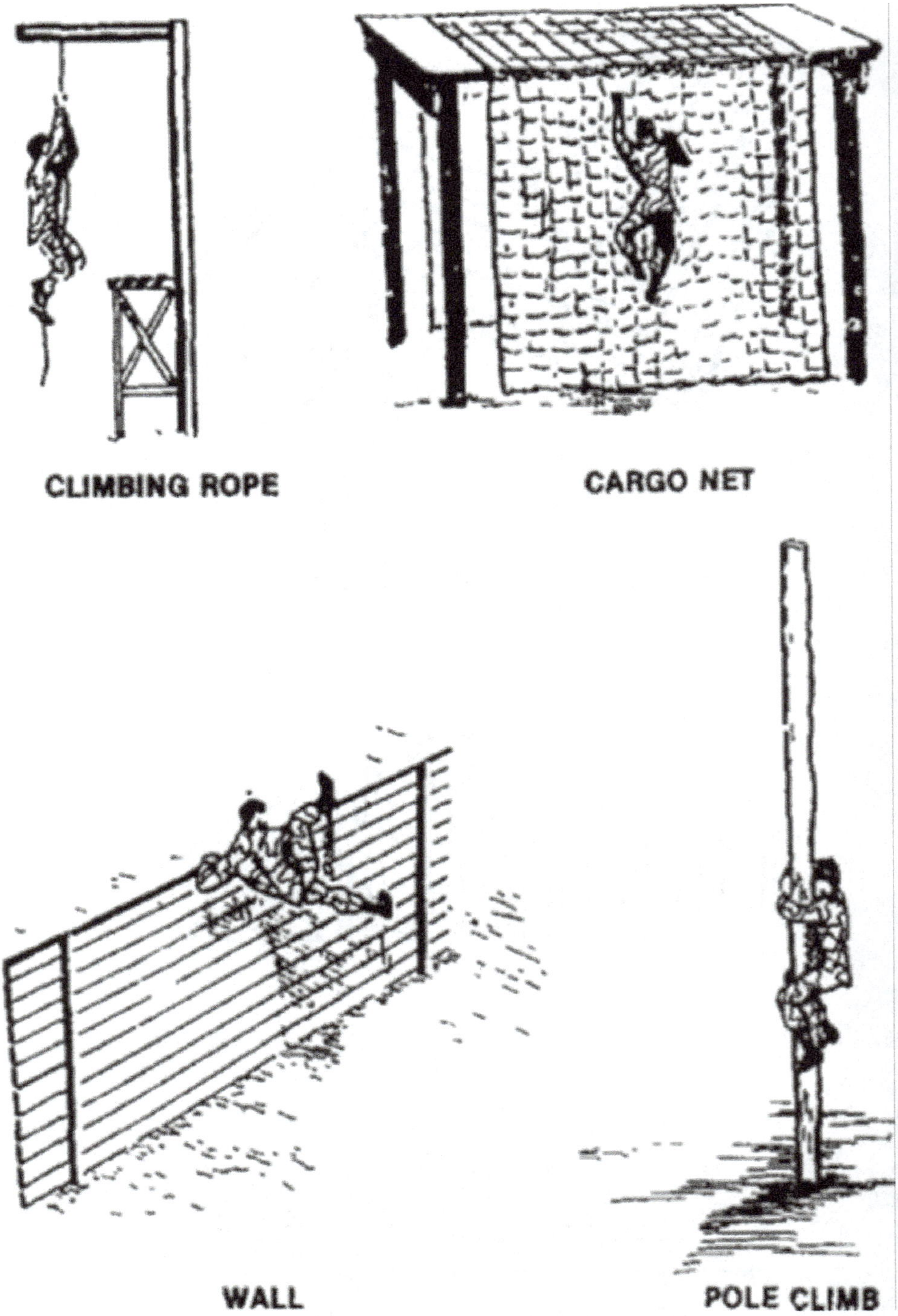

Figure 3-31. Vertical Climbing Obstacles.

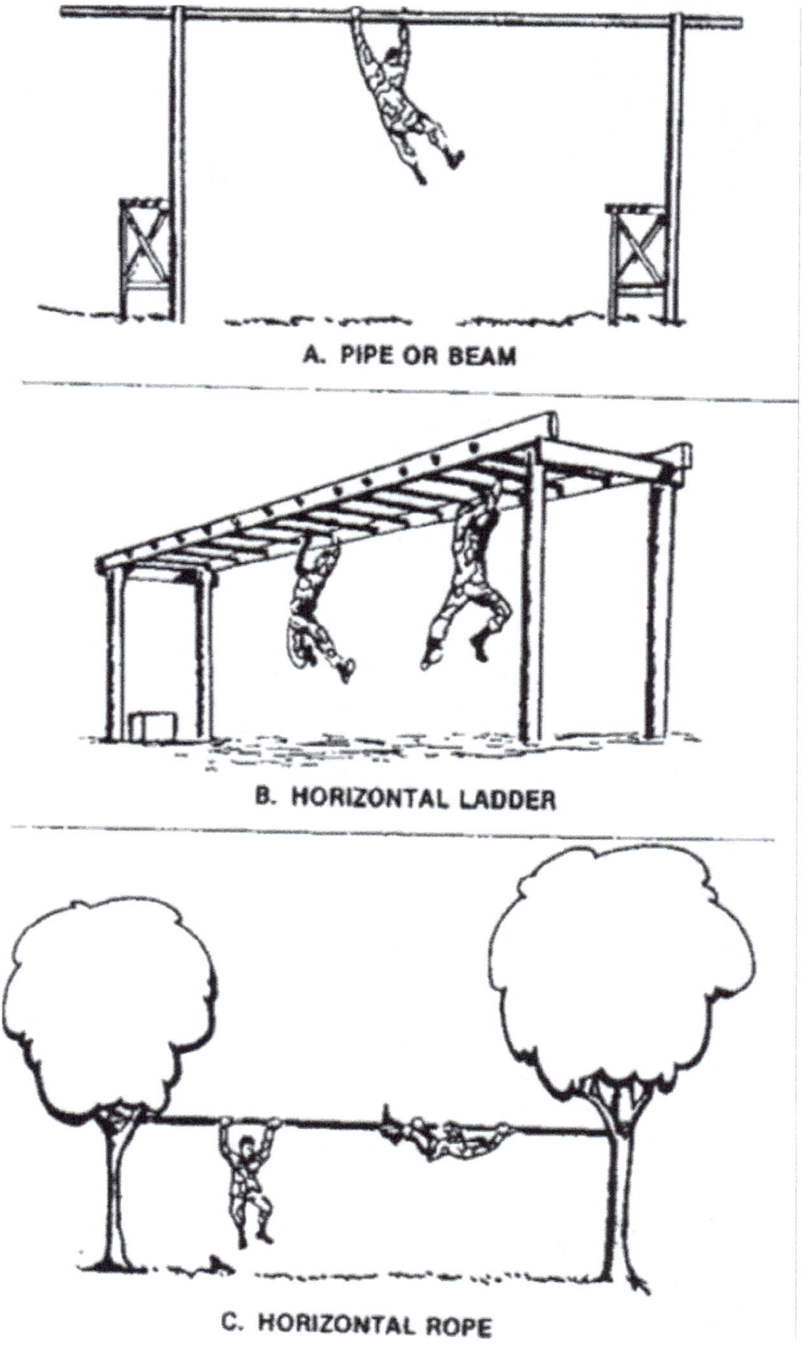

Figure 3-32.Horizontal Type Obstacles.

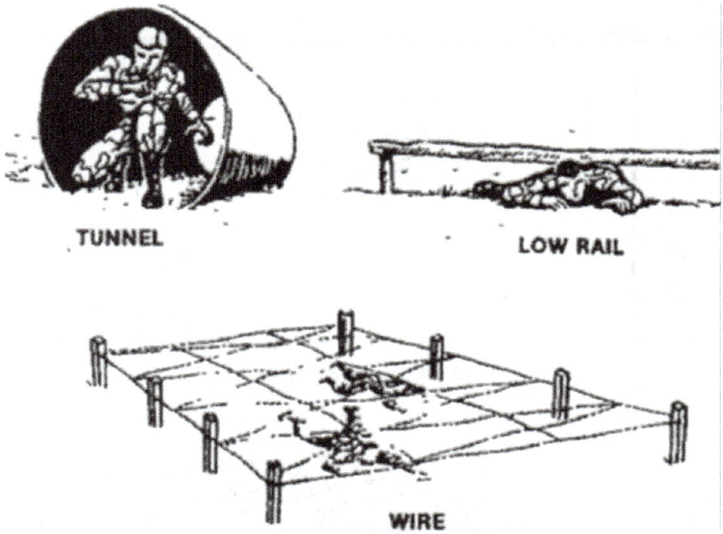

Figure 3-33.Crawling-Type Obstacles.

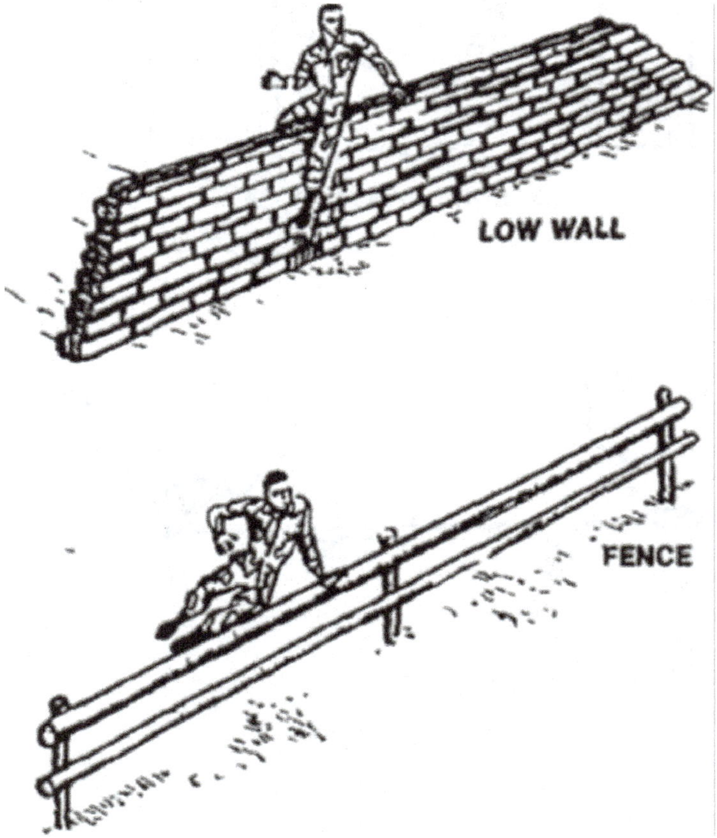

Figure 3-34.Vaulting Obstacles.

Figure 3-35.Balance-Type Obstacles.

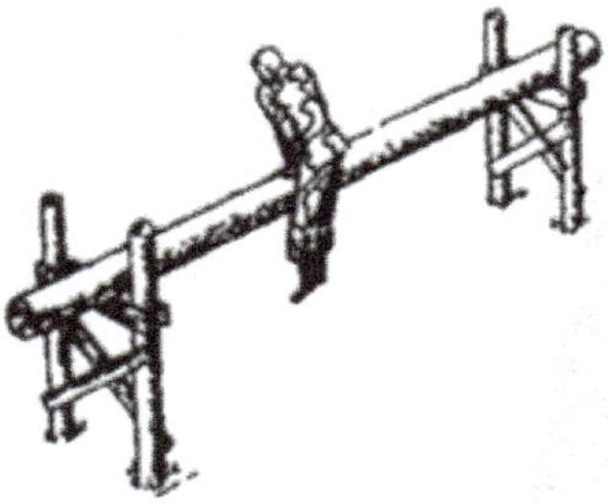

A. THE BELLY BUSTER

B. REVERSE CLIMB

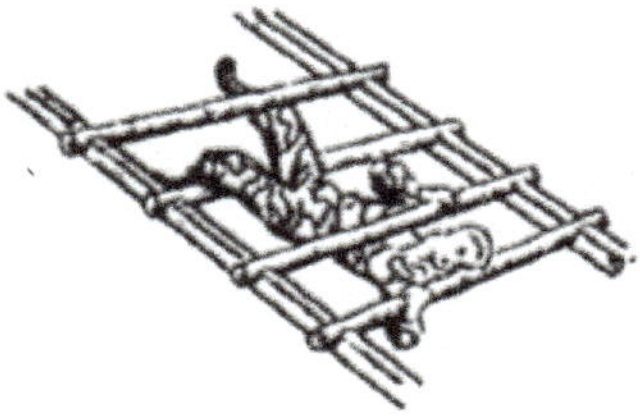

C. THE WEAVER

D. HIP—HIP

E. BALANCING LOGS

F. ISLAND HOPPER

Figure 3-36.Red Group.

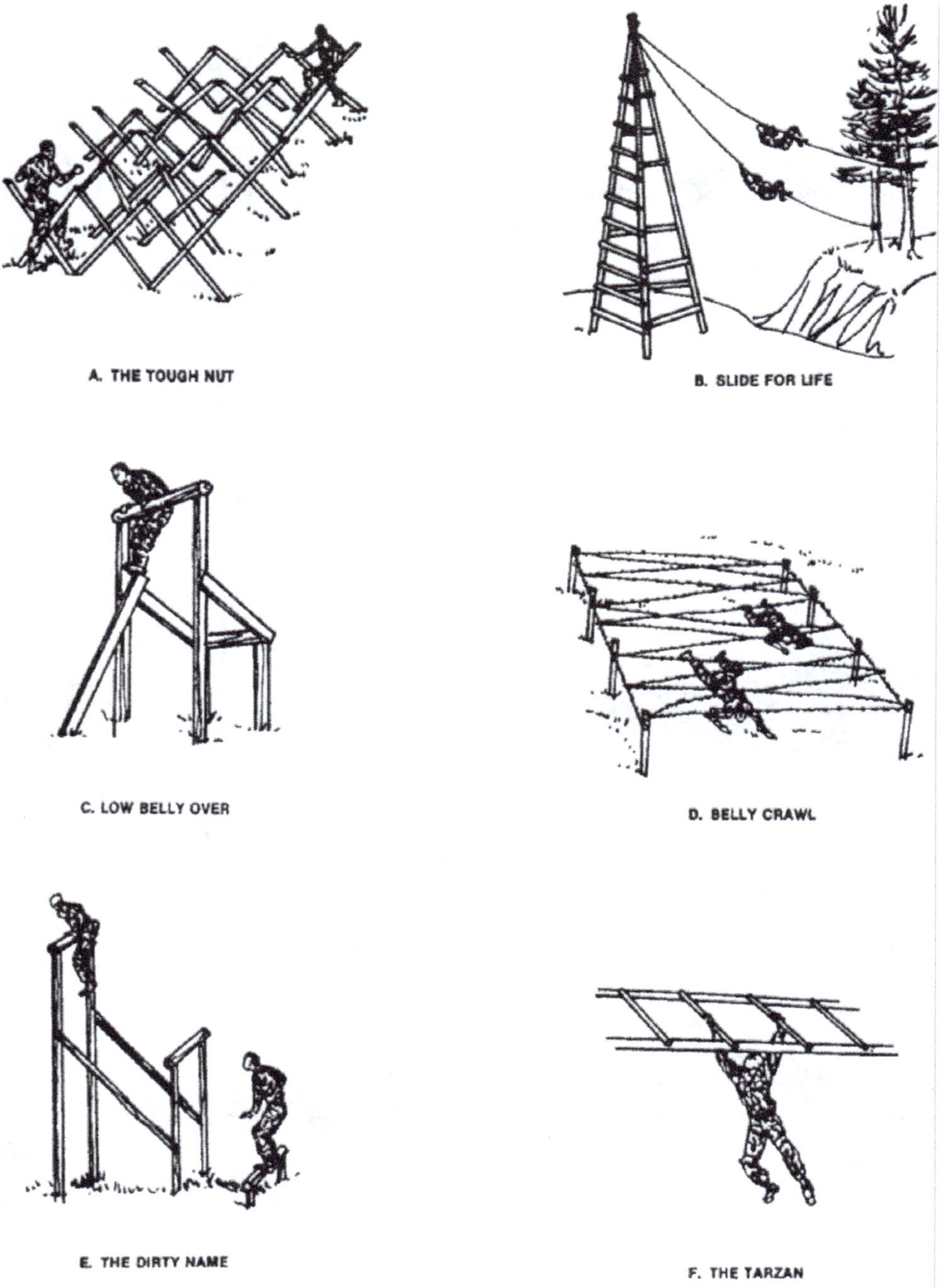

Figure 3-37.White Group.

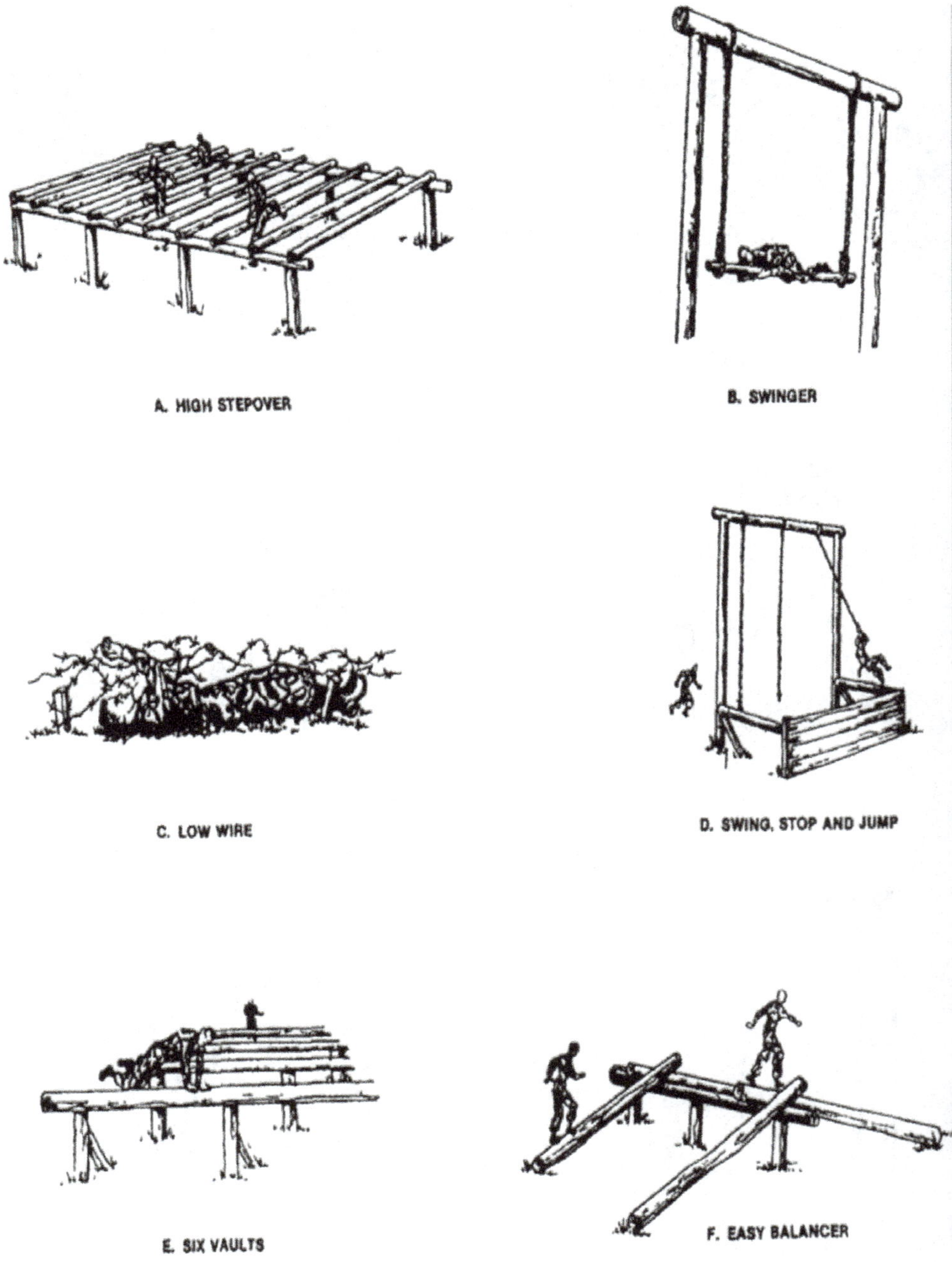

Figure 3-38.Blue Group.

A. INCLINE WALL

B. SKYSCRAPER

C. JUMP AND LAND

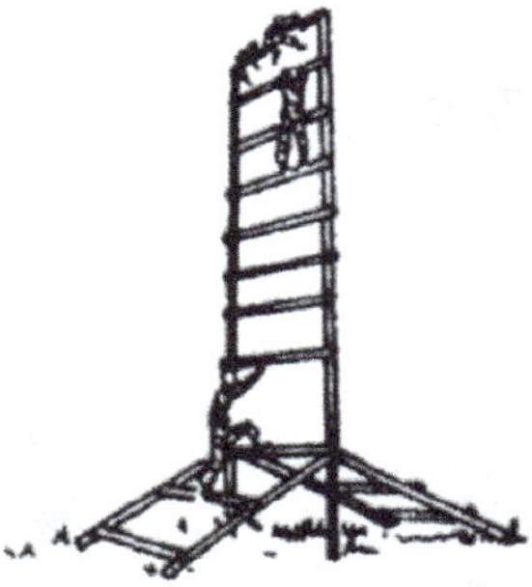

D. CONFIDENCE CLIMB

E. BELLY ROBBER

F. THE TOUGH ONE

Figure 3-39.Black Group.

Figure 3-40. Bench Conditioning, Table I.

Figure 3-41. Bench Conditioning, Table II.

PROGRESSION GUIDE

AGE GROUP	LEVEL	EXERCISES 1	2	3	4	5	6
17	A	15	18	14	15	15	250
to	B	13	16	13	13	13	235
29	C	11	14	12	11	11	[illegible]
30	A	13	14	12	13	13	200
to	B	11	13	11	11	11	185
39	C	9	12	10	9	9	165
40	A	11	11	10	11	11	150
to	B	9	10	9	9	9	135
44	C	7	9	8	7	7	120
45	A	9	8	8	9	9	100
to	B	7	7	7	7	7	90
49	C	5	6	6	5	5	80
50	A	7	6	6	7	7	75
to	B	5	5	5	5	5	70
59	C	3	4	4	3	3	60
60	A	4	5	4	4	4	50
and	B	3	4	3	3	3	40
over	C	2	3	2	2	2	30
Minutes for each exercise		2	1	1	1	2	3

Table I. Progression Guide.

Table I. Exercises.

PROGRESSION GUIDE

AGE GROUP	LEVEL	EXERCISES 1	2	3	4	5	6
17	A	17	17	17	9	19	300
to	B	15	15	15	8	17	270
29	C	13	13	13	7	15	245
30	A	15	15	15	8	17	235
to	B	13	13	13	7	15	210
39	C	11	11	11	6	13	190
40	A	13	13	13	7	15	175
to	B	11	11	11	6	13	155
44	C	9	10	9	5	11	135
45	A	11	11	11	6	13	125
to	B	9	9	9	5	11	110
49	C	7	7	7	4	9	100
50	A	9	9	9	5	11	95
to	B	7	7	7	4	9	85
59	C	5	5	5	3	7	75
60	A	6	7	7	4	9	70
and	B	5	5	5	3	7	60
over	C	4	4	4	2	5	50
Minutes for each exercise		1	1	1	1 1/2	1 1/2	6

Table II. Progression Guide.

Table II. Exercises.

PROGRESSION GUIDE

AGE GROUP	LEVEL	EXERCISES 1	2	3	4	5	6
17	A	10	19	19	16	10	350
to	B	9	17	17	15	9	315
29	C	8	15	15	14	8	280
30	A	9	17	17	14	9	270
to	B	8	15	15	13	8	240
39	C	7	13	13	12	7	210
40	A	8	15	15	12	8	200
to	B	7	13	13	11	7	180
44	C	6	11	11	10	6	160
45	A	7	13	13	10	7	150
to	B	6	11	11	9	6	135
49	C	5	9	9	8	5	120
50	A	6	11	11	8	6	115
to	B	5	9	9	7	5	105
59	C	4	7	7	6	4	95
60	A	5	9	9	7	5	90
and	B	4	7	7	6	4	80
over	C	3	5	5	4	3	70
Minutes for each exercise		1 1/2	1	1	1 1/2	1	6

Table III. Progression Guide.

Table III. Exercises.

PROGRESSION GUIDE

AGE GROUP	LEVEL	EXERCISES 1	2	3	4	5	6
17	A	12	9	12	24	25	400
to	B	11	8	11	22	23	380
29	C	10	7	10	21	21	360
30	A	11	8	11	23	23	305
to	B	10	7	10	21	21	290
39	C	9	6	9	20	20	275
40	A	10	7	10	20	21	225
to	B	9	6	9	18	18	215
44	C	8	5	8	16	16	205
45	A	8	6	8	16	16	175
to	B	7	5	7	14	14	165
49	C	6	4	6	12	12	155
50	A	6	5	6	13	13	135
to	B	5	4	5	11	11	130
59	C	4	3	4	10	10	120
60	A	5	4	5	10	10	100
and	B	4	3	4	9	9	95
over	C	3	2	3	8	8	90
Minutes for each exercise		1	2	1	1	1	6

Table IV. Progression Guide.

Table IV. Exercises.

PROGRESSION GUIDE

AGE GROUP	LEVEL	EXERCISES 1	2	3	4	5	6
17	A	14	13	28	14	30	450
to	B	13	12	27	13	28	430
29	C	12	11	26	12	26	410
30	A	12	12	25	12	26	350
to	B	11	11	24	11	24	330
39	C	10	10	23	10	22	310
40	A	11	11	23	11	23	250
to	B	10	10	21	10	21	240
44	C	9	9	19	9	19	230
45	A	9	9	20	9	20	200
to	B	8	8	18	8	18	190
49	C	7	7	16	7	16	180
50	A	7	7	16	7	16	170
to	B	6	6	14	6	14	155
59	C	5	5	12	5	12	140
60	A	6	6	12	6	12	115
and	B	5	5	11	5	10	110
over	C	4	4	9	4	9	105
Minutes for each exercise		2	1	1	2	1	5

Table V. Progression Guide.

Table V. Exercises.

PROGRESSION GUIDE

AGE GROUP	LEVEL	EXERCISES 1	2	3	4	5	6
17	A	17	15	32	32	35	500
to	B	16	14	30	30	33	480
29	C	15	13	28	28	31	460
30	A	15	13	30	30	31	400
to	B	14	12	28	28	29	380
39	C	13	11	26	26	27	360
40	A	13	10	27	27	27	310
to	B	12	9	25	25	25	285
44	C	11	8	23	23	23	265
45	A	11	9	23	23	23	250
to	B	10	8	21	21	21	230
49	C	9	7	19	19	19	210
50	A	9	8	19	19	19	200
to	B	8	7	17	17	17	190
59	C	7	6	15	15	15	175
60	A	8	7	15	15	17	140
and	B	7	6	13	13	15	130
over	C	5	5	10	10	12	120
Minutes for each exercise		2	1	1	1	1	6

Table VI. Progression Guide.

Table VI. Exercises.

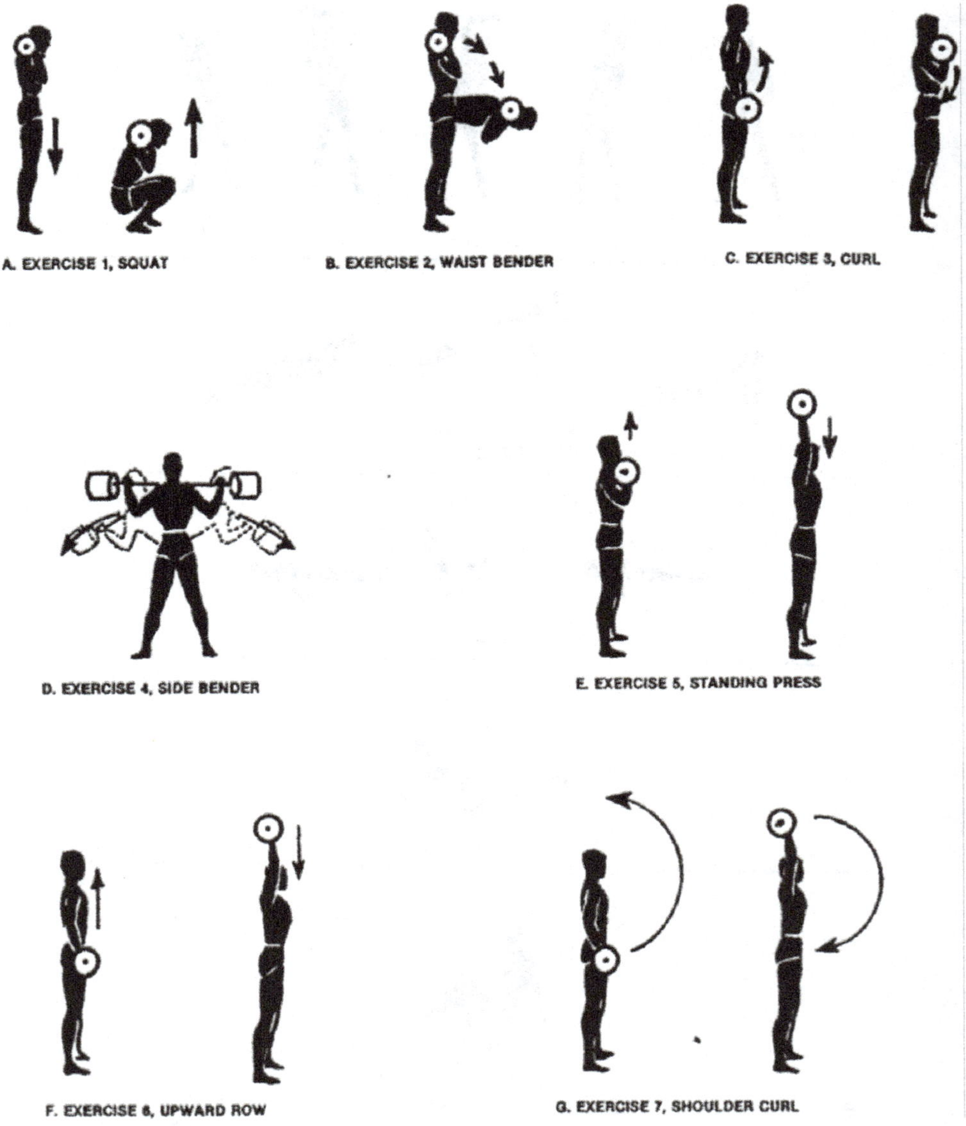

Figure 3-42. Weight Training.

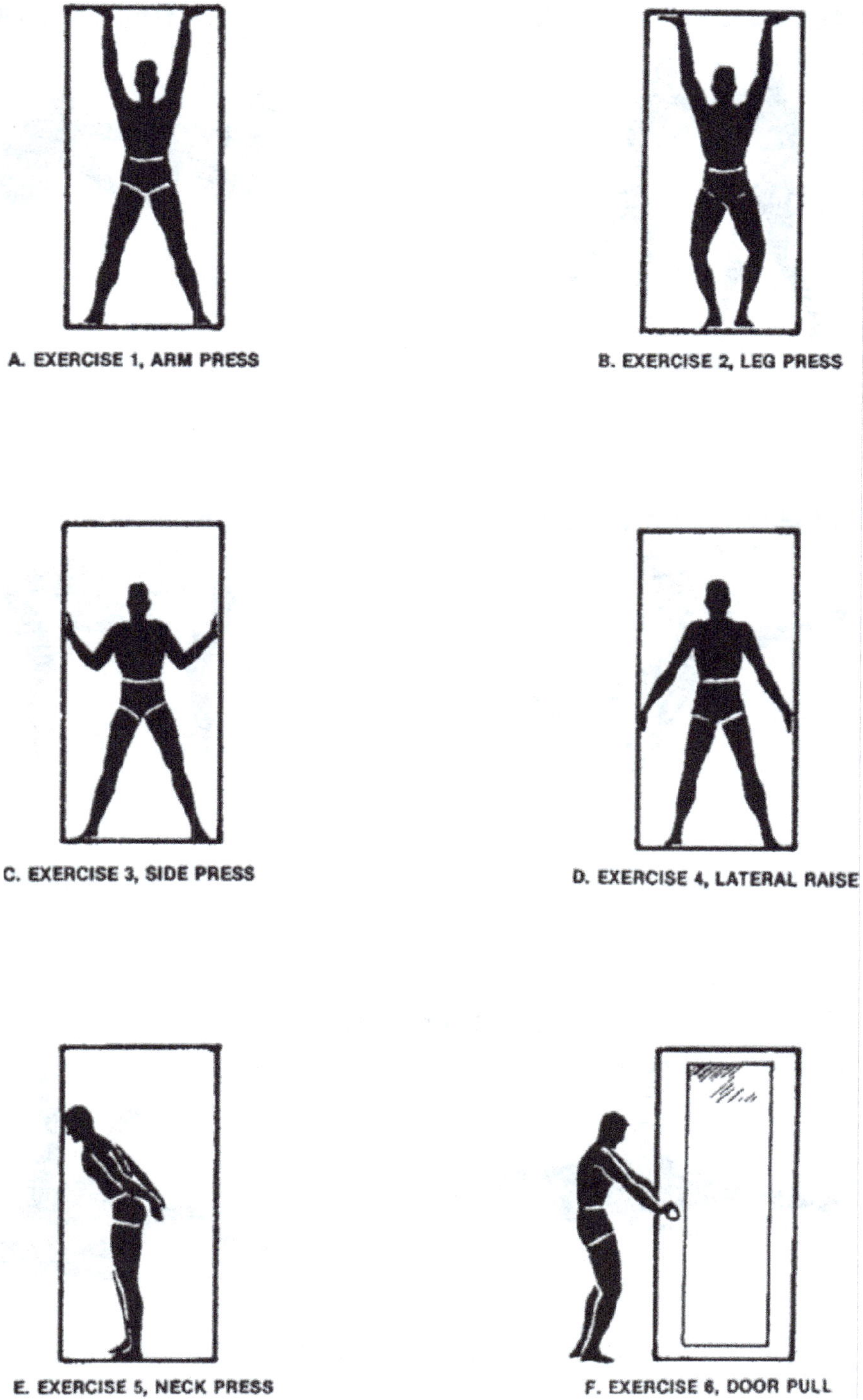

Figure 3-43. Isometric Exercises.

Figure 5-1.Combatives Table I.

Figure 5-2.Combatives Table II.

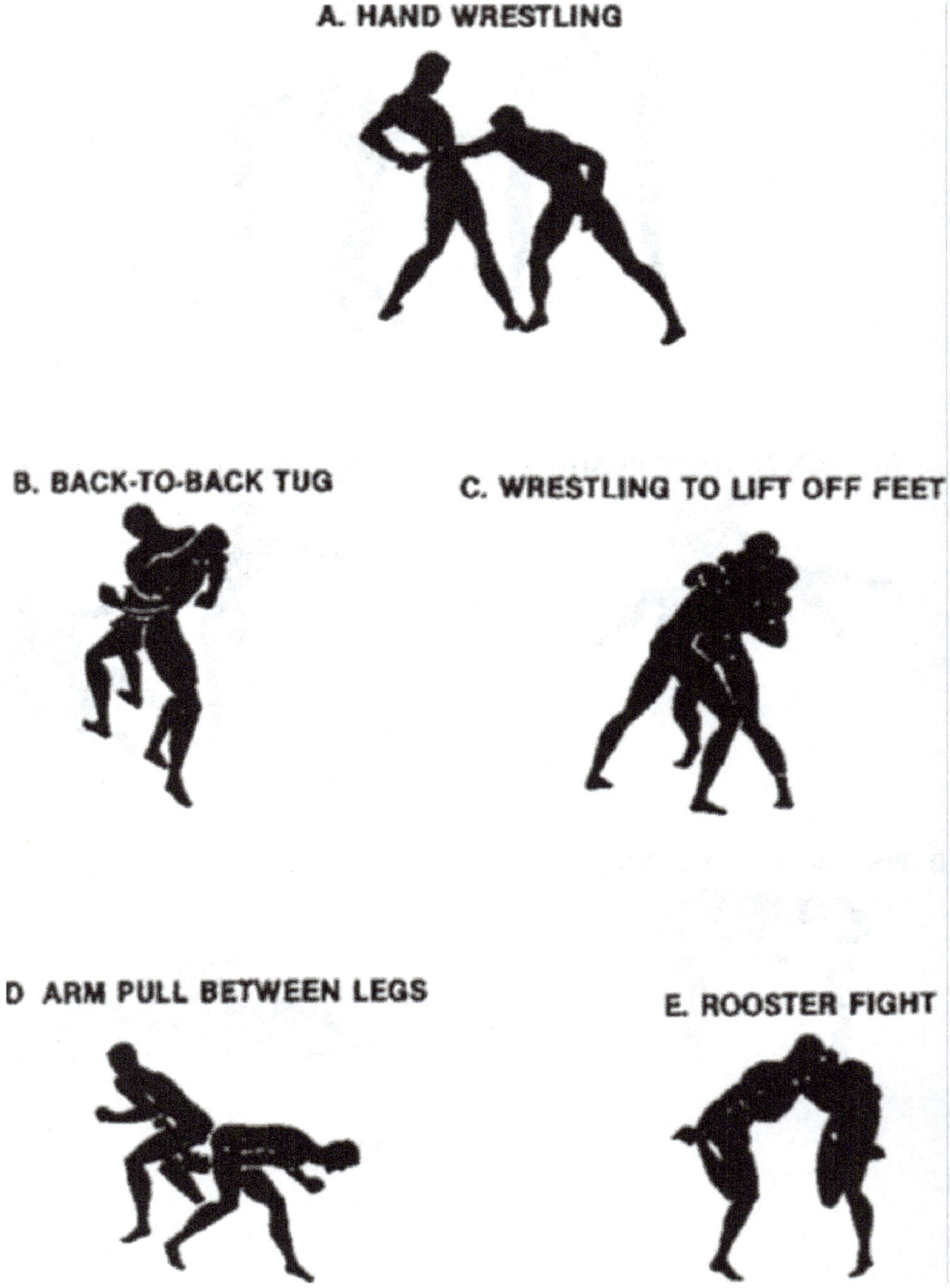

Figure. 5-3.Combatives Table III.

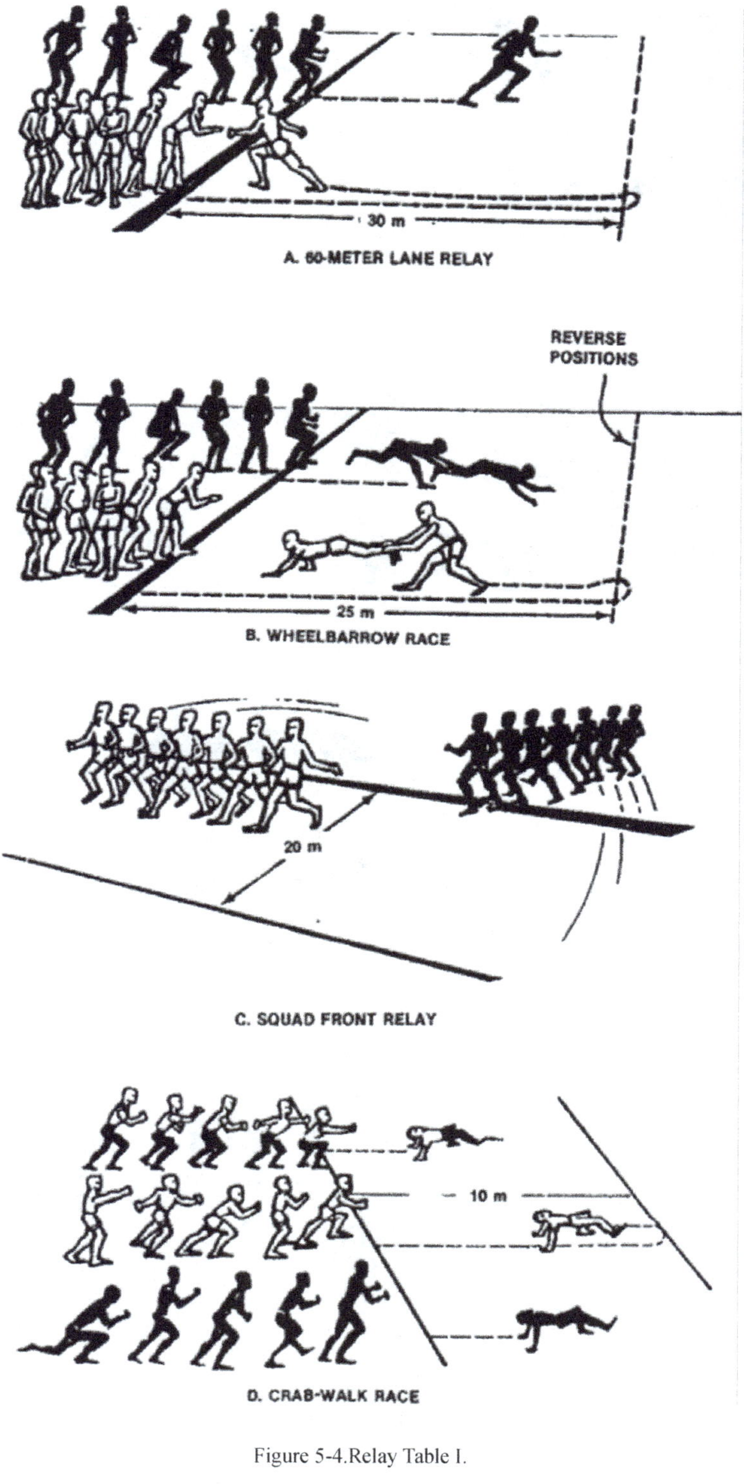

Figure 5-4.Relay Table I.

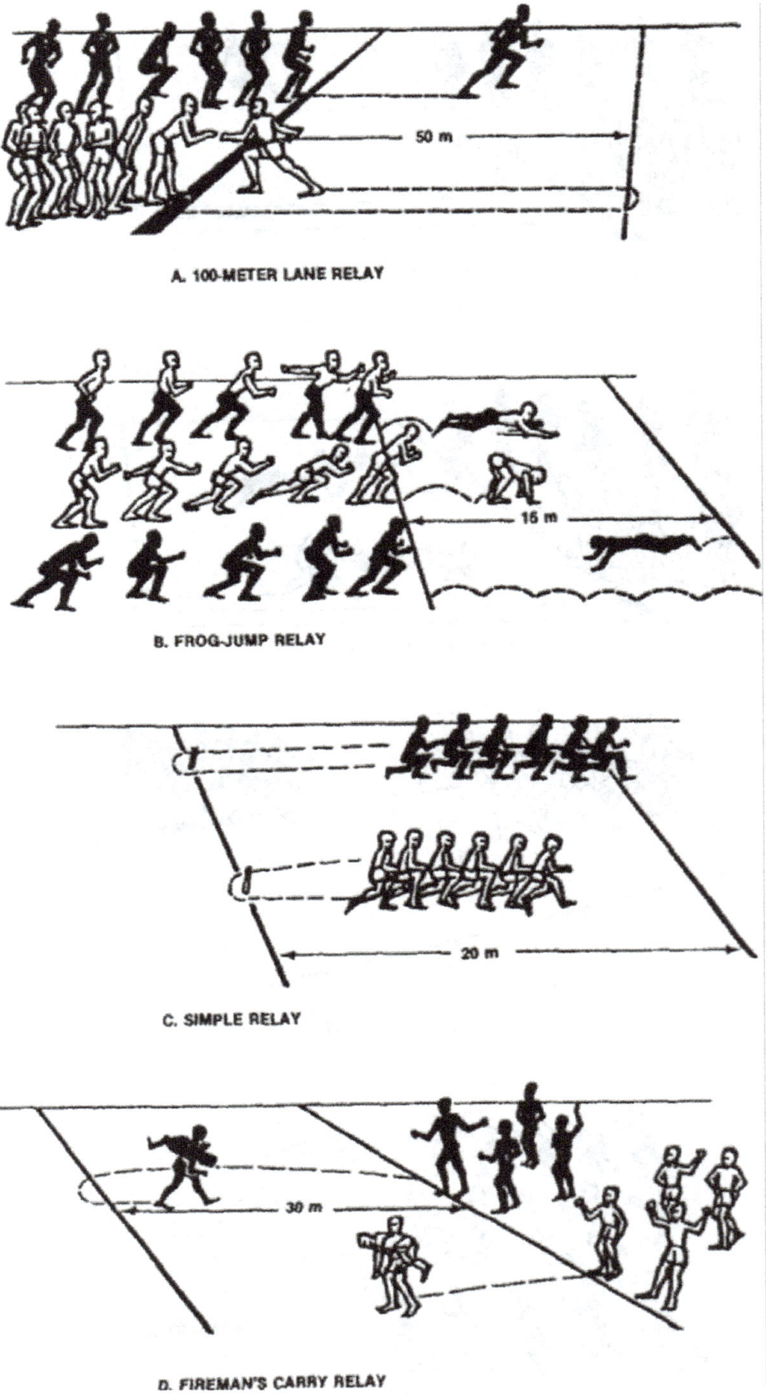

Figure 5-5.Relay Table II.

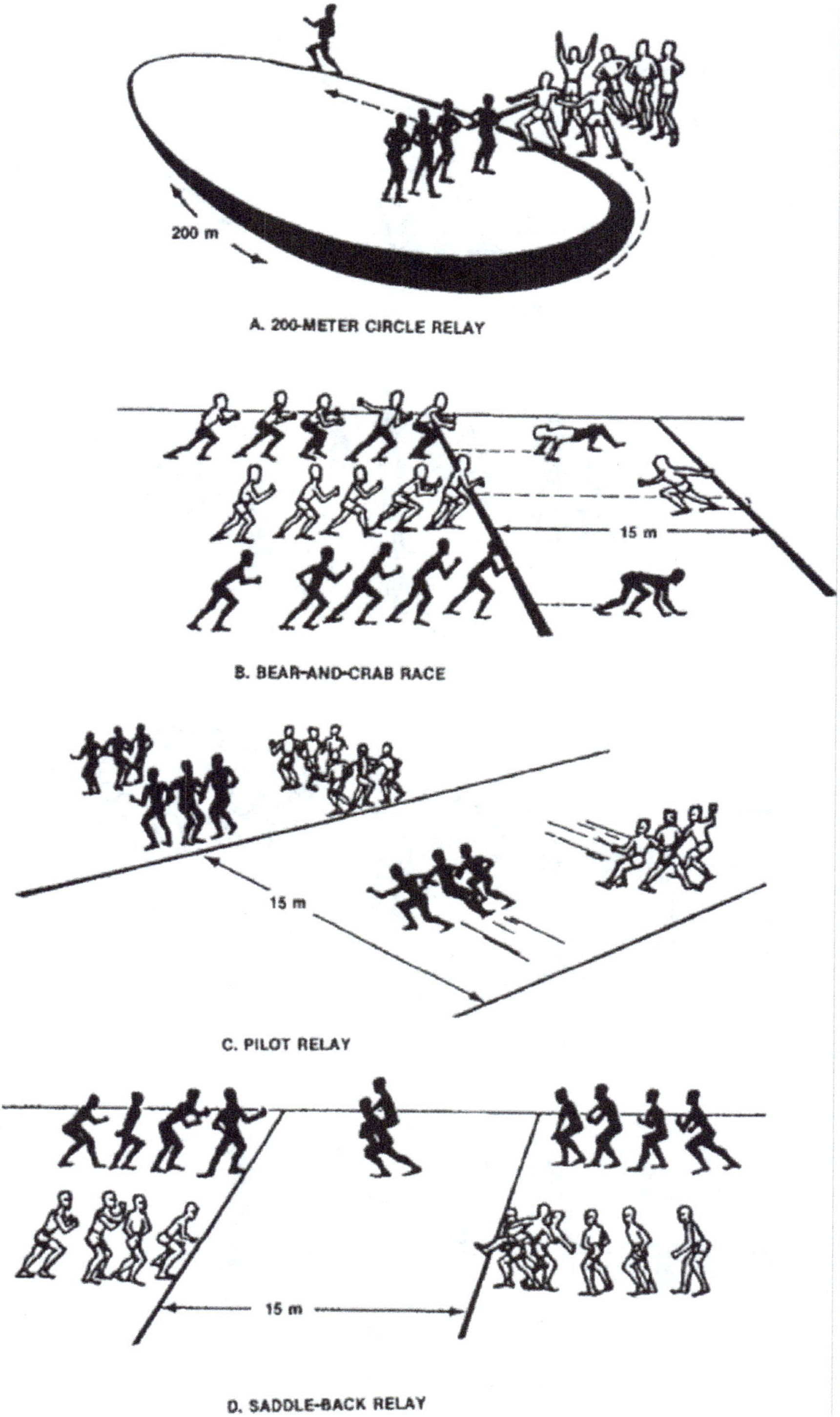

Figure 5-6.Relay Table III.

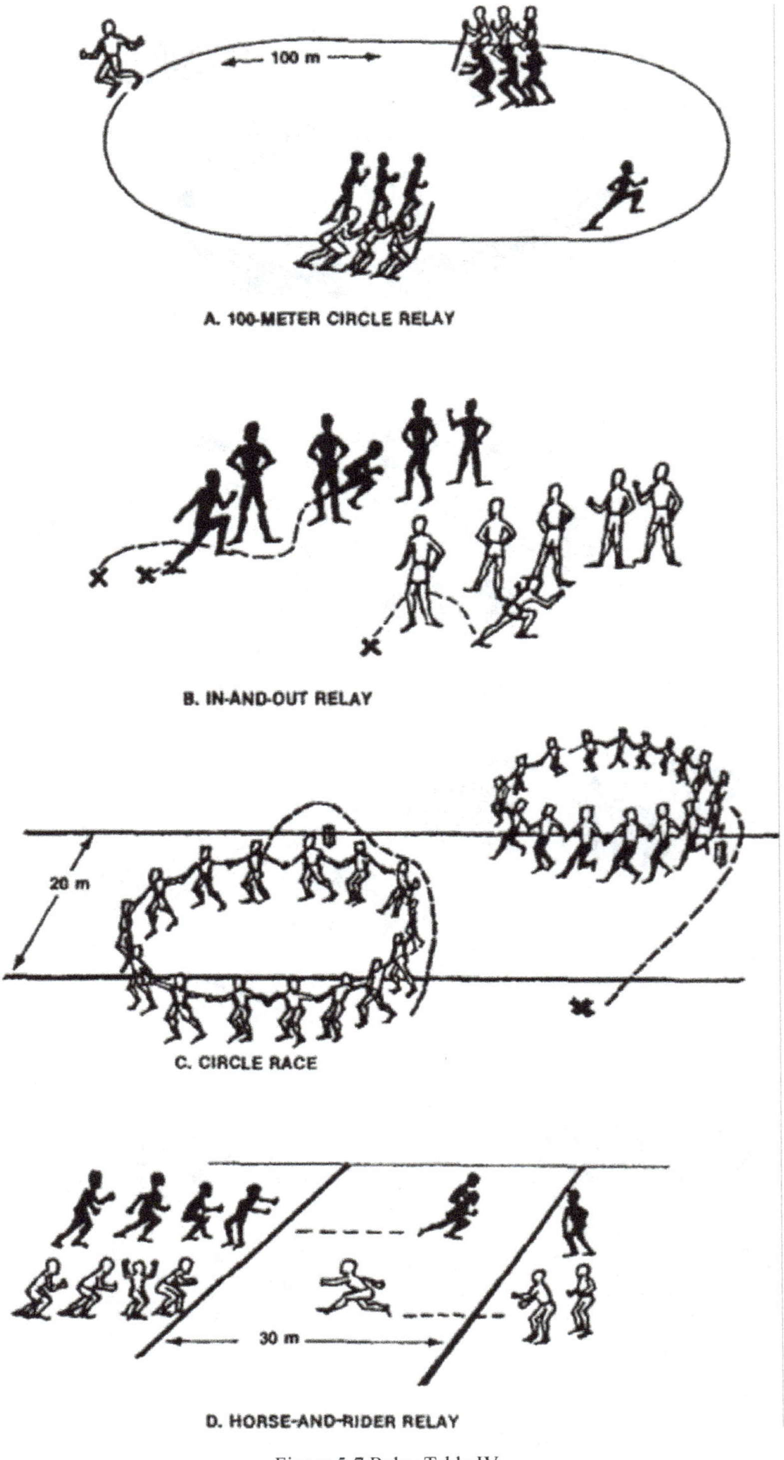

Figure 5-7.Relay Table IV.

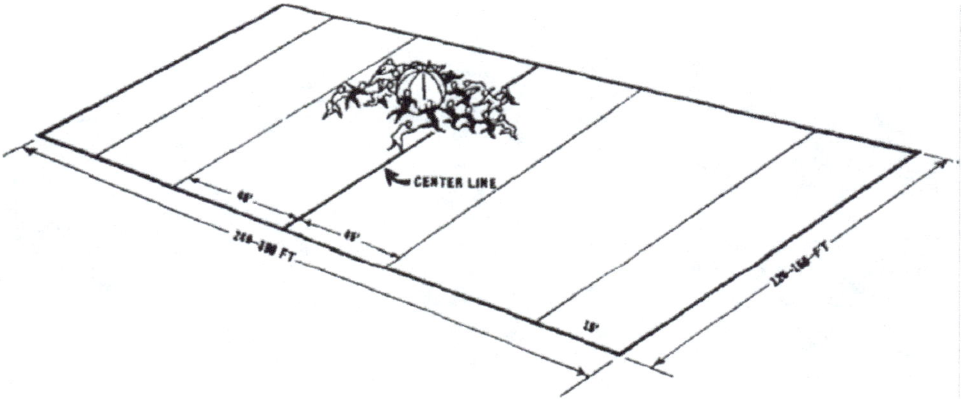

Figure 5-8.Pushball.

Figure 5-9.Line Rush.

Figure. 5-10. Human Tug-of-War.

Figure 5-11. Master of the Ring.

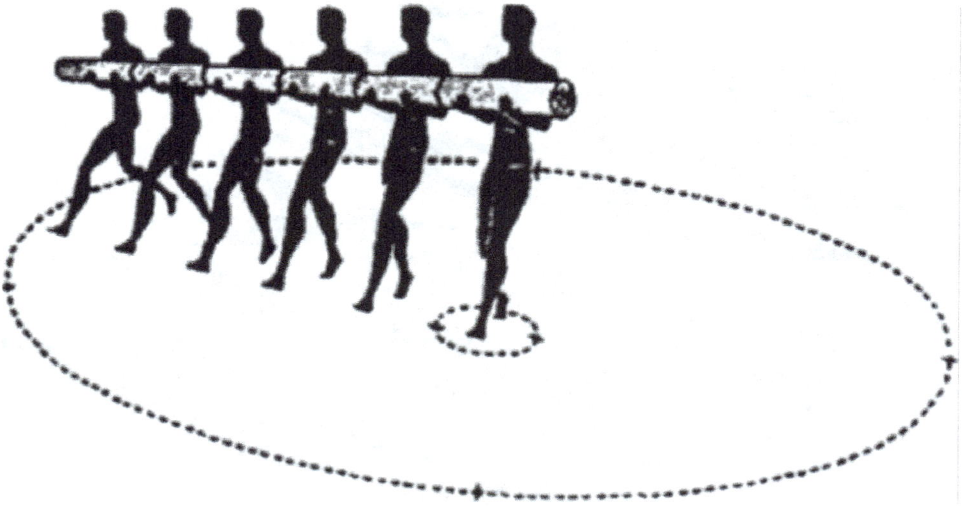

Figure 5-12.Log Pivot Circle.

Figure 5-13.Log-Rolling Race.

Figure 5-14. Prone Pushing Contest.

Figure 5-15. Shuttle Relay Race.

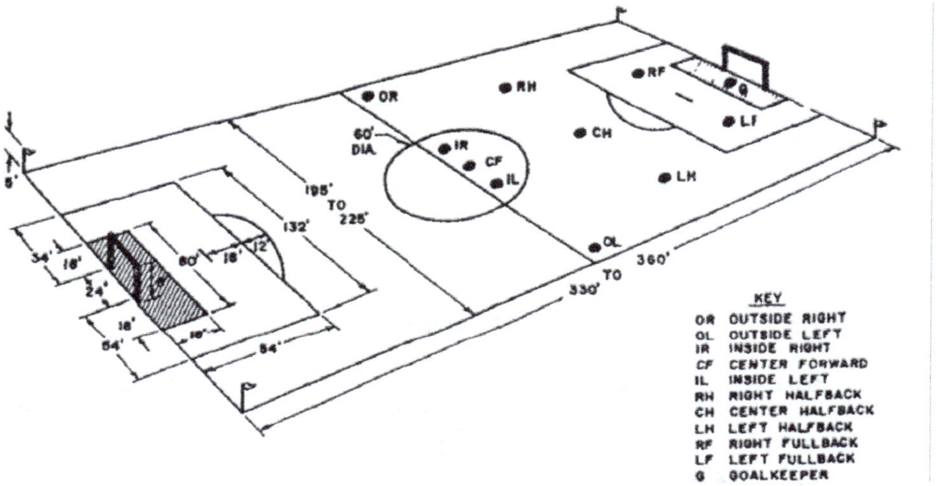

Figure 5-16. Soccer.

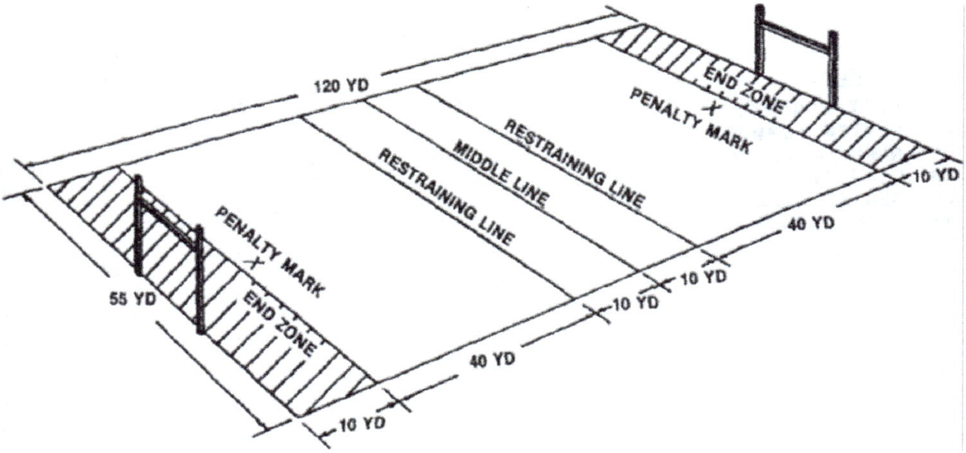

Figure 5-17.Speedball Field.

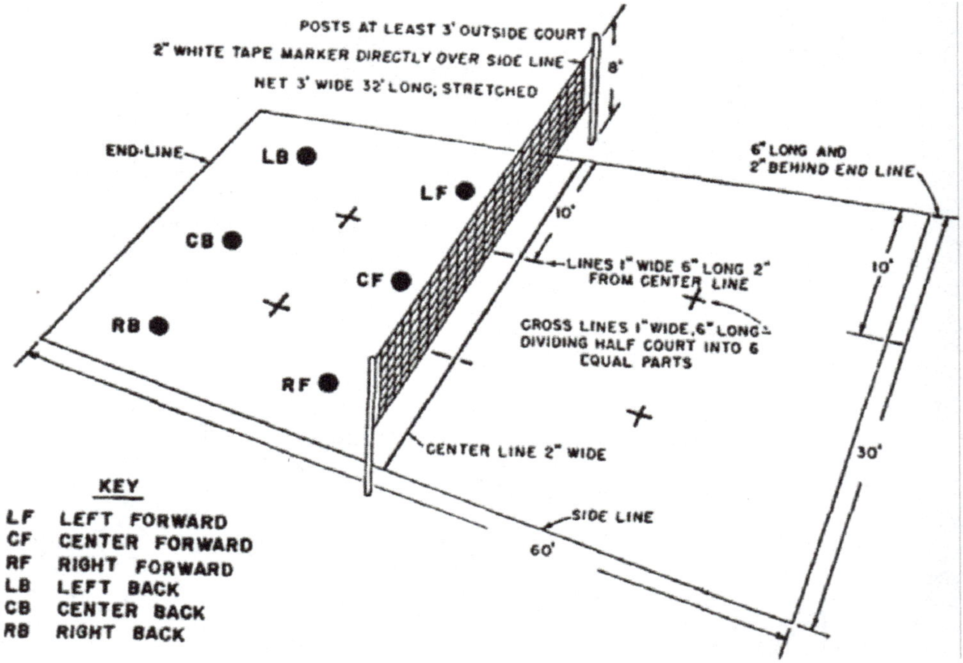

Figure 5-18. Volleyball.

PTS	UPHILL CLIMB	ROPE CLIMB	EVACUATION	FIRE AND MANEUVER	FORCED MARCH
100	1:20-BELOW	0:10-BELOW	0:25-BELOW	1:20-BELOW	26:00-BELOW
99	1:21-1:22	0:11		1:21-1:22	26:01-26:12
98	1:23-1:24	0:12	0:26	1:23-1:24	26:13-26:24
97	1:25-1:26	0:13		1:25-1:26	26:25-26:36
96	1:27-1:28	0:14	0:27	1:27-1:28	26:37-26:48
95	1:29-1:30	0:15		1:29-1:30	26:49-27:00
94	1:31-1:32	0:16	0:28	1:31-1:32	27:01-27:12
93	1:33-1:34	0:17		1:33-1:34	27:13-27:24
92	1:35-1:36	0:18	0:29	1:35-1:36	27:25-27:36
91	1:37-1:38	0:19		1:37-1:38	27:37-27:48
90	1:39-1:40	0:20	0:30	1:39-1:40	27:49-28:00
89	1:41-1:42	0:21	0:31	1:41-1:42	28:01-28:12
83	1:43-1:44	0:22	0:32	1:43-1:44	28:13-28:24
87	1:45-1:46	0:23	0:33	1:45-1:46	28:25-28:36
86	1:47-1:48	0:24	0:34	1:47-1:48	28:37-28:48
85	1:49-1:50	0:25	0:35	1:49-1:50	28:49-29:00
84	1:51-1:52	0:26	0:36	1:51-1:52	29:01-29:12
83	1:53-1:54	0:27	0:37	1:53-1:54	29:13-29:24
82	1:55-1:56	0:28	0:38	1:55-1:56	29:25-29:36
81	1:57-1:58	0:29	0:39	1:58-1:58	29:37-29:48
80	1:59-2:00	0:30	0:40	1:59-2:00	29:49-30:00
79	2:01-2:02	0:31	0:41	2:01-2:02	30:01-30:12
78	2:03-2:04	0:32	0:42	2:03-2:04	30:13-30:24
77	2:05-2:06	0:33	0:43	2:05-2:06	30:25-30:36
76	2:07-2:08	0:34	0:44	2:07-2:08	30:37-30:48
75	2:09-2:10	0:35	0:45	2:09-2:10	30:49-31:00
74	2:11-2:12	0:36	0:46	2:11-2:12	31:01-31:12
73	2:13-2:14	0:37	0:47	2:13-2:14	31:13-31:24
72	2:15-2:16	0:38	0:48	2:15-2:16	31:25-31:36
71	2:17-2:18	0:39	0:49	2:17-2:18	31:37-31:48
70	2:19-2:20	0:40	0:50	2:19-2:20	31:49-32:00
69	2:21-2:22	0:41	0:51	2:21-2:22	32:01-32:12
68	2:23-2:24	0:42	0:52	2:23-2:24	32:13-32:24
67	2:25-2:26	0:43	0:53	2:25-2:26	32:25-32:36
66	2:27-2:28	0:44	0:54	2:27-2:28	32:37-32:48
65	2:29-2:30	0:45	0:55	2:29-2:30	32:49-33:00

Figure 6-1. Physical Readiness Test Scoring Table for Male.

	EXAMPLE 1	EXAMPLE 2	EXAMPLE 3
Event #1	100 pts	100 pts	99 pts
Event #2	100 pts	99 pts	86 pts
Event #3	85 pts	75 pts (F)	80 pts
Event #4	90 pts	79 pts (F)	100 pts
Event #5	79 pts	85 pts	90 pts
Pts	454 pts	438 pts	455 pts
Final Score	79.0% (F)	69.0% (F)	91.0%
(F) = Failure			

Figure 6-2.Scoring Examples.

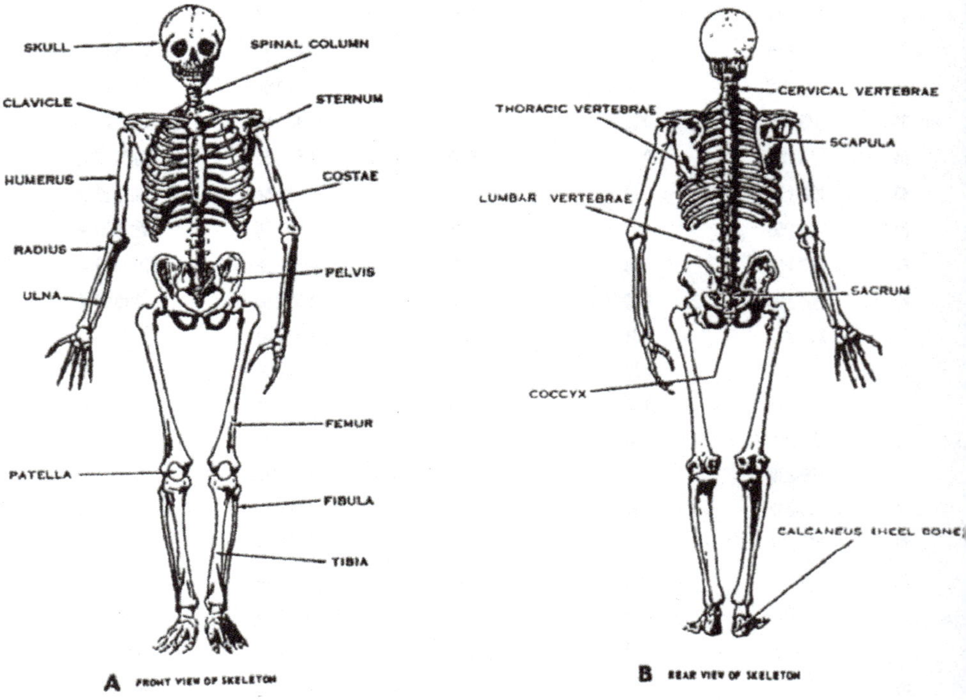

Figure 7-1.Front and Rear Views of Skeleton.

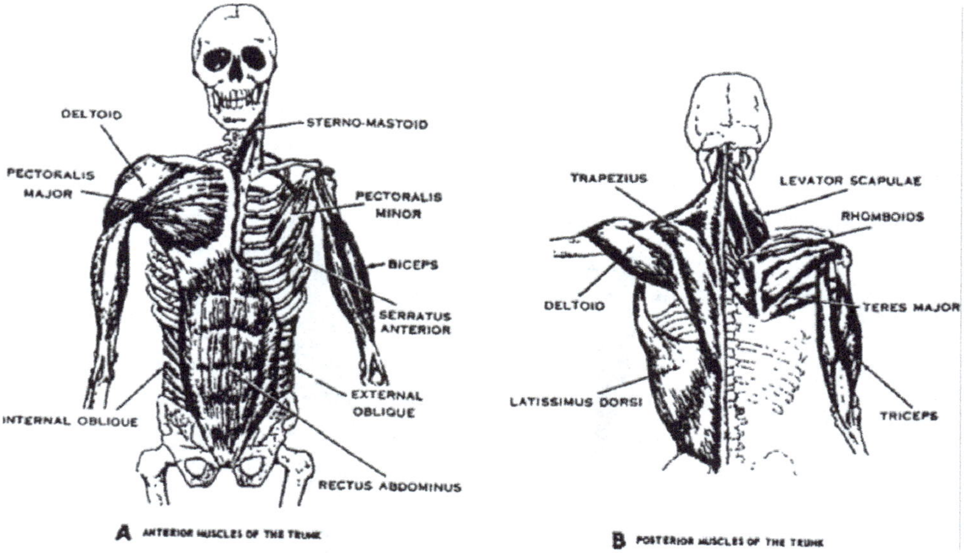

Figure 7-2.Muscles of the Trunk.

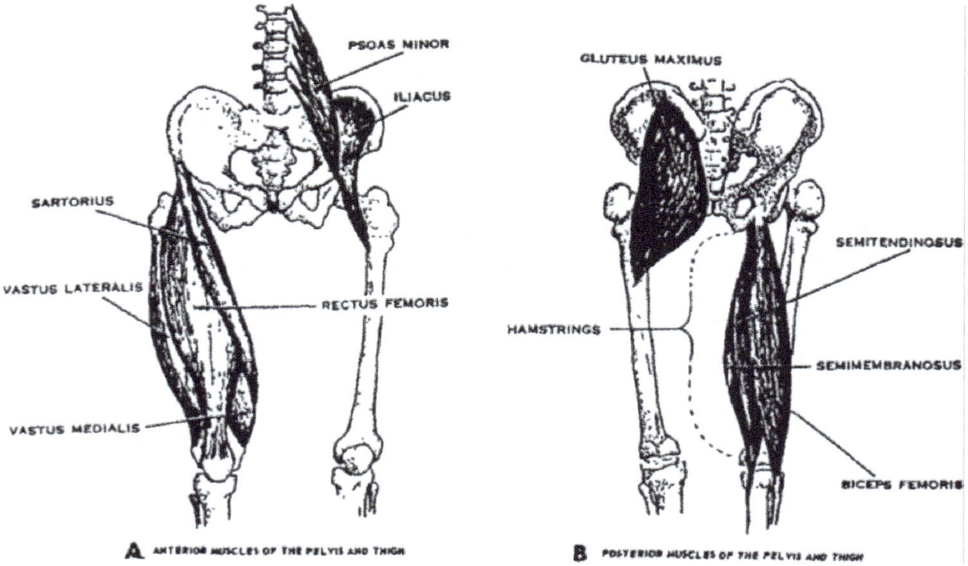

Figure 7-3.Anterior and Posterior Muscles of the Pelvisand Thigh.

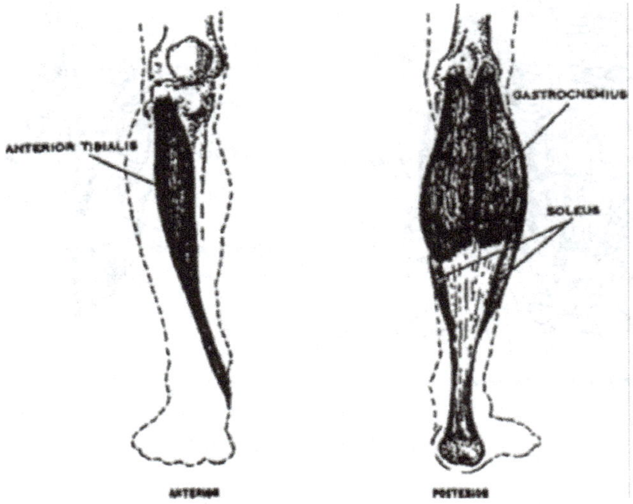

Figure 7-4. Muscles of Lower Leg.

Figure 7-5.Circulation of Blood from Heart to the Body.

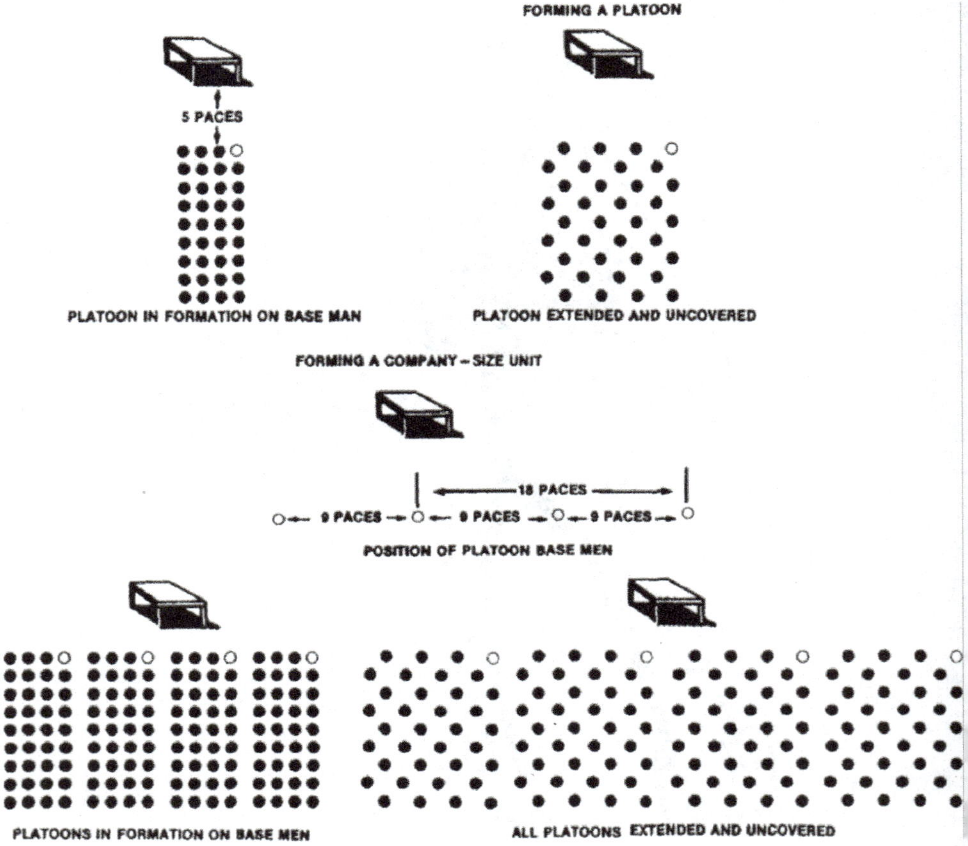

Figure A-l. Forming the Extended Rectangular Formation.

Figure A-2.Circle Formation.

TECHNICAL MANUAL
ARMY NO. 9-1005-319-23&P
AIR FORCE TO 11W3-5-5-42

***ARMY TM 9-1005-319-23&P**
AIR FORCE TO 11W3-5-5-42

DEPARTMENTS OF THE ARMY
AND AIR FORCE
Washington, DC, 1 May 1991

Technical Manual for 5.56MM,

M16A2, M4 and M4A1 Carbine

DISTRIBUTION STATEMENT. APPROVED FOR PUBLIC RELEASE; DISTRIBUTION IS UNLIMITED.

REPORTING ERRORS AND RECOMMENDING IMPROVEMENTS

You can help improve this manual. If you find any mistakes or if you know of a way to improve the procedures please let us know.

Army users mail your letter, DA Form 2028 (recommended Changes to Publications and Blank Forms), or DA Form 2028-2 located in the back of this manual direct to: Director, Armament and Chemical Acquisition and Logistics Activity, ATTN: AMSTA-AC-NML, Rock Island, IL 61299-7630.

Air Force users submit AFTO Form 22, Technical Order System Publication Improvement Report and Reply, to: WR-ALC/MMDET, Robins AFB, GA 31098-5609.

A reply will be furnished to you.

*This manual supersedes ARMY TM 9-1005-319-23&P dated 28 August 1987, including all changes.

WARNING

ALL WARNINGS in this technical manual pertain to both the rifle and the carbines unless otherwise specified.

Before starting an inspection, be sure to clear the rifle. Do not pull the trigger until the rifle has been cleared. Inspect the chamber to ensure that it is empty and no ammunition is in position to be chambered.

Do not keep live ammunition near work area.

To avoid injury to your eyes, use care when removing and installing spring-loaded parts.

All M1 6A2 rifles and M4 carbines must be inspected and gaged at least once annually for safety and serviceability. Initial gaging is required 1 year from receipt of the weapons. Air Force users refer to inspection requirements in Air Force Manual (AFM) 36-2227, Volume 1.

All Army Reserve and Army National Guard M16A2 rifles and M4 carbines must be inspected and gaged at least once every 2 years, after the initial inspection/gaging procedures have been accomplished. This initial gaging procedure is required 1 year from receipt of the weapons. This 2 year interval may be maintained unless preventive maintenance checks and services (PMCS) or other physical evidence indicates that an individual unit's M16A2 rifles and M4 carbines require inspection/ gaging at a more frequent interval. If it is determined that a yearly inspection is necessary for an individual unit, only that unit will be affected. This will not affect other units in regard to the interval of inspection.

It is recommended that training units inspect/gage all rifles and carbines at the end of each training cycle. Training units will inspect/gage all rifles and carbines at least once annually.

Below direct support maintenance, **DO NOT** interchange bolt assemblies from one rifle/carbine to another. Doing so may result in injury to, or death of, personnel.

Bolt cam pin must be installed or rifle/carbine will blow up while firing the first round. If the bolt cam pin is not installed, injury to or death of, personnel may result.

Dry cleaning solvent is flammable and toxic and should be used in a well ventilated area. The use of rubber gloves is necessary to protect the skin when washing rifle parts.

When using solid film lubricant or dichloromethane, be sure the area is well ventilated.

When using carbon removing compound (item 8, app D), avoid skin contact. If carbon removing compound comes In contact with the skin, wash thoroughly with running water. using a good lanolin base cream after exposure to the compound is helpful. Using gloves and protective equipment is required.

The lock plate prevents the selector lever from being placed in BURST and will be installed at the discretion of the unit commander. It is mandatory for use in civil disturbance (riot control).

Only blank cartridge M200 is to be used when the blank firing attachment is attached to the rifle/carbine.

WARNING (CONT)

Do not fire blank ammunition at a representative enemy at distances of less than 20 feet (6.10 Om). The unburned propellant grains can cause injury within this distance.

For further information on safety, care, and handling of ammunition: Army and Air Force users refer to M16A2 Rifle Operator's Manual.

For additional first aid data, see Field Manual (FM) 21-11.

HOW TO USE THIS MANUAL

Read this manual carefully before performing required maintenance. This manual will be referred to for Inspection/Maintenance and Repair procedures.

GENERAL

There are several things you need to know to use this manual efficiently.

1. All references in the manual are to pages only. Reference to maintenance procedures is to the page where the respective initial setup appears.

2. Illustrations for the maintenance procedures show only those parts affected by the operation being performed.

3. Whenever the male gender is mentioned in the manual (i.e., crewman, repairman), it also pertains to females.

4. When the term "evacuate to support maintenance" is used, the entire rifle must be evacuated.

5. When a procedure is common to M16A2 rifle, and M4/M4A1 carbine, **ONLY** the M16A2 configuration will be depicted. If a procedure is not common to both weapons, the procedure will be incorporated.

6. When the word rifle is referenced in text, it will reference the rifle and the carbines.

INDEXES

This manual is organized to help you find the information you need quickly. There are several useful indexes.

1. **Table of Contents**. Lists in order all chapters, sections, and appendixes. Gives page references.

2. **Nomenclature Cross-References List.**

3. **Chapter Overviews**. Summarize material covered in the chapter. Are located at the beginning of each chapter.

4. **Symptom Index**. Located just before the troubleshooting table in each maintenance chapter. Lists, in alphabetical order, parts of the rifle with possible malfunctions. References pages of the troubleshooting table.

5. **Alphabetical Index**. Located at the end of the manual. An extensive subject index for everything in the manual. Gives page references.

MAINTENANCE PROCEDURES

There are two maintenance chapters:

Army personnel use chapter two for unit maintenance procedures and chapter three for direct support maintenance procedures.

Air Force personnel: Only Air Force Specialty Code 753XX Combat Arms Training and Maintenance (CATM) specialists, technicians, and gunsmiths are authorized to perform maintenance procedures contained in this manual.

Each maintenance task has an initial setup containing a list of the following things you will need in order to do your maintenance task.

1. **Tools and Special Tools**. For standard and special tools, see appendixes B and C. Army users are to use the Tool Set, Gage Set, and/or Shop Set listed in the initial setup.

2. **Materials/Parts**. Lists expendable materials and 100 percent replaceable parts. Each material or part is followed by a part number or appendix reference.

3. **References**. Lists other publications containing necessary information.

4. **Equipment Condition**. Lists conditions to be met before starting the procedure. The reference on the left of the condition is a page reference to instructions for setting up the condition.

5. **General Safety Instructions**. Lists safety instructions to follow before performing maintenance procedures.

EXTERNAL VIEW OF 5.56 MM RIFLE M16A2

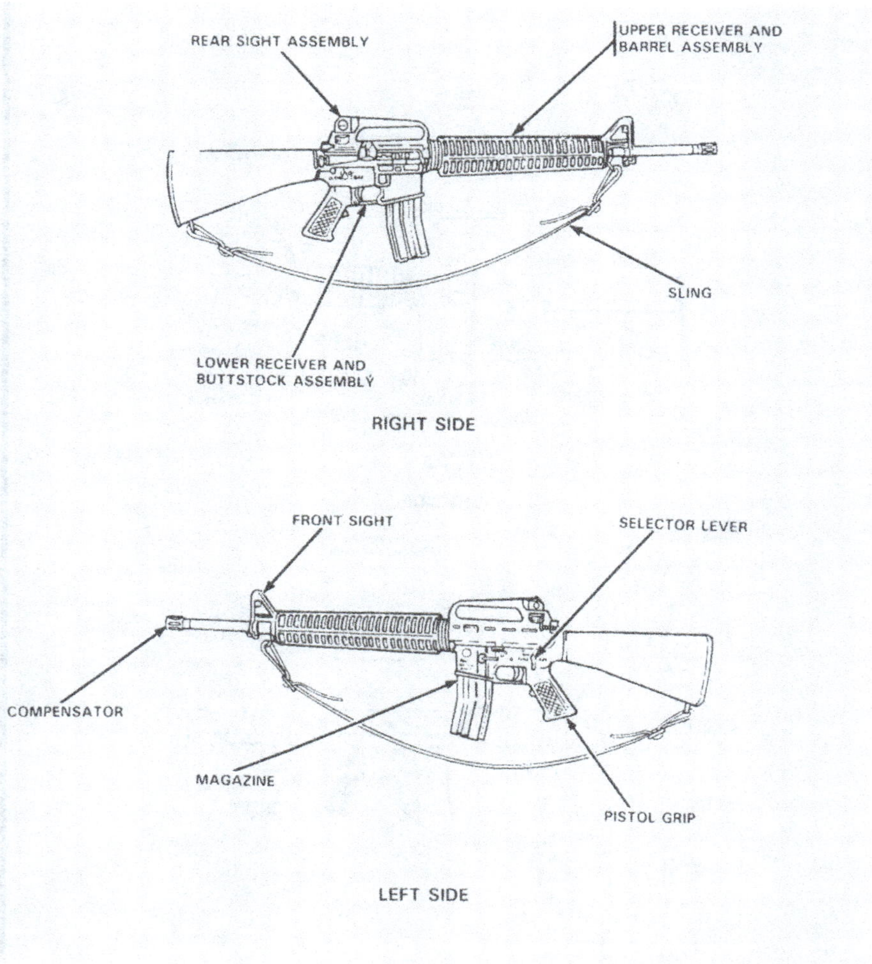

EXTERNAL VIEW OF 5.56MM CARBINE, M4/M4A1

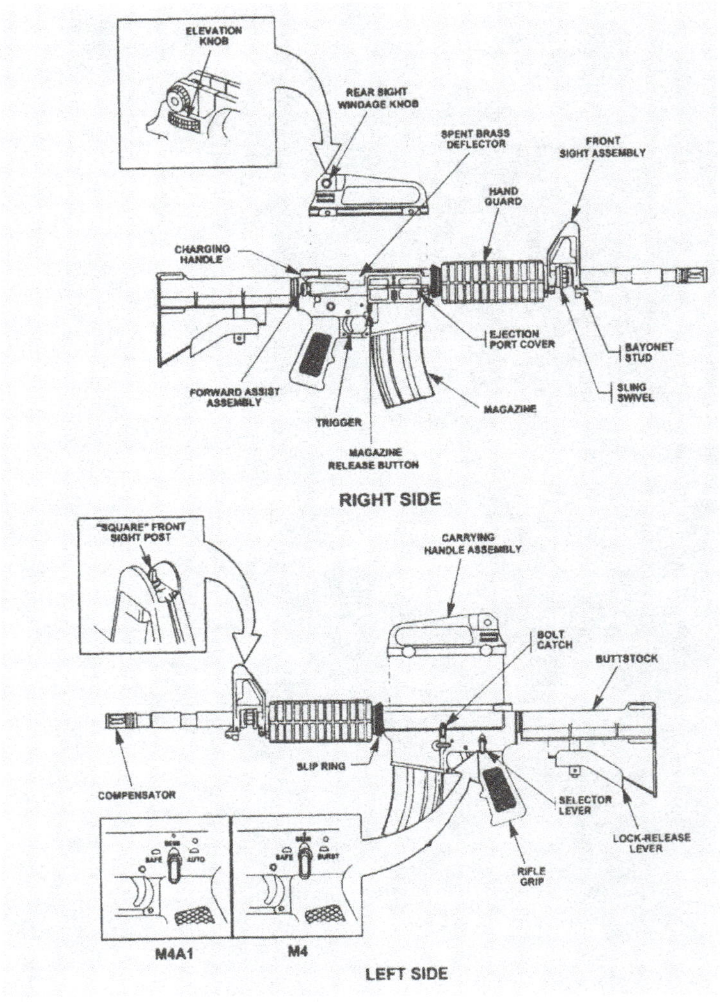

CHAPTER 1
INTRODUCTION

CHAPTER OVERVIEW

This chapter contains general information, equipment description and data, and principles of operation for the M16A2 rifle and M4/M4A1 carbines.

Section I. GENERAL INFORMATION

1-1. SCOPE.

a. Type of Manual: Unit and Direct Support Maintenance.

b. Model Number and Equipment Name: 5.56mm Rifle M16A2, M4 and M4A1 Carbines.

c. Purpose of Equipment. Provides personnel an offensive/defensive capability to engage targets with small arms fire.

1-2. MAINTENANCE FORMS, RECORDS, AND REPORTS. Department of the Army forms and procedures used for equipment maintenance will be those prescribed by DA PAM 738-750, The Army Maintenance Management System.

Air Force users refer to TO 11W-1-10 for applicable forms and records.

1-3. DESTRUCTION OF ARMY MATERIEL TO PREVENT ENEMY USE. See TM 750-244-7.

1-4. PREPARATION FOR STORAGE OR SHIPMENT. Refer to page 2-70.

Air Force users refer to Special Package Instruction (SPI) 00-856-6885.

1-5. OFFICIAL NOMENCLATURE, NAMES AND DESIGNATIONS.

NOMENCLATURE CROSS-REFERENCE LIST

Common Name	Official Nomenclature
Action Spring	Compression Helical Spring
Ball Bearing	Bearing Ball
Bolt Catch Spring	Compression Helical Spring
Bolt Carrier Key Tool	Machine Key
Burst Disconnector	Lock-Release Lever
Cam Clutch Spring	Helical Spring
Carbine	M4/M4A1 Carbine
Charging Handle Assembly	Handle Assembly

1-5. OFFICIAL NOMENCLATURE, NAMES AND DESIGNATIONS (CONT).

NOMENCLATURE CROSS-REFERENCE LIST

Common Name	**Official Nomenclature**
Disconnector Springs	Compression Helical Spring
Ejector Spring	Helical Spring
Extractor Spring Assembly	Spring Assembly
Hammer Spring	Torsion Helical Spring
Lower Receiver Extension	Spring Receiver Holder
Magazine	Cartridge Magazine
Magazine Catch Spring	Compression Helical Spring
Peel Washer	Shim
Pistol Grip	Rifle Grip
Pivot Pin Detent	Takedown Pin Detent
Rifle	Rifle, 5.56mm, M16A2
Rifle Barrel Assembly	Barrel Assembly
Selector Lever	Fire Control Selector
Semiautomatic Disconnector	Lock-Release Lever
Sling	Small Arms Sling
Trigger Spring	Torsion Helical Spring
Upper Receiver	Upper Cartridge Receiver

1-6. REPORTING EQUIPMENT IMPROVEMENT RECOMMENDATIONS (EIR). If your M16A2 rifle needs improvement, let us know. Send us an EIR. You, the user, are the only one who can tell us what you don't like about your equipment. Let us know why you don't like the design.

Army users submit SF 368 (Product Quality Deficiency Report) to: Commander, U.S. Army Armament Research, Development and Engineering Center, ATTN: AMSTA-AR-QAW (R)/Customer Feedback Center, Rock Island, IL 61299-7300.

Air Force users submit Materiel Deficiency Report (MDR) to: DIR MAT MGT ROBINS AFB GA//MMIBTC// and Product Quality Deficiency Report to: DIR MAT MGT ROBINS AFB GA//MMQA// IAW Technical Order 00-35D-54.

A reply will be sent to you.

1-7. CORROSION PREVENTION AND CONTROL (CPC). CPC of Army materiel is a continuing concern. It is important that any corrosion problems with this item be reported so that the problem can be corrected and improvements can be made to prevent the problem in future items,

While corrosion is typically associated with rusting of metals, it can also include deterioration of other materials such as rubber and plastic. Unusual cracking, softening, swelling, or breaking of these materials may be a corrosion problem.

If a corrosion problem is identified, it can be reported using Standard Form 368, Product Quality Deficiency Report. Use of key words such as "corrosion", "rust", " deterioration", or "cracking" will assure that the information is identified as a CPC problem.

Army users submit Product Quality Deficiency Report (SF 368) to:

Commander
U.S. Army Armament Research, Development and Engineering Center
ATTN: AMSTA-AR-QAW (R)
Rock Island, IL 61299-7300

Air Force users submit Materiel Deficiency Report (MDR) to:

DIR MAT MGT
ATTN: MMIBTC
Robins AFB, GA

and Product Quality Deficiency Report to:

DIR MAT MGT
ATTN: MMQA
Robins AFB, GA

Section II. EQUIPMENT DESCRIPTION AND DATA

1-8. EQUIPMENT CHARACTERISTICS, CAPABILITIES, AND FEATURES.

a. Characteristics.

(1) Light weight
(2) Air-cooled
(3) Gas-operated
(4) Magazine-fed
(5) Semiautomatic or burst fire

b. Capabilities. Provides personnel an offensive/defensive capability to engage targets with direct small arms fire.

c. Features.

(1) Receivers are made of light-weight aluminum alloys; however, the safety , durability, and function of the rifles are in no way reduced. The portability and logistical values are greatly increased, particularly when air transport is used.

(2) The bolt locking action is one of the mechanical features of the rifle. The bolt assembly and barrel extension contain locking lugs which engage and lock the bolt assembly firmly in the barrel extension. The initial force of the explosion of the cartridge is absorbed by the barrel, barrel extension, and bolt assembly.

1-8. EQUIPMENT CHARACTERISTICS, CAPABILITIES, AND FEATURES (CONT).

(3) The trigger guard is easily adaptable to winter operations. A spring-loaded retaining pin is depressed to allow ready access to the trigger when wearing arctic mittens.

(4) The ejection port cover prevents dirt or sand from getting into the ejection port. The ejection port cover must be closed during periods when firing is not anticipated. It opens automatically by the forward or rearward movement of the bolt carrier assembly.

1-9. LOCATION AND DESCRIPTION OF MAJOR COMPONENTS.

(A) **MAGAZINE**. 30 cartridge capacity.

(B) **SLING**. The sling is adjustable and provides a means to carry the weapon.

(C) **BOLT CARRIER ASSEMBLY**. Carries bolt assembly to chamber and fires the weapon. Contains the firing pin, cartridge extractor, bolt assembly, cartridge ejector, and bolt cam pin.

(D) **CHARGING HANDLE ASSEMBLY**. Provides a means of charging the weapon.

(E) **M16A2 UPPER RECEIVER AND BARREL ASSEMBLY**. Upper receiver contains rear sight assembly, ejection port, ejection port cover, and a housing for the key and bolt carrier assembly and bolt assembly. Rifle barrel assembly is air-cooled, contains compensator and front sight assembly, and holds the two handguard assemblies and the sling swivel.

(F) **LOWER RECEIVER AND BUTTSTOCK ASSEMBLY**. Lower receiver contains the trigger assembly, sear, hammer assembly, selector lever, rifle grip, bolt catch, and buttstock assembly. The buttstock assembly houses the action spring, buffer assembly, and extension assembly.

(G) **M4/M4A1 CARRYING HANDLE**. Provides a means of carrying carbine.

(H) **M4/M4A1 UPPER RECEIVER AND BARREL ASSEMBLY**. Upper receiver contains, ejection port, ejection port cover, a housing for key and bolt carrier assembly and bolt assembly, and mounting surface for the carrying handle assembly. Carbine barrel assembly is air-cooled, contains compensator and front sight assembly, and holds the two handguard assemblies and the sling swivel.

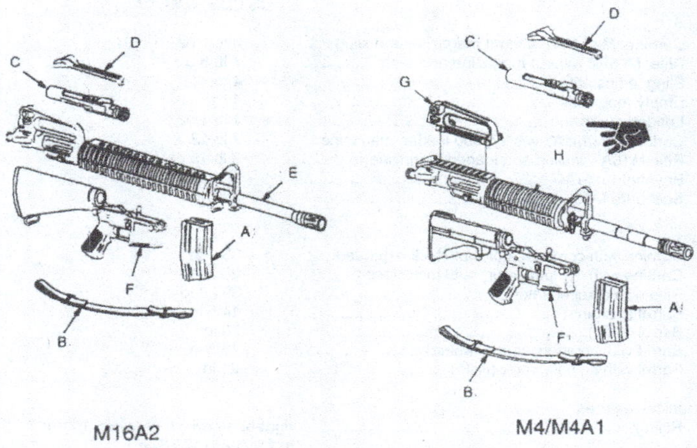
D
C
E
A
F
B
M16A2
D
C
G
H
A
F
B
M4/M4A1

1-10. EQUIPMENT DATA.

	US CUSTOMARY	METRIC
Weight:		
Carbine, M4/M4A1 without magazine and sling	6 lb 7 oz	2.91 kg
Rifle, MI 6A2 without magazine and sling	7 lb 8 oz	3.40 kg
Sling, adjustable	4 oz	0.11 kg
Empty magazine	4 oz	0.11 kg
Loaded magazine	1 lb 1 oz	0.48 kg
Carbine, M4/M4A1 w/sling and loaded magazine ...	7 lb 12 oz	3.51 kg
Rifle M16A2 w/sling and loaded magazine	8 lb 13 oz	4.00 kg
Bayonet-Knife M7	10.5 oz	0.30 kg
Scabbard M10 o	5 oz	0.14 kg
Length:		
Carbine with compensator, buttstock extended.......	33.0 in	83.82 cm
Carbine with compensator, buttstock closed	29.75 in	75.57 cm
Rifle with compensator	39.63 in	100.66 cm
Barrel (Carbine)	14.5 in	36.83 cm
Barrel (Rifle)...	20 in	50.8 cm
Barrel with compensator (Carbine)	15.5 in	39.37 cm
Barrel with compensator (Rifle)	21 in	53.34 cm
Mechanical features:		
Rifling	right-hand twist 6 grooves, 1 turn	
....	in 7 inches (17.78 cm)	
Method of operation	direct gas	
Type of breech mechanism	rotating bolt	
Method of feeding	magazine	
Cooling	air	
Trigger pull (M16A2 & M4).......................................	5.5 to 9.5 lb	2.49 to 4.31kg
Trigger pull (M4A1)	5.5 to 8.5 lb	2.49 to 3.86 kg
Ammunition:		
Caliber	223	5.56mm
Type	ball, blank, dummy, and tracer	
Firing characteristics:		
Muzzle velocity (Carbine) (approximate)	2,970 fps	905.85 mps
Muzzle velocity (Rifle) (approximate).......................	3,100 fps	945.5 mps
Chamber pressure	52,000 psi	358,540 kPa
Cyclic rate of fire (Carbine) (approximate)..............	700-970 rds/m	
Cyclic rate of fire (Rifle) (approximate)	700-900 rds/m	

1-10. EQUIPMENT DATA (CONT).

	US CUSTOMARY	METRIC
Maximum rate of fire:		
Semiautomatic	45 rds/m	
Burst	90 rds/m	
Sustained rate of fire	12/15 rds/m	
Maximum range	3,938 yards	Approximately 3,600 meters
Maximum effective range:		
Individual/point targets (Carbine)	547 yards	500 meters
Individual/point targets (Rifle)	602 yards	550 meters
Area targets (Carbine)	650 yards	600 meters
Area targets (Rifle)	875 yards	800 meters

Section III. PRINCIPLES OF OPERATION

1-11. GENERAL. The 5.56mm M16A2 and M4/M4A1 carbine:

a. Is gas-operated. It fires in either the semiautomatic or burst mode.

b. Has positive locking of the bolt. Firing pin is part of the bolt carrier assembly and cannot strike the primer until the bolt assembly is fully locked.

Section III. PRINCIPLES OF OPERATION (CONT).

1-12. PRINCIPLES OF OPERATION.

(A) **MAGAZINE**. Holds cartridges ready for feeding and provides a guide for positioning cartridges for stripping. Provides quick reload capabilities for sustained firing.

(B) **SLING**. Provides the means for carrying the weapon.

(C) **BOLT CARRIER ASSEMBLY**. Provides stripping, chambering, locking, firing, extraction, and ejection of cartridges using the drive springs and projectile propelling gases for power.

(D) **CHARGING HANDLE ASSEMBLY**. Provides initial charging of the weapon. The handle latch locks the charging handle assembly in the forward position during sustained fire to prevent injury to the operator.

(E) **M16A2 UPPER RECEIVER AND BARREL ASSEMBLY**. Provides support for the bolt carrier assembly. The barrel chambers the cartridge for firing and directs the projectile.

(F) **LOWER RECEIVER AND BUTTSTOCK ASSEMBLY**. Provides firing control for the rifle and carbine. M16A2 ONLY provides storage for basic cleaning materials.

(G) **M4/M4A1 CARRYING HANDLE ASSEMBLY**. Provides a means of carrying the carbine, contains rear sight assembly, and can be removed for mounting various optics.

(H) **M4/M4A1 UPPER RECEIVER AND BARREL ASSEMBLY**. Provides support for the bolt carrier assembly. The barrel chambers the cartridge for firing and directs the projectile.

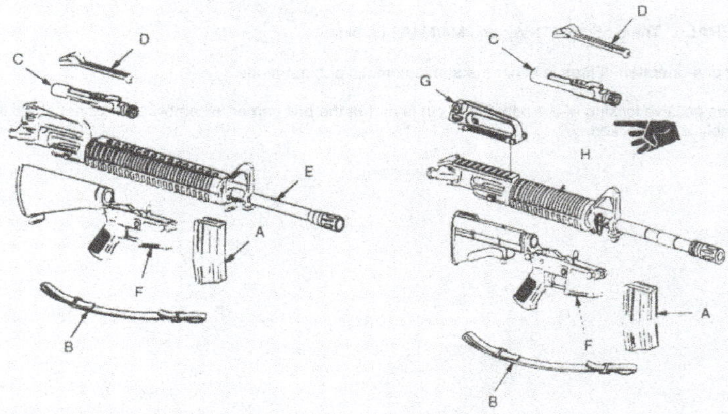

M16A2 M4/M4A1

CHAPTER 2
UNIT MAINTENANCE INSTRUCTIONS

CHAPTER OVERVIEW

This chapter provides information and Instructions to help keep the rifle in good repair and contains the following sections:

a. Repair Parts, Special Tools, TMDE, and Support Equipment
b. Service Upon Receipt
c. Preventive Maintenance Checks and Services (PMCS)
d. Troubleshooting
e. Maintenance Procedures

Section I. REPAIR PARTS, SPECIAL TOOLS, TMDE, AND SUPPORT EQUIPMENT

2-1. COMMON TOOLS AND EQUIPMENT. For authorized common tools and equipment, refer to the Modified Table of Organization and Equipment (MTOE) applicable to your unit.

Air Force users must maintain the following common tools:

3-ounce soft-brass hammer
Vise
Flat tip screwdriver
Punch
Tweezers/round nose pliers
Hammer
Needle nose pliers

2-2. SPECIAL TOOLS, TMDE, AND SUPPORT EQUIPMENT. Special tools required for unit support are listed in appendixes B and C. Fabricated tools are listed and illustrated in appendix E.

2-3. REPAIR PARTS. Repair parts are listed and illustrated in appendix C of this manual.

Section II. SERVICE UPON RECEIPT

2-4. GENERAL.

a. Inspect the rifle for damage Incurred during shipment. If rifle has been damaged, report the damage on SF 364, Report of Discrepancy (ROD).

b. Check the rifle against the packing slip to see If shipment is complete. Army users report all discrepancies in accordance with DA PAM 738-750.

Air Force users submit Materiel Deficiency Report (MDR) to: DIR MAT MGT ROBINS AFB GA//MMIBTC// and Product Quality Deficiency Report to: DIR MAT MGT ROBINS AFB GA//MMQA//. IAW Technical Order 00-35D-54.

c. Check to see whether the equipment has been modified.

2-5. SERVICE UPON RECEIPT OF MATERIEL.

WARNING

Before starting an inspection, be sure to clear the rifle. Do not actuate the trigger before clearing the rifle. Inspect the chamber to make sure it is empty and free of obstructions. Check to see there are no obstructions in the barrel and no ammunition is in position to be chambered.

SERVICE UPON RECEIPT

LOCATION	ITEM	ACTION	REMARKS
1. Container	a. M16A2 rifle or M4/M4A1 carbine	a. Remove rifle from containers.	
		b. Inspect the equipment for damage incurred during shipment	If the equipment has been damaged, report the damage on SF Form 364, Report of Discrepancy (ROD).
		c. Check the equipment against the packing list to see if the shipment is complete	Report all discrepancies in accordance with the instructions of DA PAM 738-750.
	b. Basic issue items	Check for missing items	Refer to TM 9-1005-319-10 (operator's manual).
2. M16A2 rifle or M4/M4A1 carbine	a. Barrel assembly	If volatile corrosion inhibitor (VCI) is in barrel, remove and discard.	
	b. All parts	a. Field-strip rifle and inspect for missing, damaged, and rusted or corroded parts.	Refer to operator's manual.
		b. Clean and lubricate	Refer to operator's manual.
		c. Reassemble	Refer to operator's manual.
		d. Function check	Refer to page 2-69.

SERVICE UPON RECEIPT (CONT)

LOCATION	ITEM	ACTION	REMARKS
		e. Check to see whether the equipment has been modified	Refer to DA PAM 25-30.
	c. Magazine	Check for positive retention and functioning of bolt catch	Refer to operator's manual.

Section III. PREVENTIVE MAINTENANCE CHECKS AND SERVICES (PMCS)

2-6. GENERAL This section contains the procedures and instructions necessary to perform unit preventive maintenance checks and services. These services are to be performed by unit maintenance personnel with the assistance of the operator where practical.

2-7. PREVENTIVE MAINTENANCE CHECKS AND SERVICES

WARNING

Before starting an inspection, be sure to clear the rifle. Do not keep live ammunition near the work area.

a. General. The PMCS procedures are contained in the table following. They are arranged in logical sequence requiring a minimum amount of time and motion on the part of the persons performing them and are arranged so that there will be minimum interference between persons performing checks simultaneously on the same end item.

b. Item No. Column. Checks and services are numbered in disassembly sequence. This column shall be used as a source of item numbers for the "TM Number" column on DA Form 2404, Equipment Inspection and Maintenance Worksheet, in recording results of PMCS.

c. Interval Column. This column gives the designated interval when each check is to be performed.

d. Item To Be Checked Or Serviced Column. This column lists the items to be checked or serviced.

e. Procedure Column. This column contains a brief description of the procedure by which the check is to be performed. It contains all the information required to accomplish the checks and services. Information marked SH Indicates a specific equipment shortcoming and the procedure needed to correct the shortcoming.

NOTE

For the purpose of this technical manual, the following definition is supplied. This definition is not intended to apply to any other document

Shortcoming (SH): A fault that requires maintenance or supply action on a piece of equipment, but does not render equipment Not Mission Capable

f. Not Fully Mission Capable If: Column. This column contains a brief statement of the condition (e.g., malfunction, shortage) that would cause the covered equipment to be less than fully ready to perform its assigned mission.

2-7. PREVENTION MAINTENANCE CHECKS AND SERVICES (CONT).

PREVENTIVE MAINTENANCE CHECKS AND SERVICES FOR M16A2 RIFLE (CONT)

Item No.	Interval	Item to be Checked or Serviced	Procedure	Not Fully Mission Capable if:

WARNING

Before starting an inspection, be sure to clear the weapon. Do not pull the trigger until the weapon has been cleared. Inspect the chamber to ensure that it is empty and no ammunition is in position to be chambered. Do not keep live ammunition near work area.

NOTE

An inactive weapon is a weapon which has been stored in an arms room for a period of 90 days without use. The weapon may or may not have been assigned to an individual.

Inactive weapons shall receive quarterly PMCS unless inspection reveals more frequent servicing is necessary.

Normal cleaning (PMCS) of an inactive weapon will be performed every 90 days. Should the unit armorer detect corrosion on a weapon prior to the end of the 90-day period, the PMCS should be performed immediately.

Solid Film Lubricant (SFL) is the authorized touch up for the M16A2 Rifle and M4/M4A1 Carbine and may be used on up to one third of the exterior finish of the weapon.

FOR ARMY CONUS USE ONLY AND AIR FORCE TRAINING WEAPONS ONLY: Solid Film Lubricant may be used as a touch up without limitation on the upper receiver and barrel assembly. This is to say that units which **DO NOT** fall under the category of Divisional Combat Units or rapid deployment type units may have up to 100 percent of the exterior surface of the upper receiver and barrel assembly protected with SFL.. Prior to application of SFL, the surface must be thoroughly cleaned and inspected for corrosion and/or damage. If corroded or damaged, the part must be repaired or replaced prior to application of SFL. Continued use under combat conditions would result in an unprotected surface when the SFL wears off. This would result in a large light reflecting surface and accelerated deterioration of the unprotected surface. Therefore, Divisional Combat Units and units which fall under the definition of Rapid Deployment type must adhere to the limitation of **NOT** over one third of their exterior surface covered by SFL.

When determining mission capability, deadline if it is a deficiency.

PREVENTIVE MAINTENANCE CHECKS AND SERVICES FOR M16A2 RIFLE (CONT)

Item No.	Interval	Item to be Checked or Serviced	Procedure	Not Fully Mission Capable if:
1	Quarterly	Magazine (serviceability check)	Disassemble as In TM 9-1005-319-10 (operator's manual). Inspect tube (1) for bulges, dents, or damaged feeder lips (2). Inspect spring (3) and follower (4) for kinks or damage. SH-Replace the magazine if any of these conditions exist. Reassemble magazine and check for binding during operation of follower (4). SH-Replace the magazine if the follower binds.	A magazine is not available fu; use with the rifle.

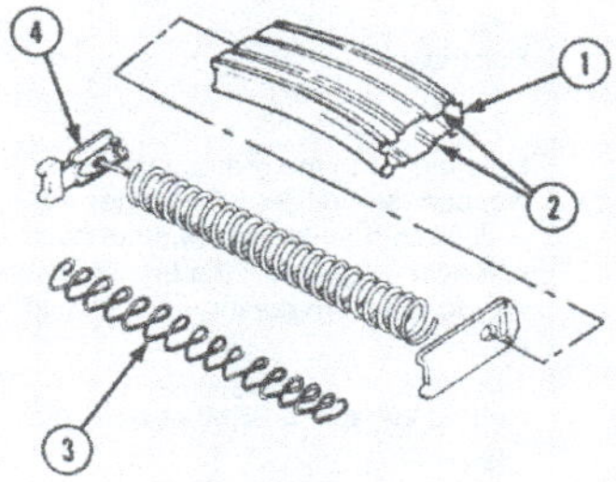

Item No.	Interval	Item to be Checked or Serviced	Procedure	Not Fully Mission Capable if:
2	Quarterly	Charging handle assembly and selector lever	**WARNING** If the rifle falls any of the following selector lever tests, evacuate It to support maintenance. Continued use of the rifle could result In injury to, or death of, personnel.	

2-7. PREVENTIVE MAINTENANCE CHECKS AND SERVICES (CONT).

PREVENTIVE MAINTENANCE CHECKS AND SERVICES FOR M16A2 RIFLE (CONT)

Item No.	Interval	Item to be Checked or Serviced	Procedure	Not Fully Mission Capable if:
2	Quarterly (cont)	Charging handle assembly and selector lever (cont)	Pull charging handle (1) to rear. Check that chamber is clear. Let bolt carrier assembly (2) close. Leave hammer in cocked position. Do not pull trigger	Charging handle does not lock in place when in the forward position.
			Place selector lever (3) in SAFE position Pull trigger. Place selector lever (3) in SEMI position.	Hammer falls.
			NOTE For the purpose of the following test, "SLOW" is defined as 1/4 to 1/2 the normal rate of trigger release.	
			Pull trigger	Hammer does not fall.
			Hold trigger to the rear, charge weapon, and release the trigger with a slow, smooth motion, without hesitations or stops, until the trigger is fully forward (an audible click should be heard).	Hammer falls.
			Repeat the above SEMI position test five times	The weapon malfunctions during any of these five tests.

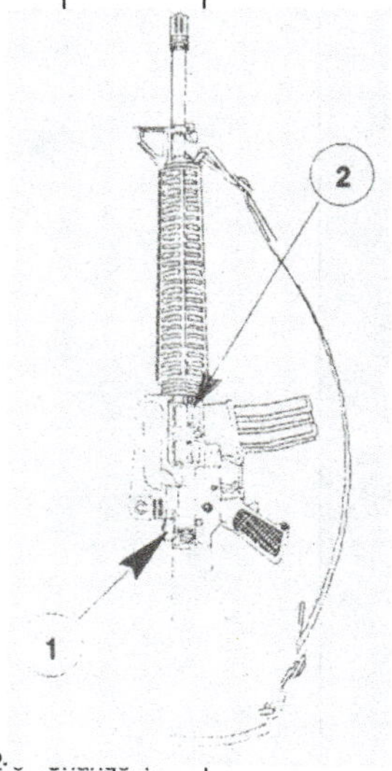

PREVENTIVE MAINTENANCE CHECKS AND SERVICES FOR M16A2 RIFLE (CONT)

Item No.	Interval	Item to be Checked or Serviced	Procedure	Not Fully Mission Capable if:
2	Quarterly (cont)	Charging handle assembly and selector lever (cont)	**M16A2 and M4 ONLY** Place selector lever (3) in BURST position. Charge weapon and squeeze trigger.	Hammer does not fall.
			Hold trigger to the rear, pull the charging handle to the rear and release it three times. Release trigger. Squeeze trigger.	Hammer does not fall.
			NOTE The burst disconnector should have held the hammer to the rear when it engaged the deep notch of the burst cam.	
			M4A1 ONLY Place selector lever (3) in AUTO position Charge carbine and squeeze trigger Hammer should fall.	Hammer does not fall.
			Hold trigger to the rear, charge carbine, and release trigger. Squeeze trigger. Hammer should not fall.	Hammer falls.

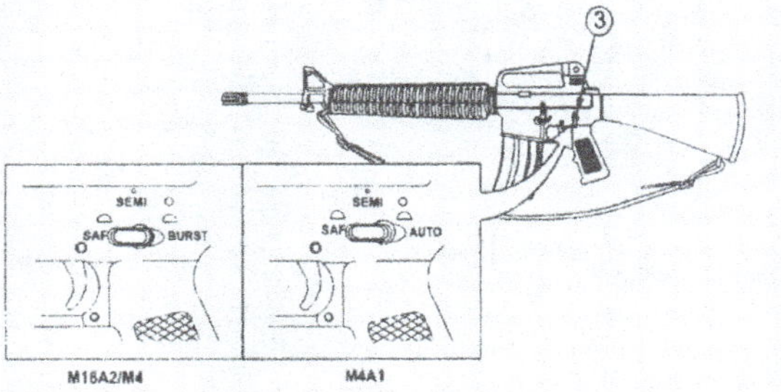

2-7. PREVENTIVE MAINTENANCE CHECKS AND SERVICES (CONT).

PREVENTIVE MAINTENANCE CHECKS AND SERVICES FOR M16A2 RIFLE (CONT)

Item No.	Interval	Item to be Checked or Serviced	Procedure	Not Fully Mission Capable if:
2	Quarterly (cont)	Charging handle assembly and selector lever (cont)	**NOTE** Automatic sear should have released hammer while holding trigger in the squeezed position before releasing and resqueezing the trigger.	
			All weapons With hammer in forward position, using moderate finger/thumb pressure attempt to place the selector lever (3) in SAFE position	Moderate finger/ thumb pressure moves selector lever to SAFE position.

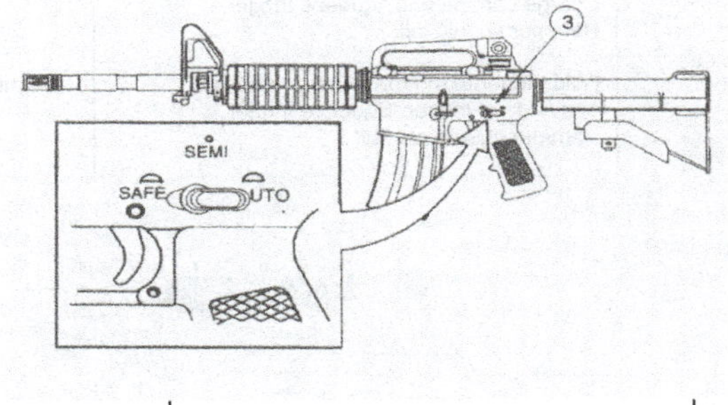

2-7. PREVENTIVE MAINTENANCE CHECKS AND SERVICES (CONT).

PREVENTIVE MAINTENANCE CHECKS AND SERVICES FOR M16A2 RIFLE (CONT)

Item No.	Interval	Item to be Checked or Serviced	Procedure	Not Fully Mission Capable if:
3	Quarterly	Upper receiver and barrel assembly (handguard assemblies) 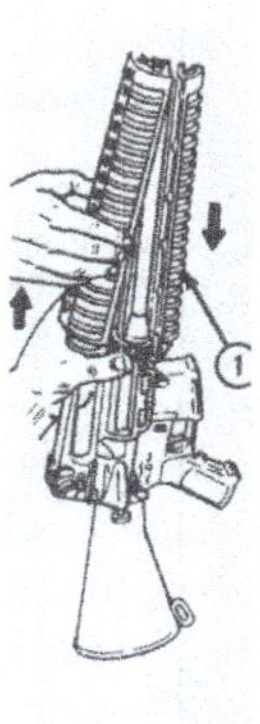	**CAUTION** Do not use screwdriver or any other tool when removing the handguard assemblies, doing so may damage the handguard assemblies and/or slip. **NOTE** Refer to operator's manual for "buddy system" procedure on removing handguard assemblies. Remove and Inspect handguard assemblies (1) internally and externally for cracks and/or damage. Cracks are acceptable providing they do not extend into the handguard retaining flange, or adversely affect rifle operation or operator safety or proper retention of handguard assembly. Discard and replace the handguard assembly (1) If the heatshield is loose enough to rattle when Installed on rifle.	Handguard missing or unserviceable.

PREVENTIVE MAINTENANCE CHECKS AND SERVICES FOR M16A2 RIFLE (CONT)

<table>
<tr><th>Item No.</th><th>Interval</th><th>Item to be Checked or Serviced</th><th>Procedure</th><th>Not Fully Mission Capable if:</th></tr>
<tr><td>4</td><td>Quarterly</td><td>Upper receiver and barrel assembly (serviceability check)

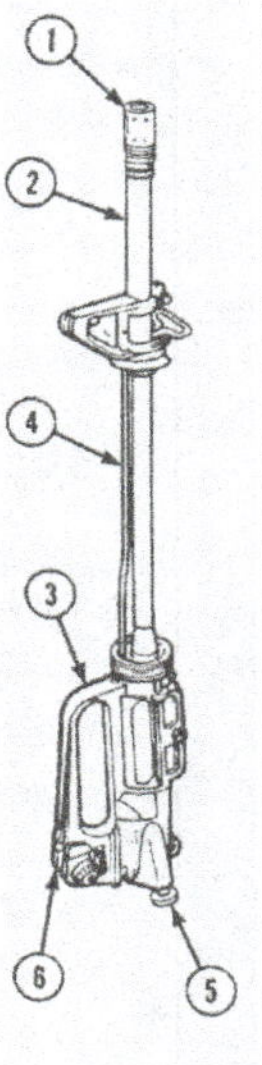

</td><td>

WARNING

Dry cleaning solvent is flammable and toxic and should be used in a well ventilated area The use of rubber gloves is necessary to protect the skin when washing rifle parts

CAUTION

Damage may occur if excessive force is used to release takedown pin or pivot pin. Use hand pressure ONLY.

Release takedown pins and open and separate receivers Hand check compensator (1) for looseness on barrel (2), then hand check barrel (2) for looseness on upper receiver (3). Check center slot of compensator for alignment (p 2-50). If compensator or barrel is loose, evacuate to support maintenance

Check gas tube (4), forward assist assembly (5), and rear sight assembly (6) for damage, The rear sight spring should retain the rear sight assembly (6) In either position with firmness SH - If damaged, evacuate to support maintenance.
</td><td>Compensator or barrel is loose</td></tr>
</table>

2-7. PREVENTIVE MAINTENANCE CHECKS AND SERVICES (CONT).

PREVENTIVE MAINTENANCE CHECKS AND SERVICES FOR M16A2 RIFLE (CONT)

Item No.	Interval	Item to be Checked or Serviced	Procedure	Not Fully Mission Capable if:
4	Quarterly (cont)	Upper receiver and barrel as sembly (service ability check) (cont)	**NOTE** If front or rear sight is moved, return to original position Check front sight post, de tent, and spring (7) for dam age and corrosion Clean and lubricate Check charg ing handle (8) and election port cover (9) for defects and proper function Check sling swivel (10) and rivet (11) for damage and proper function SH Other corn ponents are defective, re place as necessary.	Charging handle (8) is defective.

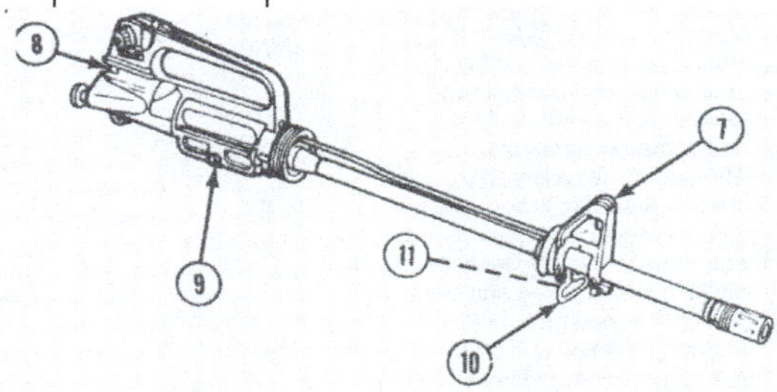

CAUTION

Do not use a wire brush to roughen surfaces Use a well ventilated area during cleaning and application of solid film lubricant. If solid film lubricant comes in contact with moving parts or functioning surfaces of the rifle, remove lubricant immediately by washing with dry cleaning solvent.

NOTE

Shiny metal exterior surfaces of the rifle should be recoated with solid film lubricant (Item 21, app D). Clean surface with dry cleaning solvent (item 16, app D); dry, roughen with abrasive cloth (Item 13, app D) and apply solid film lubricant.

PREVENTIVE MAINTENANCE CHECKS AND SERVICES FOR M16A2 RIFLE (CONT)

Item No.	Interval	Item to be Checked or Serviced	Procedure	Not Fully Mission Capable if:
4	Quarterly (cont)	Upper receiver and barrel assembly (serviceability check) (cont) 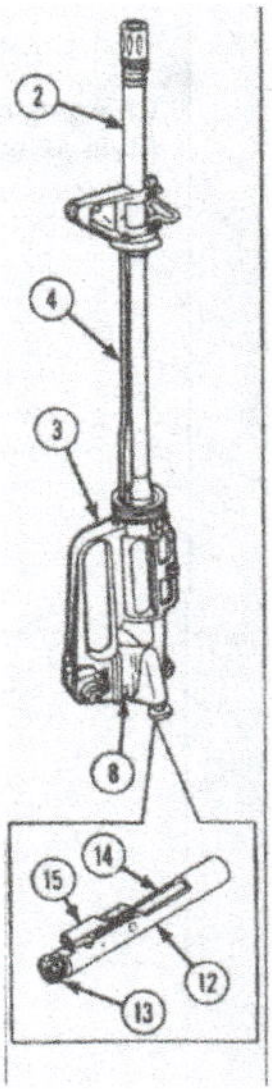	Inspect upper receiver (3) finish for scratches or worn shiny spots. If scratched or worn shiny in spots, disassemble and remove all lubricant from surface with dry cleaning solvent (item 16, app D). Wear rubber gloves (item 18, app D) and use a wash pan (item 24, app D) to apply solvent. Let parts dry thoroughly. Roughen the surface using abrasive cloth (item 1 3, app D) and apply solid film lubricant (item 16, app D). Allow 16 to 24 hours to dry before handling. Hold barrel (2) at 40-degree angle (muzzle down). Pull charging handle (8) to rear. Hold bolt carrier assembly (12) to rear and push charging handle forward. Release bolt carrier assembly (12). The bolt carrier assembly should close and lock under its own weight. If It does not, remove the bolt assembly (13) from the key and bolt carrier assembly (14) and slide the key and bolt carrier assembly (without bolt) back and forth In the upper receiver and barrel assembly. If the gas tube (4) hits the carrier key (1 5), or If the gas tube binds in the carrier key, try to correct the malfunction by adjusting (slightly bending) the gas tube In the area of the handguard assemblies. If unable to adjust, evacuate to support maintenance.	 Adjustment does not correct the malfunction.

2-7. PREVENTIVE MAINTENANCE CHECKS AND SERVICES (CONT).

PREVENTIVE MAINTENANCE CHECKS AND SERVICES FOR M16A2 RIFLE (CONT)

Item No.	Interval	Item to be Checked or Serviced	Procedure	Not Fully Mission Capable if:
4	Quarterly (cont)	Upper receiver and barrel assembly (serviceability check) (cont)	**M4/M4A1** Inspect carrying handle assembly (16) and mounting surface (17) of upper receiver for damage. If the carrying handle is missing or can not be correctly mounted, repair as authorized or evacuate to support maintenance. Inspect carrying handle assembly (16) to insure unit applied identification (ID) code matches unit applied ID code on carbine. If it doesn't match locate correct carrying handle assembly and match up to correct M4/ M4A1 carbine. If a match can not be found, the weapon should be re-zeroed by the operator.	

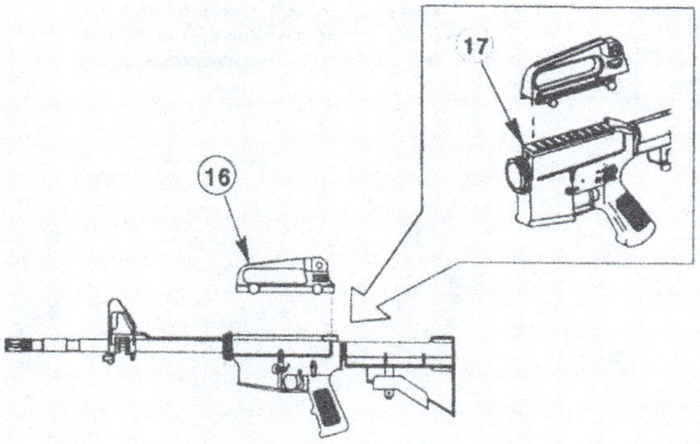

2-7. PREVENTIVE MAINTENANCE CHECKS AND SERVICES (CONT).

PREVENTIVE MAINTENANCE CHECKS AND SERVICES FOR M16A2 RIFLE (CONT)

WARNING

Below direct support maintenance, do not interchange bolt assemblies from one rifle to another. Doing so may result in injury to, or death of, personnel.

Item No.	Interval	Item to be Checked or Serviced	Procedure	Not Fully Mission Capable if:
5	Quarterly	Key and bolt carrier assembly and bolt assembly (serviceability check)	Remove and disassemble. Visually inspect bolt assembly (1) for cracks, especially in the area of the bolt cam pin hole (2). Check for cracks on locking lugs (3), for a cluster of pits or chipped bolt face (4), and for an elongated firing pin hole (5). If cracked or broken, evacuate to support maintenance for repair.	Defects are found.
			Check for worn or missing bolt rings (6) Check for proper staggering of bolt rings Insert the bolt assembly (1) into the key and bolt carrier assembly (7). Turn key and bolt carrier assembly (7) so the bolt assembly (1) points down. The bolt assembly must not drop out. Remove bolt assembly (p 2-35). Check for broken or missing firing pin retaining pin (8) and bolt cam pin (9); replace as necessary.	The bolt assembly drops out of the key and bolt carrier assembly due to its own weight. Missing or broken firing pin retaining pin or bolt cam pin.

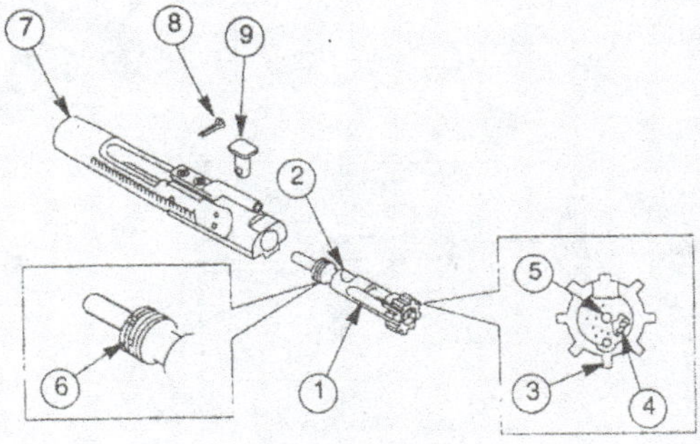

PREVENTIVE MAINTENANCE CHECKS AND SERVICES FOR M16A2 RIFLE (CONT)

Item No.	Interval	Item to be Checked or Serviced	Procedure	Not Fully Mission Capable if:
5	Quarterly (cont)	Key and bolt carrier assembly and bolt assembly (serviceability check) (cont)	Check cartridge extractor (10), extractor spring assembly (11), cartridge ejector (12), and ejector spring (13) for dirt and serviceability. If dirty, clean, lubricate and assemble. If unserviceable, replace as necessary.	Parts are missing or unserviceable
			Check key and bolt carrier assembly (7) and carrier key (14) for damage and looseness. If damaged or loose, evacuate to support maintenance	Key and bolt carrier assembly or carrier key is damaged, or carrier key is loose.
			NOTE If carrier key Is dented, evacuate to support maintenance.	
			Check firing pin 15) for chips or breaks If damaged, evacuate to support maintenance.	Firing pin is damaged.
			Pits or wear in area Illustrated (16) is permissable	

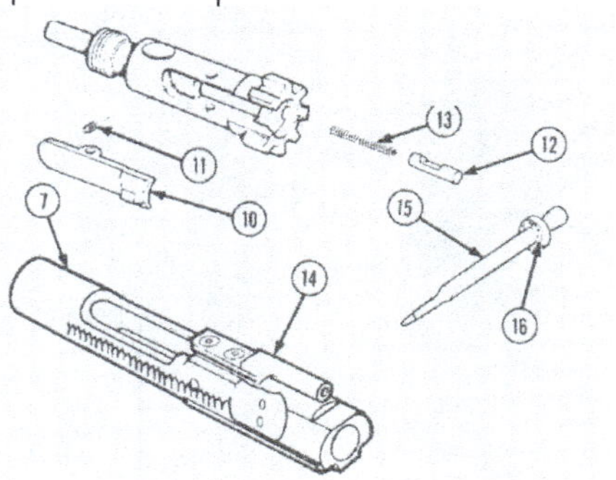

2-7. PREVENTIVE MAINTENANCE CHECKS AND SERVICES (CONT).

PREVENTIVE MAINTENANCE CHECKS AND SERVICES FOR M16A2 RIFLE (CONT)

Item No.	Interval	Item to be Checked or Serviced	Procedure	Not Fully Mission Capable if:
6	Quarterly	Lower receiver and buttstock assembly (serviceability check)	Remove buffer assembly (1) and action spring (2) Check buffer assembly for cracks SH - Buffer assembly is cracked Check action spring (2) for kinks and free length Free length should be **RIFLE**: 11 3/4 Inches (29 85 cm) minimum to 13 1./2 inches (34 29 cm) maximum SH - If action spring is kinked or does not meet free length requirements **CARBINE**: 10 1/16 inches (25.56 cm) minimum to 11 1/4 inches (28.58 cm) maximum. Do not attempt to adjust spring length SH - If action spring is kinked or does not meet free length requirements	

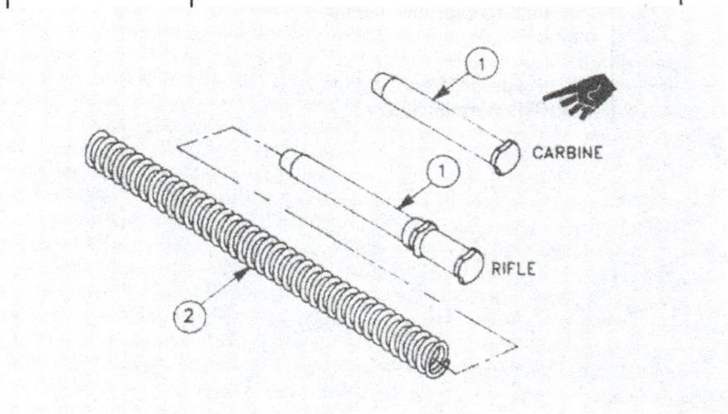

PREVENTIVE MAINTENANCE CHECKS AND SERVICES FOR M16A2 RIFLE (CONT)

Item No.	Interval	Item to be Checked or Serviced	Procedure	Not Fully Mission Capable if:
6	Quarterly (cont)	Lower receiver and buttstock assembly (serviceability check) (cont)	Remove pistol grip screw (3), lockwasher (4), pistol grip (5), helical spring (6), safety detent (7), pivot pin (8), pivot pin detent (9), and helical spring (10). Clean and lubricate metal components. Also clean and generously lubricate pivot pin holes and spring/detent holes. Replace defective/damaged components as necessary.	Components are defective/damaged.
			Disengage takedown pin (11) and pull out, push back in to re-engage takedown pin (an audible click should be heard). If an audible click is not heard, see page 2-57 for repair.	Components are defective/damaged.

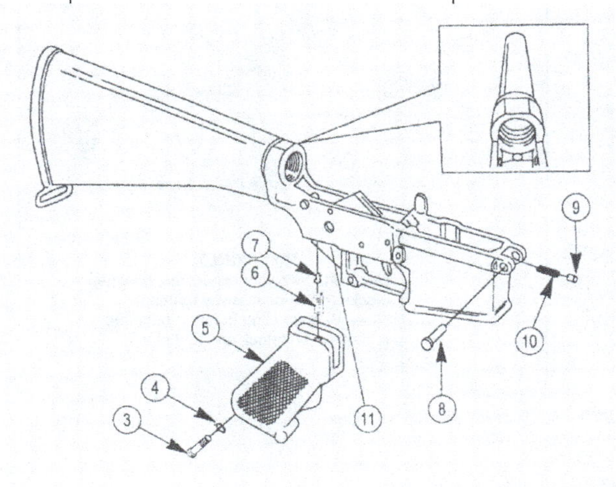

2-7. PREVENTIVE MAINTENANCE CHECKS AND SERVICES (CONT).

PREVENTIVE MAINTENANCE CHECKS AND SERVICES FOR M16A2 RIFLE (CONT)

Item No.	Interval	Item to be Checked or Serviced	Procedure	Not Fully Mission Capable if:
6	Quarterly (cont)	Lower receiver and buttstock assembly (serviceability check) (cont)	Lubricate helical compression spring and takedown pin detent (11) by placing one drop of lubricant on takedown pin detent and lowering the buttstock assembly (12) to vertical position. Allow the lubricant to work Its way around the helical compression spring and takedown pin detent (11). Check buttstock assembly(12) components for damage.	

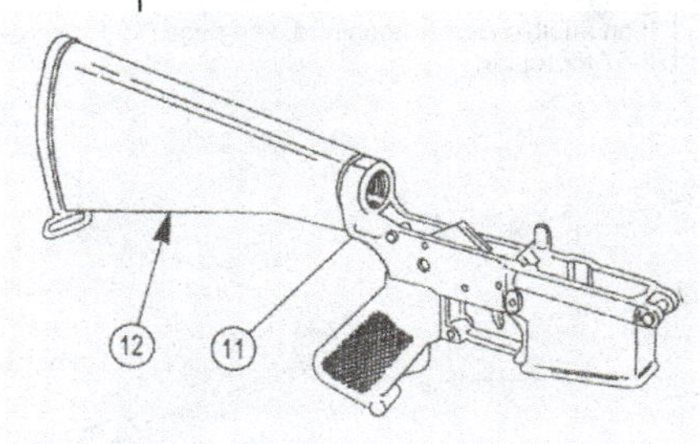

Item No.	Interval	Item to be Checked or Serviced	Procedure	Not Fully Mission Capable if:
			RIFLE ONLY Under the following conditions, hairline cracks (no chipped away material allowed) originating from the buttplate end of the buttstock are acceptable.	Components are damaged.

PREVENTIVE MAINTENANCE CHECKS AND SERVICES FOR M16A2 RIFLE (CONT)

Item No.	Interval	Item to be Checked or Serviced	Procedure	Not Fully Mission Capable if:
6	Quarterly (cont)	Lower receiver and buttstock assembly (serviceability check) (cont)	**RIFLE ONLY** a. One hairline crack, not to exceed 1 in. (2.54 cm) in length, per side of buttstock	The buttstock is cracked in the critical area or does not meet the crack criteria.
			b. Two additional hairline cracks up to 0.25 in. (0.64 cm) in length, per side of buttstock.	
			c. A total of three cracks per side of the buttstock, originating from the buttplate end, are allowable.	
			Cracks in the critical area at the front end of the buttstock are not acceptable.	
			Check buttstock assembly (12) for forward to rear movement and/or a 1/32 in. (0.079 cm) gap between the buttstock assembly (12) and the lower receiver (13). If forward to rear movement and/or a 1/32 in. (0.079 cm) gap appears, tighten self-locking screw. If still not tight, remove buttstock assembly and check for loose lower receiver extension. If loose evacuate to support maintenance. If not loose, replace buttplate (14).	Lower receiver extension cannot be tightened.

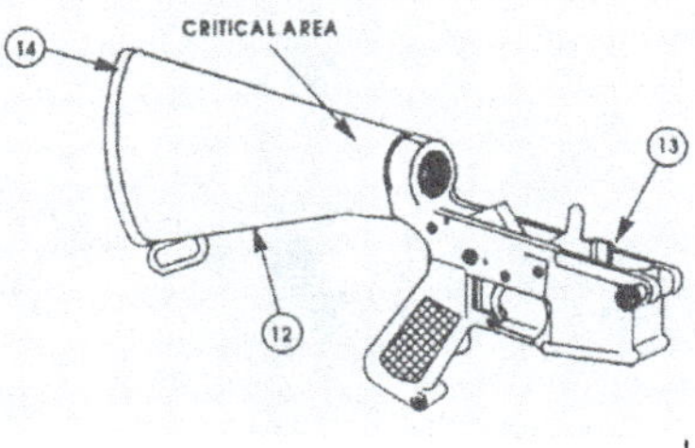

2-7. PREVENTIVE MAINTENANCE CHECKS AND SERVICES (CONS).

PREVENTIVE MAINTENANCE CHECKS AND SERVICES FOR M16A2 RIFLE (CONT)

Item No.	Interval	Item to be Checked or Serviced	Procedure	Not Fully Mission Capable if:
6	Quarterly (cont)	Lower receiver and buttstock assembly (serviceability check) (cont)	**RIFLE ONLY** Small amounts of side-to-side, up-and-down or rotational movement of the buttstock assembly is acceptable. (1) Cracks visible around the butt-plate mounting holes while screws are mounted SH - Cracks are visible around mounting holes when installed on rifle. (2) Cracks or separations around the door assembly are visible when the door assembly is closed SH - Cracks are visible when door assembly is closed. (3) If buttplate is cracked in excess of 0.25 in (O 64 cm) in length and extends through the buttplate (14). see pg. 2-64 for repair SH - If cracked in excess of 0 25 in. extends thru buttplate (4) The buttplate (14) should not be removed other than for repair or replacement of parts at which time a new self-locking screw, NSN 5305-01-147-8585, must be used.	

PREVENTIVE MAINTENANCE CHECKS AND SERVICES FOR M16A2 RIFLE (CONT)

Item No.	Interval	Item to be Checked or Serviced	Procedure	Not Fully Mission Capable if:
6	Quarterly (cont)	Lower receiver and buttstock assembly (serviceability check) (cont)	**NOTE** If a weapons lower receiver is missing one third or more of its exterior protective finish, resulting in an unprotected/light reflecting surface, it is candidate for overhaul. This missing finish will be considered a shortcoming. This shortcoming requires action to obtain a replacement weapon. Once a replacement has been received, evacuate the original weapon to depot for overhaul. If scratched or worn shiny in spots, repair in the same manner as outlined for upper receiver (see item 4 above).	
7	Quarterly	M16A2 Rifle	Assemble as in TM 9-1005-319-10 (operator's manual). Check sling for damage. If damaged, replace. Check for improperly assembled, broken, missing, or damaged parts. Check over all general appearance. Replace parts as required and authorize evacuation to support maintenance for repair.	

2-7. PREVENTIVE MAINTENANCE CHECKS AND SERVICES (CONT).

PREVENTIVE MAINTENANCE CHECKS AND SERVICES FOR M16A2 RIFLE (CONT)

Item No.	Interval	Item to be Checked or Serviced	Procedure	Not Fully Mission Capable if:
8	Quarterly	Annual DS safety and serviceability inspection and gaging	Check to ensure annual DS safety and serviceability inspection and gaging has been done and that the next gaging and inspection is scheduled. If annual gaging has not been performed within the last year, notify support maintenance.	Annual gaging has not been performed.

Section IV. TROUBLESHOOTING

2-8. GENERAL.

a. This section contains unit level troubleshooting information for locating and correcting most of the operating troubles which may develop in the M16A2 rifle, M4/M4A1 carbine. Each malfunction for the individual part or assembly is followed by a list of tests or inspections which will help you to determine the corrective actions in the order listed.

b. This manual cannot list all malfunctions that may occur, nor all tests or inspections and corrective actions. If a malfunction is not listed or is not corrected by listed corrective actions, see individual repair sections in the maintenance procedures on each major assembly.

2-9. TROUBLESHOOTING PROCEDURES. Refer to troubleshooting table for malfunctions, tests, and corrective actions. The symptom index is provided for a quick reference of the malfunctions covered in the table.

SYMPTOM INDEX

TROUBLESHOOTING

MALFUNCTION
TEST OR INSPECTION
CORRECTIVE ACTION

1. FAILURE OF MAGAZINE TO LOCK IN RIFLE.

Step 1. Dirty or corroded magazine catch (1).

Disassemble and clean.

Step 2 Defective magazine catch spring (2)

Evacuate to support maintenance.

Step 3 Worn or broken magazine catch (1).

Evacuate to support maintenance

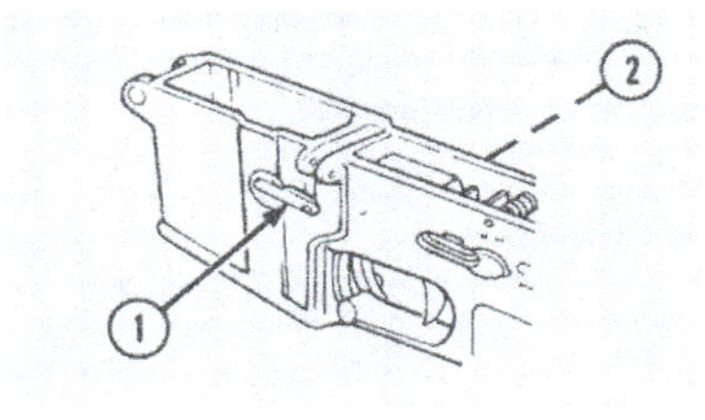

2-9. TROUBLESHOOTING PROCEDURES (CONT).

TROUBLESHOOTING (CONT)

MALFUNCTION
TEST OR INSPECTION
CORRECTIVE ACTION

2. FAILURE TO FEED

Step 1. Magazine catch spring weak or broken

Evacuate to support maintenance

Step 2. Magazine catch (1) defective.

Evacuate to support maintenance

Step 3. Magazine catch (1) out of adjustment (will not retain magazine)

Refer to operator's manual.

Step 4. Short recoil.

Refer to page 2-28

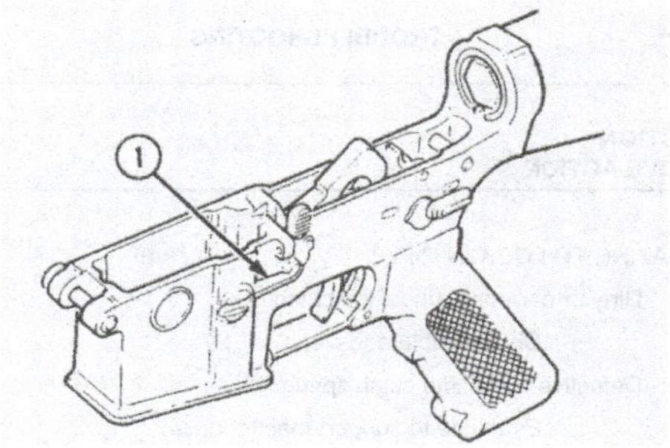

3. FAILURE TO CHAMBER.

Step 1. Weak or broken action spring (1), RIFLE ONLY (free length 11 3/4 inches (29.85 cm) minimum to 13 1.'2 Inches (34 29 cm) maximum) CARBINE ONLY (10 1 16 inches (2S 56 cm) minimum to 11 1,4 inches (28 58 cm) maximum)

Replace action spring (p 2-57)

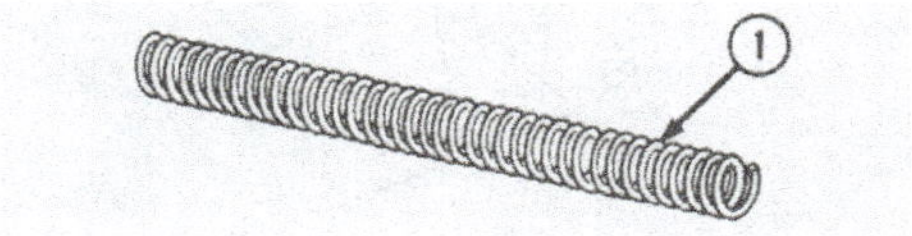

TROUBLESHOOTING (CONT)

MALFUNCTION
TEST OR INSPECTION
CORRECTIVE ACTION

Step 2. Short recoil

Refer to page 2-28

4. FAILURE TO LOCK

Step 1. Bolt cam pin (1) missing

Replace (p 2-35)

Step 2. Loose or damaged bolt carrier key (2).

a. Evacuate to support maintenance.

b. Dented bolt carrier key may be repaired (p 2 35).

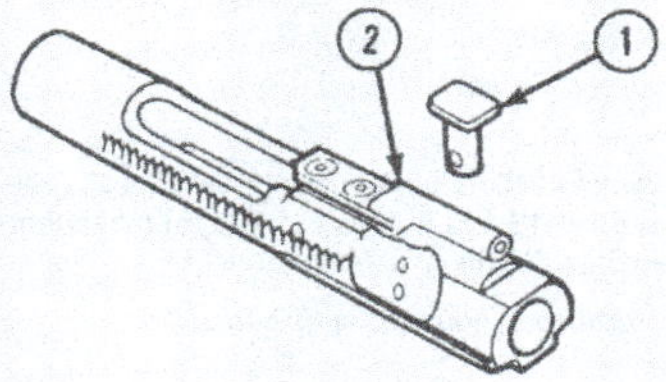

Step 3. Improperly assembled extractor spring assembly (3)

Assemble correctly (p 2-38).

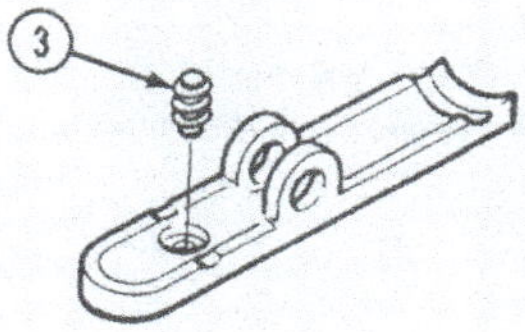

2-9. TROUBLESHOOTING PROCEDURES (CONT).

TROUBLESHOOTING (CONT)

MALFUNCTION
TEST OR INSPECTION
CORRECTIVE ACTION

4. FAILURE TO LOCK (CONT)

Step 4. Bent gas tube 14J

a Adjust to Its original configuration by bending in area of hand-guard assembly.

b If the gas tube cannot be returned to Its original configuration, evacuate the rifle to support maintenance

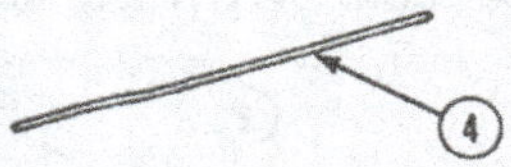

Step 5 Weak or broken action spring (5); RIFLE ONLY: (free length 11 3/4 inches (29,85 cm) minimum to 13 1,/2 Inches (34 29 cm) maximum). CARBINE ONLY (10 1/16 Inches (25 56 cm) minimum to 11 1,4 inches (28 58 cm) maximum

Replace action spring (p 2-571

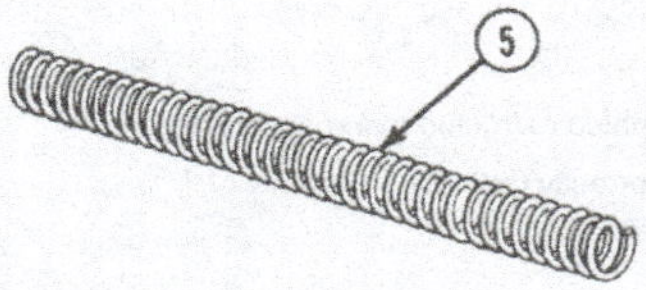

5. FAILURE TO FIRE.

Step 1 Broken or chipped firing pin (1).

Evacuate to support maintenance.

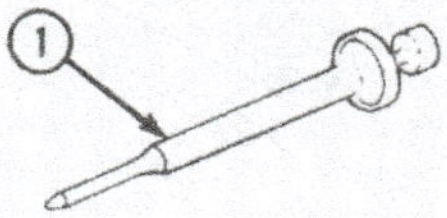

TROUBLESHOOTING (CONT)

MALFUNCTION
TEST OR INSPECTION
CORRECTIVE ACTION

Step 2 Carbon buildup in firing pin recess inside bolt assembly

Remove cartridge extractor and clean recess with pipe cleaner (Item 11, app D), refer to operator's manual.

Step 3. Firing mechanism (2) and or lower receiver assembly (31 improperly assembled or has worn, broken, or missing parts.

Evacuate to support maintenance

Step 4. Broken, defective, or missing firing pin retaining pin (4).

Replace (p 2-35)

Step 5. Selector lever (5) frozen on SAFE position.

Evacuate to support maintenance

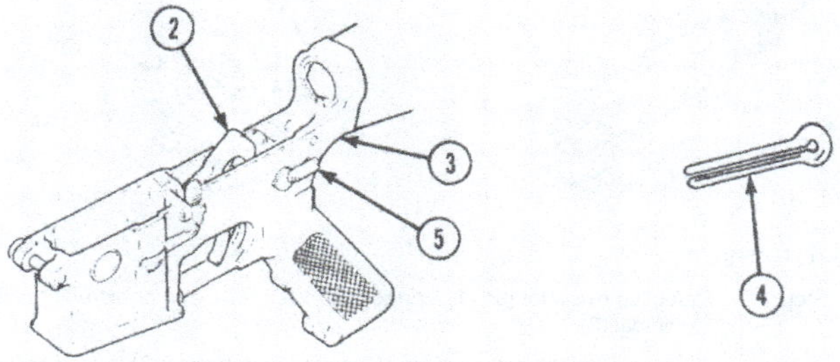

2-9. TROUBLESHOOTING PROCEDURES (CONT).

TROUBLESHOOTING (CONT)

MALFUNCTION
TEST OR INSPECTION
CORRECTIVE ACTION

6. FAILURE TO UNLOCK.

Step 1. Burred locking lugs (1) on bolt assembly.

Remove burrs.

Step 2. Burred lugs (2) on barrel extension

Remove burrs.

Step 3 Short recoil.

Refer to page 2-28.

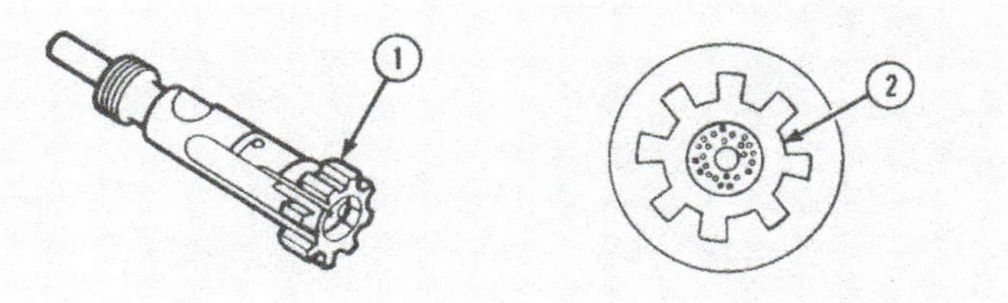

7. FAILURE TO EXTRACT.

Step 1. Defective extractor pin (1), cartridge extractor (2), and or extractor spring assembly (3).

Replace extractor pin (1), cartridge extractor (2), and or extra(:tor spring assembly (3) (p 2-381

Step 2. Short recoil

Refer to page 2-28

NOTE

Rubber insert and spring are an assembly Illustration shows Insert out of assembly for clarification only Do not remove the rubber Insert from the extractor spring assembly.

TROUBLESHOOTING (CONT)

MALFUNCTION
TEST OR INSPECTION
CORRECTIVE ACTION

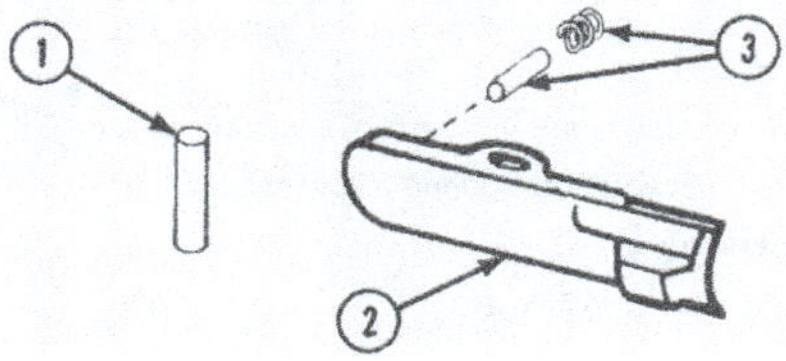

8. FAILURE TO EJECT.

Step 1. Broken cartridge ejector (1).

Replace (p 2-38).

Step 2. Cartridge ejector (1) stuck in bolt body (2).

Disassemble and clean (p 2-38).

Step 3. Weak or broken ejector spring (3).

Replace (p 2-38)

Step 4. Short recoil

Refer to page 2-28.

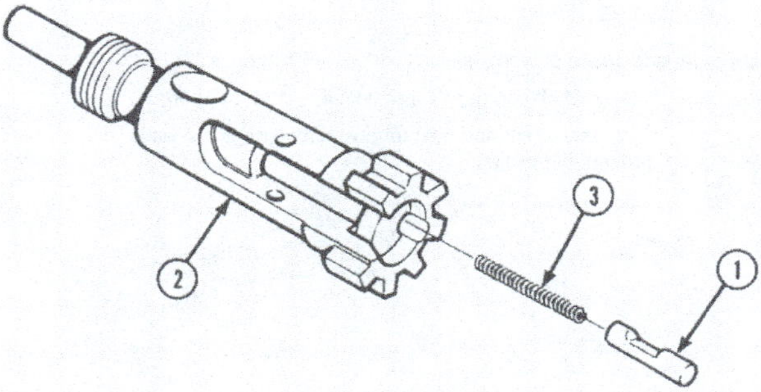

2-9. TROUBLESHOOTING PROCEDURES (CONT).

TROUBLESHOOTING (CONT)

MALFUNCTION
TEST OR INSPECTION
CORRECTIVE ACTION

9. FAILURE TO COCK

Step 1 Worn, broken, or missing parts of firing mechanism

Evacuate to support maintenance.

Step 2 Short recoil

Refer to below

10 SHORT RECOIL

Step 1 Broken or damaged action spring (1).

Replace action spring (p 2-57).

Step 2 Unlubricated or dirty action spring and receiver extension

Clean and lubricate

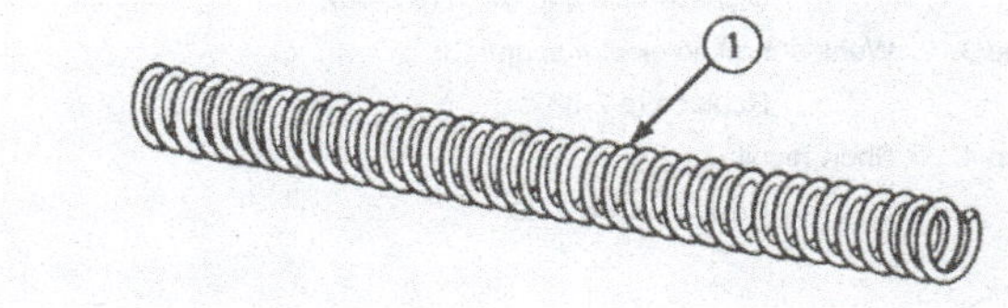

Step 3 Improper gap space or worn, missing, or broken bolt rings (2)

a. Stagger bolt ring gaps (approximately 13 turn apart}

b. Evacuate to support maintenance if bolt rings are worn, broken, or missing.

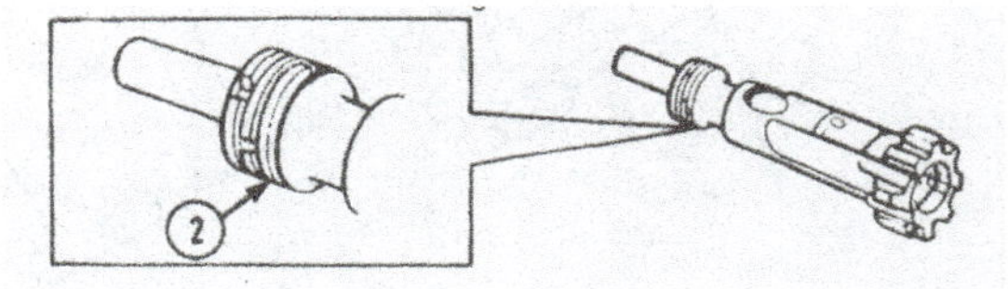

TROUBLESHOOTING (CONT)

MALFUNCTION
TEST OR INSPECTION
CORRECTIVE ACTION

Step 4. Carbon build-up or foreign matter In the narrow passage of the bolt carrier key (3).

Clean with CLP (Item 9, app D) and a pipe cleaner (Item 11, app D).

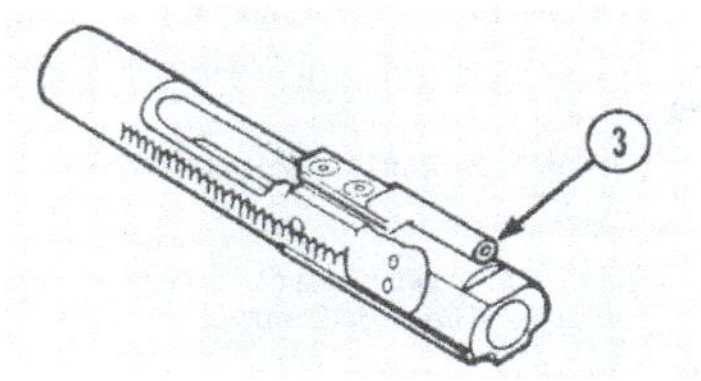

Step 5. Gas leakage caused by broken or loose gas tube (4) around front sight base.

Evacuate to support maintenance.

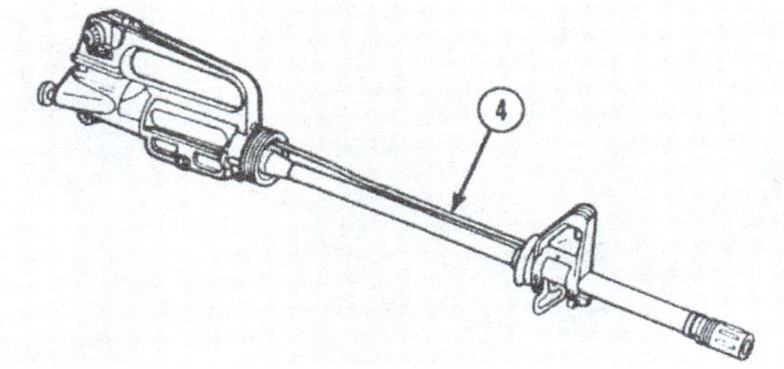

Step 6. Improper alignment of gas tube and carrier key.

a. Adjust gas tube alignment by bending in area of handguard assembly to Its original configuration

b. If gas tube cannot be returned to Its original configuration, evacuate the rifle to support maintenance.

2-9. TROUBLESHOOTING PROCEDURES (CONT).

TROUBLESHOOTING (CONT)

MALFUNCTION
TEST OR INSPECTION
CORRECTIVE ACTION

11. RIFLE CANNOT BE ZEROED.

Step 1. Defective barrel assembly (1).

Evacuate to support maintenance

Step 2. Barrel assembly out of alignment with rear sight assembly (2) on upper receiver.

Evacuate to support maintenance

Step 3. Defective front sight (3).

Remove front sight post (4), front sight detent (5), and helical spring (6). If damaged, replace.

Step 4. Defective -ear sight assembly (2).
Evacuate to support maintenance.

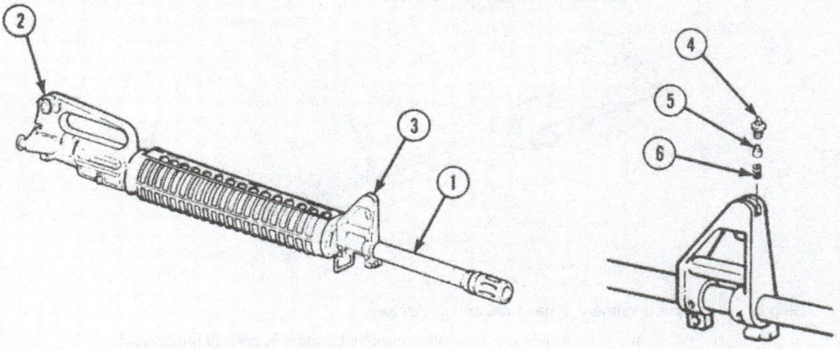

TROUBLESHOOTING (CONT)

MALFUNCTION
TEST OR INSPECTION
CORRECTIVE ACTION

12. FAILURE TO CYCLE WITH SELECTOR LEVER SET ON BURST (M16A2 AND M4).

Faulty selector lever or broken cam, cam clutch spring, or burst disconnector.

Evacuate to support maintenance.

12.1. FAILURE TO CYCLE WITH SELECTOR LEVER SET ON AUTO (M4A1).

Faulty selector lever.

Evacuate to support maintenance.

13. FIRES TWO ROUNDS WITH ONE PULL OF TRIGGER WITH SELECTOR LEVER SET ON SEMI (DOUBLE FIRING).

Perform function test.

If any part of function test (p 2-68) fails, evacuate to support maintenance.

14. FIRES WITH SELECTOR LEVER ON SAFE OR WHEN TRIGGER IS RELEASED WITH SELECTOR LEVER ON SEMI.

Worn, broken, or missing parts of firing mechanism.

Evacuate to support maintenance.

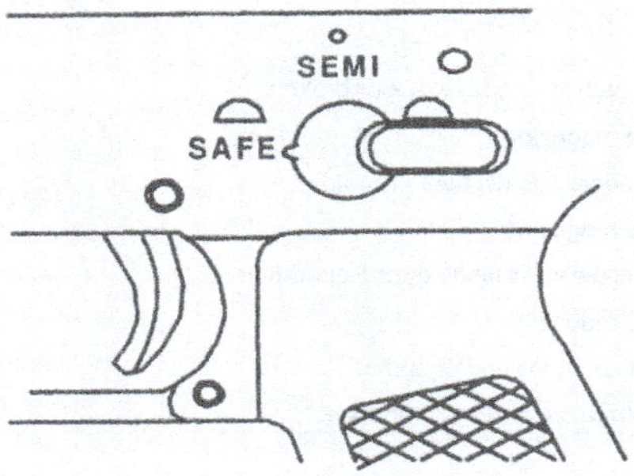

2-9. TROUBLESHOOTING PROCEDURES (CONT).

TROUBLESHOOTING (CONT)

MALFUNCTION
TEST OR INSPECTION
CORRECTIVE ACTION

15 BOLT ASSEMBLY FAILS TO LOCK TO REAR AFTER FIRING LAST ROUND.

Step 1. Magazine follower (1) worn or broken.

Replace magazine.

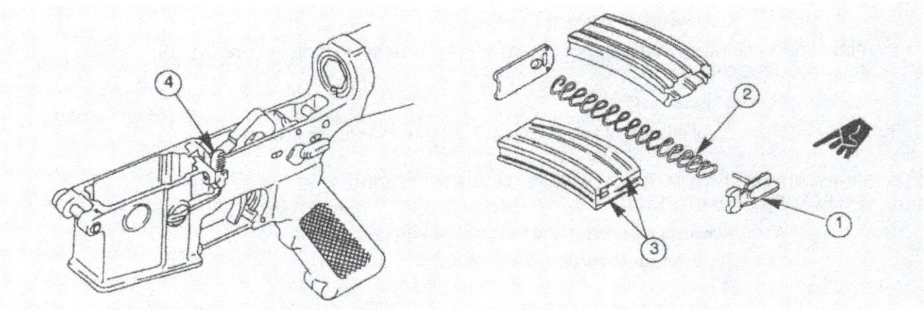

Step 2. Magazine catch spring (2) weak or broken.

Replace magazine.

Step 3. Magazine feeder lips (3) bent or broken.

Replace magazine.

Step 4. Magazine follower (1) binds during operation.

Replace magazine.

Step 5. Broken bolt catch (4) and/or spring.

Evacuate to support maintenance.

Section V. MAINTENANCE PROCEDURES

2-10. INITIAL SETUP. The following information will reduce the space required for the initial setup portion of the maintenance procedures.

a. Materials/Parts required are not listed unless they apply to the procedure

b. Personnel Required is listed only if the task requires more than one person If Personnel Required Is not listed, It means one person can do the job.

c. The normal standard equipment condition Is that the item is removed from the en(1 Item or next higher assembly and is in the assembled condition. Equipment Condition is not listed unless some other condition is required.

d. The approximate time required Is listed on the applicable Maintenance Allocation Chart (MAC).

e. When the term evacuate to support maintenance is used, the entire rifle must be evacuated.

2-11. LUBRICATION GENERAL.

a. Whenever the term "cleaner, lubricant, and preservative (CLP)" or the words "lubricant", "lube", "LSA", or "LAW" are cited In this TM, they are to be interpreted to mean CLP (Item 9, app D), weapons lubricating oil (LSA) (Item 23, app D), or weapons lubricating oil (LAW) (Item 22, app D) can be utilized as applicable The following constraints must be adhered to:

b. Under all but the coldest arctic conditions, LSA or CLP are the lubricants to use on the rifle. Either may be used at 100F (230C) and above. However, do not use both on the same rifle at the same time.

c. LAW is the lubricant to use during cold arctic conditions, + 100F (12C) and below

d. Any of the lubricants may be used from 10°F to + 100F (230C to 12°C)

e. **ARMY ONLY**: Do not mix lubricants on the same rifle The rifle must be thoroughly cleaned during change from one lubricant to another Dry Cleaning Solvent (SD) (Item 16, app D) is recommended for cleaning during change from one lubricant to another.

f. Rifle Bore Cleaner (RBC) (item 12, app D) may be used to remove carbon buildup In the bore and other portions of the rifle.

2-12. MAJOR COMPONENTS OF M16A2 RIFLE AND CARBINES.

This task covers disassembly

INITIAL SETUP:

References
TM 9-1005-319-10 (operator's manual)

Equipment Conditions
Rifle and Carbines assembled

General Safety Instructions
Before starting an inspection, be sure to clear the weapon. Do not keep live ammunition near the work area.

To avoid injury to your eyes, use care when removing and installing spring-loaded parts.

Below direct support maintenance, do not interchange bolt assemblies from one weapon to another. Doing so may result in injury to, or death of, personnel.

DISASSEMBLY

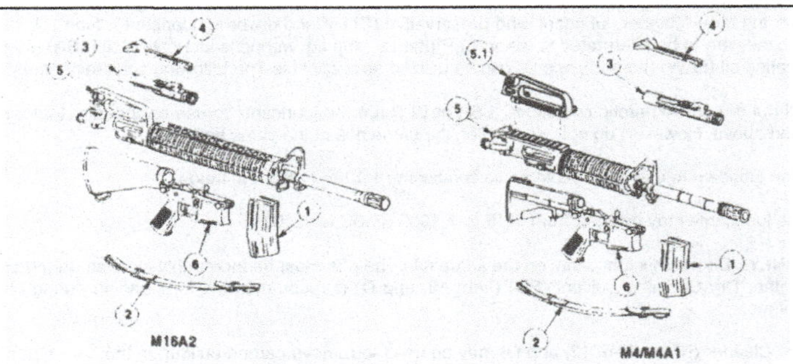

a. Refer to operator's manual.

b. Remove magazine (1), sling (2), bolt carrier assembly (3), charging handle assembly (4), and upper receiver and barrel assembly (5), carrying handle assembly (5.1), from lower receiver and buttstock assembly (6).

NOTE

Solid Film Lubricant (SFL) is the only authorized touchup for the MI 16A2 rifle and M4/M4A1 carbines and may be used on up to one third of the exterior finish of the weapon. **FOR ARMY CONUS USE ONLY AND AIR FORCE TRAINING WEAPONS ONLY**: SFL may be used as touchup without limitation on the upper receiver and barrel assembly. This is to say that units which DO NOT fall under the category of Divisional Combat Units or rapid deployment type units may have up to 100 percent of the exterior surface of the upper receiver and barrel assembly protected with SFL if necessary.

12-13. BOLT CARRIER ASSEMBLY.

This task covers:

a. **Disassembly**
b. **Cleaning**
c. **Inspection/Repair**
d. **Lubrication**
e. **Reassembly**

INITIAL SETUP:

Tools
(ARMY) Small Arms Repairman Tool Kit (item 3, app B)
Bolt Carrier Key Tool (item 11, figure C-17)

References
TM 9-1005-319-10 (operator's manual)

Equipment Conditions
2-34 Bolt carrier assembly removed

General Safety Instructions
Bolt cam pin must be installed or rifle will blow up while firing the first round. If the bolt cam pin is not installed, injury to, or death of, personnel may result.

Do not interchange bolt assemblies from one rifle to another. Doing so may result in injury to, or death of, personnel.

a. DISASSEMBLY

CAUTION

Do not spread or close legs of firing pin retaining pin (1).

1. Remove firing pin retaining pin (1). Tilt key and bolt carrier assembly (2) and catch firing pin (3) as it drops out.
2. Rotate bolt cam pin (4) one quarter turn and lift straight up to remove.
3. Remove bolt assembly (5) from key and bolt carrier assembly (2).

NOTE
For disassembly of bolt assembly (5), see page 2-38.

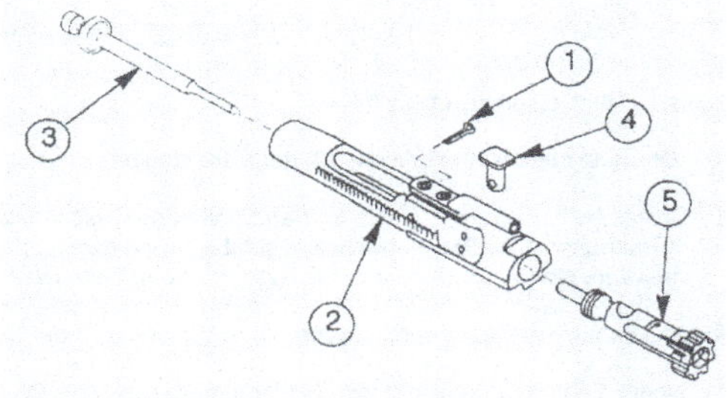

12-13. BOLT CARRIER ASSEMBLY (CONT).

b. CLEANING

Clean all items (operator's manual). Remove carbon deposits.

c. INSPECTION/REPAIR

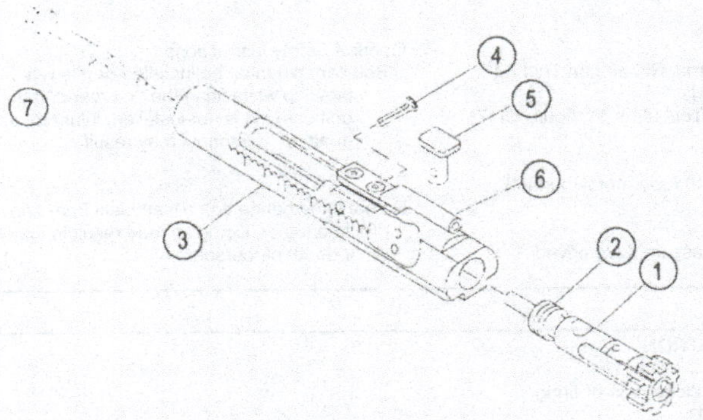

1. Inspect bolt assembly (1) as follows:

   a. Check to see that the bolt ring (2) gaps are staggered.

   b. Install bolt assembly (1) into key and bolt carrier assembly (3). Check for worn rings by holding the key and bolt carrier assembly (3) with the bolt assembly facing down. If bolt assembly falls out after firing pin retaining pin (4) and bolt cam pin (5) are removed, the rings are worn. Evacuate to support maintenance.

2.Inspect carrier key (6) for dents, distortion, or looseness. If dented or loose, evacuate to support maintenance.

3. Inspect firing pin retaining pin (4) and bolt cam pin (5) for cracks, damage, or excessive wear. Replace if unserviceable.

4. Inspect firing pin (7) for damage or If tip is chipped. Evacuate rifle to support maintenance if unserviceable

5. Inspect key and bolt carrier assembly (3) for damage or wear. If unserviceable, evacuate to support maintenance.

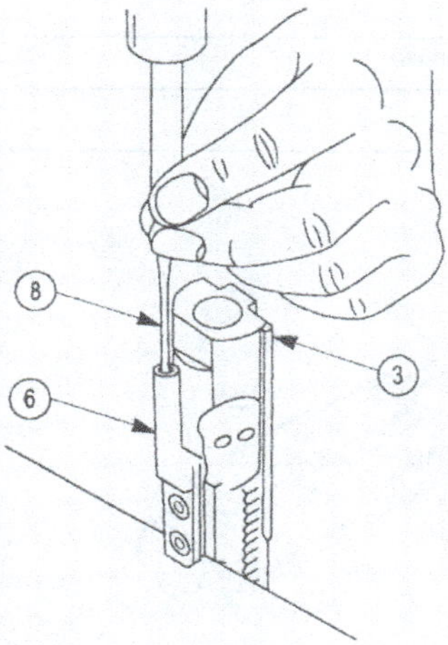

CAUTION

Extreme care must be exercised during the following procedure to assure that the striking force is not directed to the attaching screws and that the tube portion is not enlarged or flared beyond original requirement. Such enlargement would permit loss of gas pressure when the key and gas tube come together during functioning.

6. Repair small dents and/or distortions in carrier key (6) using bolt carrier key tool as follows:

a. Place the key and bolt carrier assembly (3) in a vertical position, supported In a manner that contact is made with the rear surface of the carrier key (6).

b. Insert the small end of the key tool (8) into the tube portion of the carrier key (6).

c. Strike the large end of the key tool (8) lightly with a 3 ounce, soft-brass hammer.

d. Repeat striking (gently) until carrier key (6) is reformed to original configuration.

e. If carrier key (6) cannot be reformed to original configuration, evacuate the weapon to support maintenance.

d. LUBRICATION

Lubricate all Items (p 2 33) (operator's manual)

e. REASSEMBLY

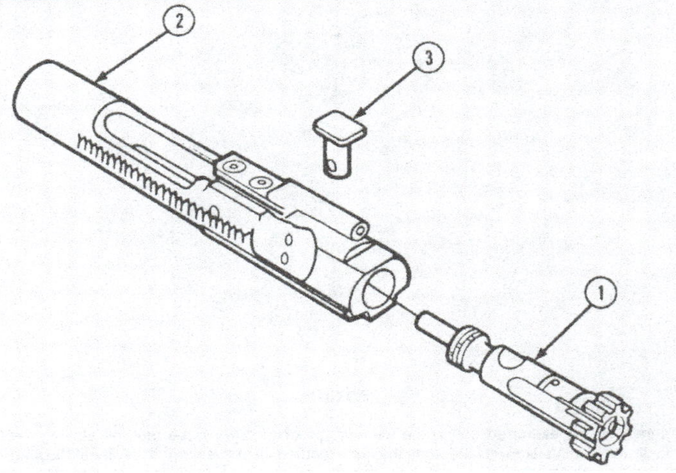

WARNING

Do not Interchange bolt assemblies from one rifle to another Doing so may result In Injury, or death of, personnel.

Bolt cam pin must be Installed or rifle will blow up while firing the first round. If the bolt cam pin is not Installed, Injury to, or death of, personnel may result

NOTE

Before installing bolt assembly, check to see that the bolt ring gaps are staggered to prevent loss of gas pressure

1. Install bolt assembly (1) In key and bolt carrier assembly (2)

2.Install bolt cam pin (3) and rotate one quarter turn

2-13. BOLT CARRIER ASSEMBLY (CONT).

e. REASSEMBLY (CONT)

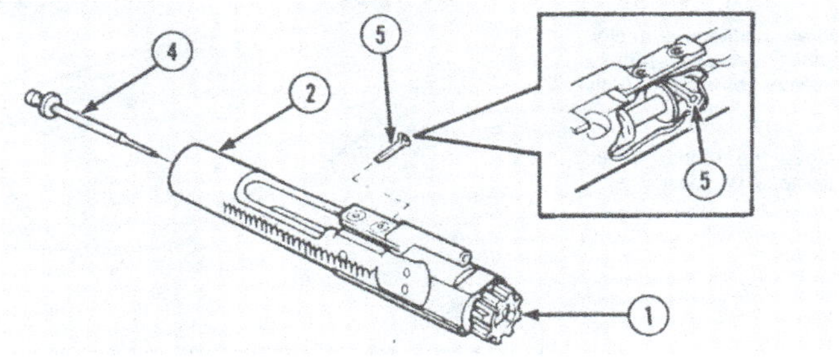

3. Hold key and bolt carrier assembly (2) with bolt assembly (1) down and drop in firing pin (4).
4. Install firing pin retaining pin (5) from left side only to ensure proper installation Check for proper installation by holding key and bolt carrier assembly (2) with bolt assembly (1) up, then attempt to shake out firing pin.
5. Reassemble rifle, refer to page 2-68.

2-14. BOLT ASSEMBLY.

This task covers:

a. **Removal**
b. **Cleaning**
c. **Inspection**
d. **Lubrication**
e. **Reassembly**

INITIAL SETUP:

Tools
(ARMY) Small Arms Repairman Tool Kit (item 3, app B)

Equipment Conditions
2-35 Bolt assembly removed

General Safety Instructions
Do not Interchange bolt assemblies from one rifle to another. Doing so may result in injury to, or death of, personnel.

To avoid injury to your eyes, use care when removing and installing spring-loaded parts.

a. DISASSEMBLY

NOTE

Do not separate cartridge extractor and extractor spring assembly unless replacement of either or both is required

Do not remove the rubber Insert from the extractor spring assembly

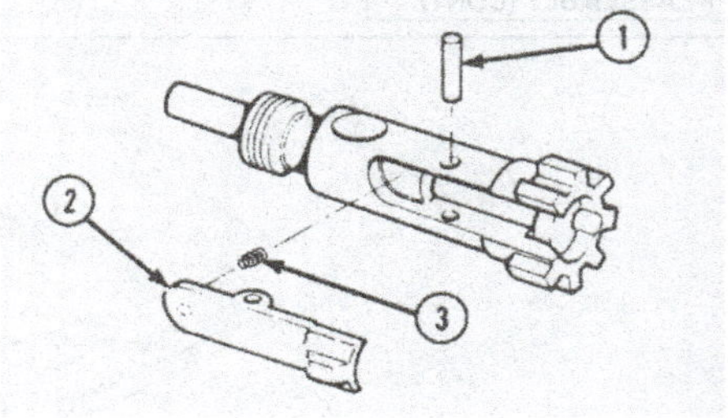

1. Push out extractor pin (1) and remove cartridge extractor (2) and extractor spring assembly (3) as a unit
2. If required, twist extractor spring assembly (3) counterclockwise to remove from cartridge extractor (2).

CAUTION

Be sure to use vise law protective caps

3. Hold bolt body In vise and remove spring pin (4) using 1 16 Inch punch and hammer

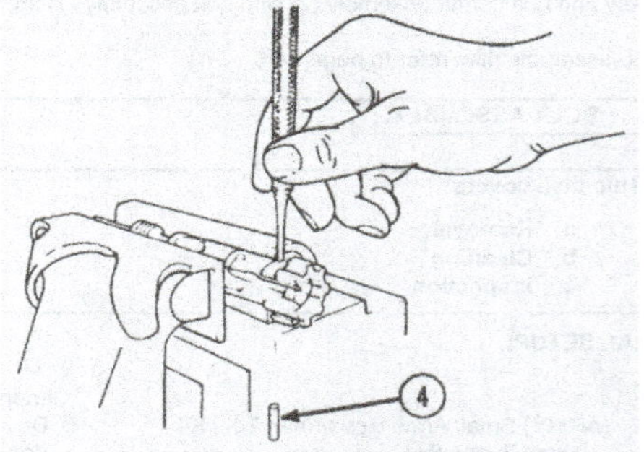

2-14. BOLT ASSEMBLY (CONT).

a. DISASSEMBLY (CONT)

WARNING

To avoid injury to your eyes, use care when removing and installing spring-loaded parts.

4. Remove punch, be sure to catch cartridge elector (5) and ejector spring (6) to prevent loss

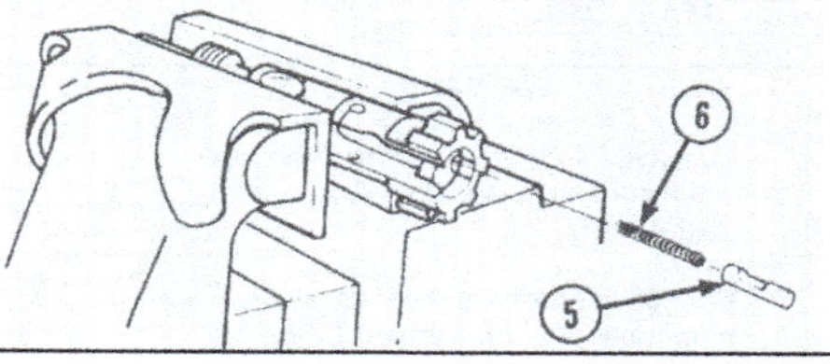

b. CLEANING

CAUTION

Do not distort extractor spring assembly during cleaning.

Clean all Items (operator's manual). Remove carbon deposits.

c. INSPECTION/REPAIR

1. Check bolt rings (1) for damage and proper stagger (gaps approximately 3 turn apart).
2. Inspect for cracks or damage, especially around bolt cam pin hole (2) and locking lugs (3). If cracked or damaged evacuate to support maintenance.
3. Inspect cartridge extractor (4), extractor spring assembly (5), and extractor pin (6) for cracks, breaks, chips, and other damage. Pay close attention to cartridge extractor lip (7). If damaged, replace.
4. Inspect cartridge ejector (8) and ejector spring (9) for cracks, breaks, and chips If damaged, replace.

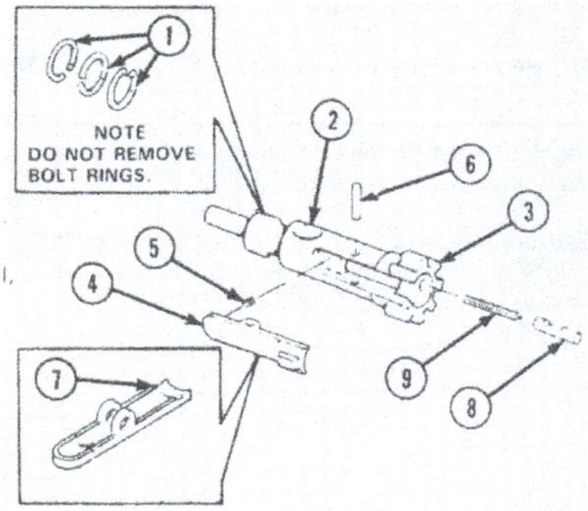

d. LUBRICATION

Lightly lubricate all items (p 2 33) (operator's manual)

e. REASSEMBLY

WARNING

To avoid Injury to your eyes, use care when removing and Installing spring loaded parts.

Do not Interchange bolt assemblies from one rifle to another Doing so may result In injury to, or death of, personnel.

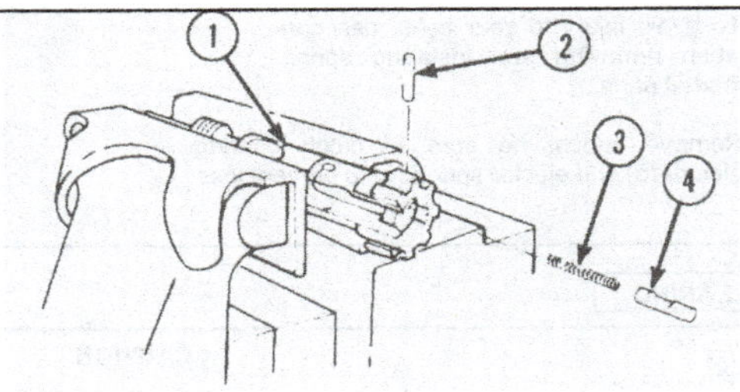

CAUTION

Be sure to use vise jaw protective caps

1. Place bolt body (1) in a vise and start spring pin (2) In hole.
2. Install ejector spring (3) and cartridge ejector (4). Align groove on cartridge ejector (4) so that spring pin 12) can be Installed
3. Compress and hold ejector spring and cartridge ejector In place with a 3 8 inch punch.
4. Using hammer and 1 16 Inch punch, complete Installation of spring pin (2) so that the ends are flush with the out side of bolt body (1).

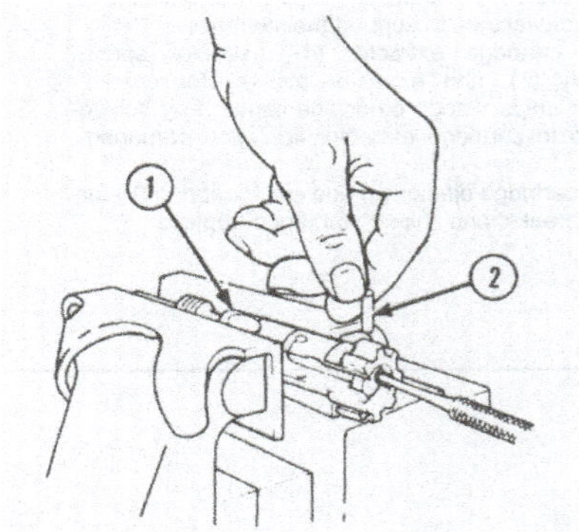

2-14. BOLT ASSEMBLY (CONT).

e. REASSEMBLY (CONT)

NOTE

Do not disassemble rubber Insert from extractor spring assembly.

5. If removed, Insert large end of extractor spring assembly (5) into cartridge extractor (6) and seat by pushing and turning clockwise.

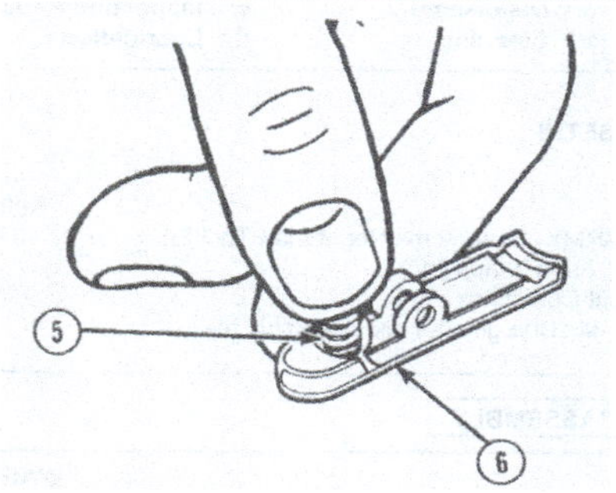

6 Position cartridge extractor (6) and ex tractor spring assembly (5) on bolt body (1).

7 Compress extractor spring assembly (5) and cartridge extractor (6) to align holes.

8. Install extractor pin (7) by hand.

9. Reassemble rifle, refer to page 2-68.

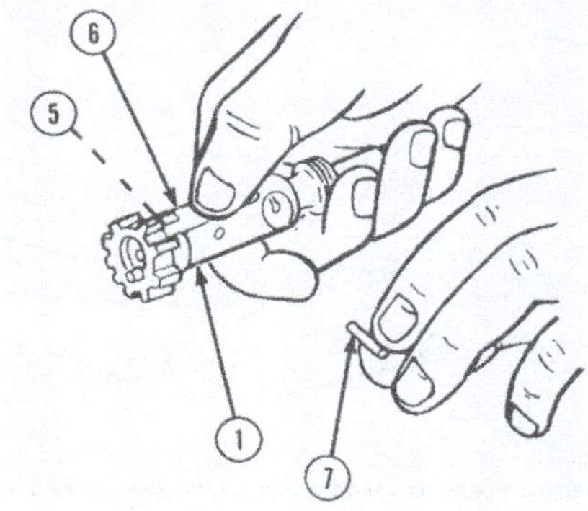

2-15. CHARGING HANDLE ASSEMBLY.

This task covers:

a. Disassembly **c. Inspection/Repair** **e. Reassembly**
b. Cleaning **d. Lubrication**

INITIAL SETUP:

Tools
(ARMY) Small Arms Repairman Tool Kit (item 3, app B)

Equipment Conditions
2-34 Charging handle assembly removed

References
TM 9-1005-319-10 (operator's manual)

a. DISASSEMBLY

WARNING

To avoid injury to eyes, use care when removing and installing spring-loaded parts.

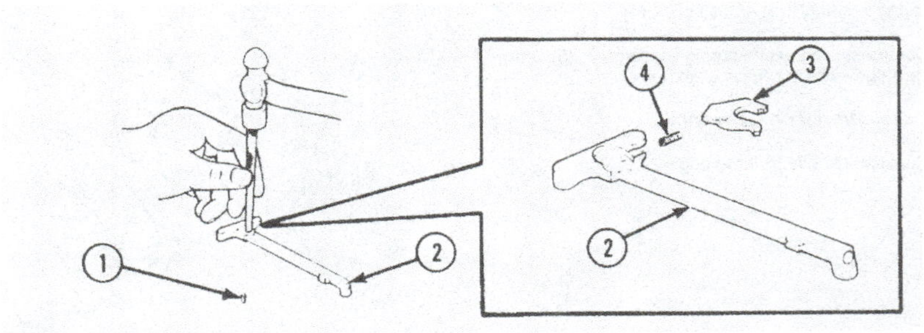

1. Remove spring pin (1) from charging handle (2) using a hammer and 1/16 inch punch.
2. As punch is withdrawn, catch charging handle latch (3) and helical spring (4) to prevent loss.

2-15. CHARGING HANDLE ASSEMBLY (CONT).

b. CLEANING

Clean all Items (operator's manual) Remove carbon deposits.

c. INSPECTION/REPAIR

Inspect all Items for breaks, cracks, or damage Replace all unserviceable Items

d. LUBRICATION

Lightly lubricate all Items (p 2-33) (operator's manual)

e. REASSEMBLY

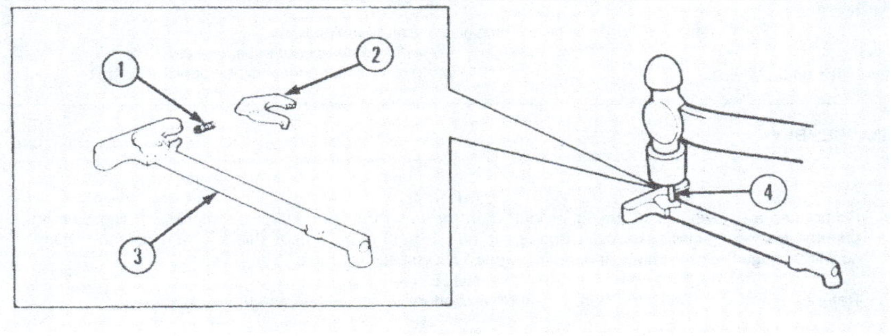

1. Position helical spring I1) and charging handle latch (2) in chdrgiin(J handle 13) Align holes and hold In position
2. Install spring pin (4) using a hammer Make sure spring pin is flush
3. Reassemble rifle, refer to page 2 68

2-16. UPPER RECEIVER AND BARREL ASSEMBLY, RIFLE BARREL ASSEMBLY, AND UPPER RECEIVER ASSEMBLY.

This task covers:

a. Disassembly
b. Cleaning
c. Inspection
d. Repair
e. Lubrication
f. Reassembly

INITIAL SETUP

Tools
(ARMY) Small Arms Repairman Tool Kit (item 3, app B)
Front sight post removal and installation tool (fig. E-2, app E)
Front sight detent depressor (fig. E-1, app E)

Materials/Parts
Lubricants (app D)

References
TM 9-1005-319-10 (operator's manual)

Equipment Conditions
2-34 Upper receiver and barrel assembly removed from lower receiver

General Safety Instructions
To avoid injury to your eyes, use care when removing and installing spring-loaded parts.

a. DISASSEMBLY

CAUTION

Do not use a screwdriver or any other tool when removing the handguard assemblies. Doing so may damage the handguard assemblies and/or slip ring. Do not remove heat shield for any reason. Doing so will damage the heat shield and the handguard assemblies will have to be replaced.

NOTE

Refer to operator's manual for "buddy system" procedure on removing handguard assemblies.

2-16. UPPER RECEIVER AND BARREL ASSEMBLY, RIFLE BARREL ASSEMBLY, AND UPPER RECEIVER ASSEMBLY (CONT).

a. DISASSEMBLY (CONT)

1. Push down on handguard slip ring (1) and lift upper handguard assembly (2) up and out.

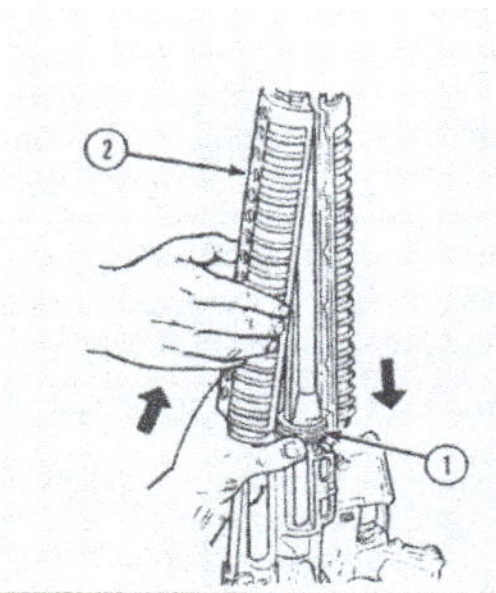

2. Push down on handguard slip ring (1) and lift the lower handguard assembly (3) up and out.

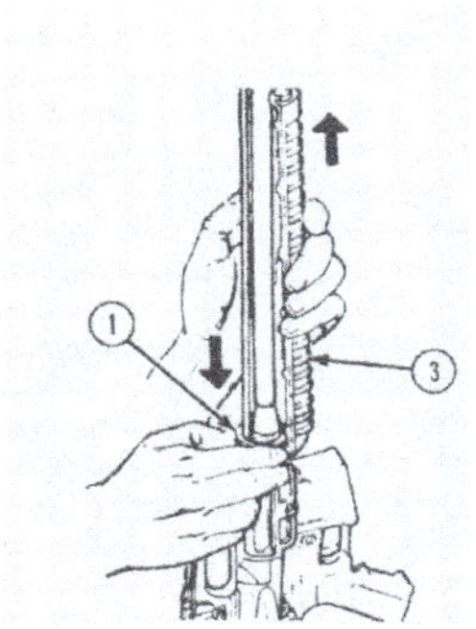

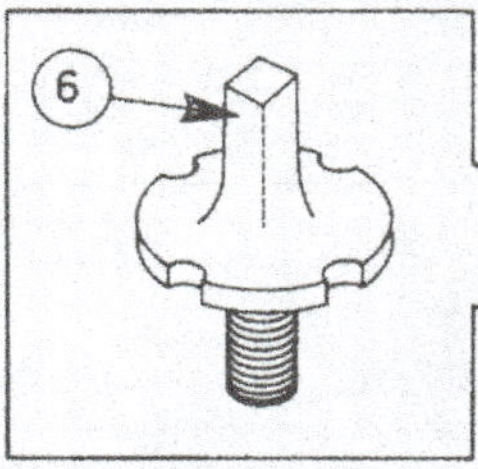

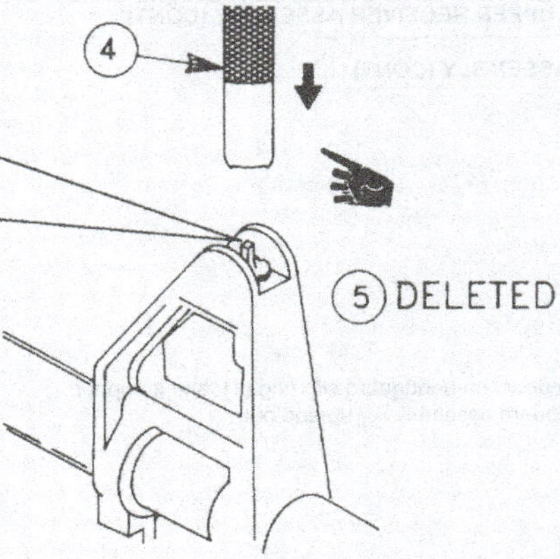

WARNING

To avoid injury to your eyes, use care when removing and installing spring-loaded parts.

NOTE

Remove front sight posts only if parts are damaged.

3. Using a dummy cartridge or a font sight post removal and installation tool (4).removed front sight post (6) by turning counter clockwise.

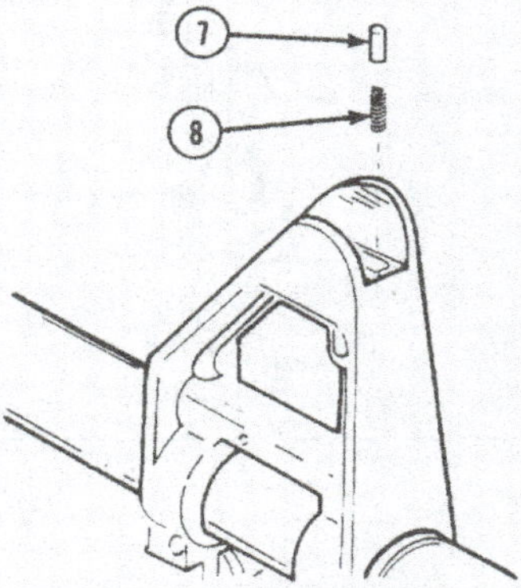

4. Catch front sight detent(7) and helical spring (8) to prevent loss.

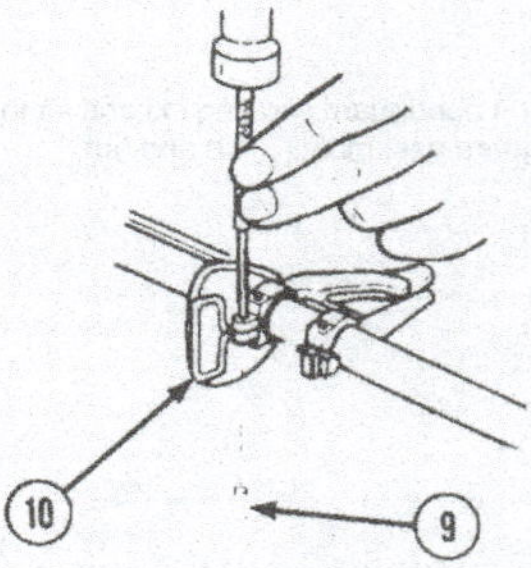

NOTE

RIFLE ONLY

Disassemble small sling swivel only if repair is necessary.

5. If necessary, knock out tubular rivet (9) with a hammer and punch and remove small sling swivel (10). Discard tubular rivet (9).

2-16. UPPER RECEIVE AND BARREL ASSEMBLY, RIFLE BARREL ASSEMBLY, AND UPPER RECEIVER ASSEMBLY (CONT).

a. DISASSEMBLY (CONT)

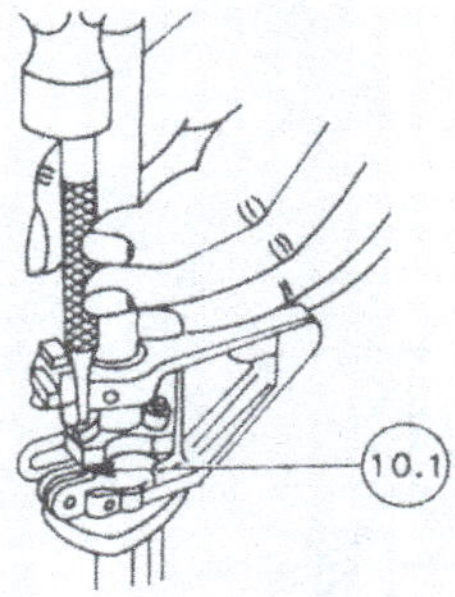

CARBINE ONLY

5A. Remove two spring pins (10 1)

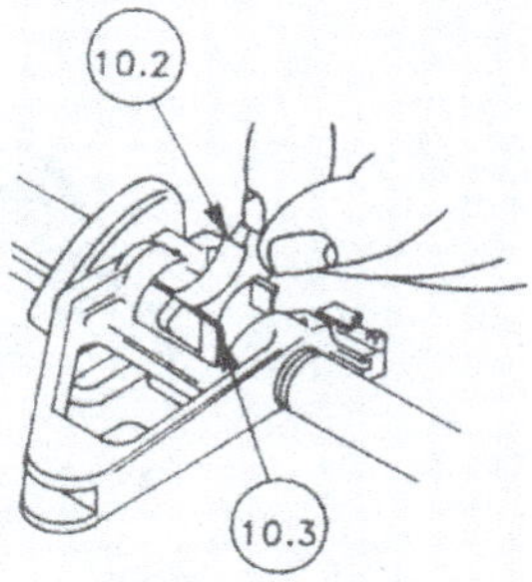

CARBINE ONLY

5B. Lift swivel locking bar (10 2) up and out of swivel mount (10 3)

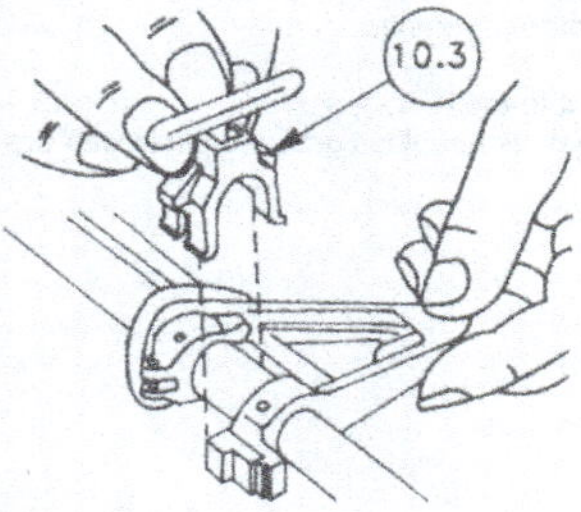

5C. Remove swivel mount (10.3).

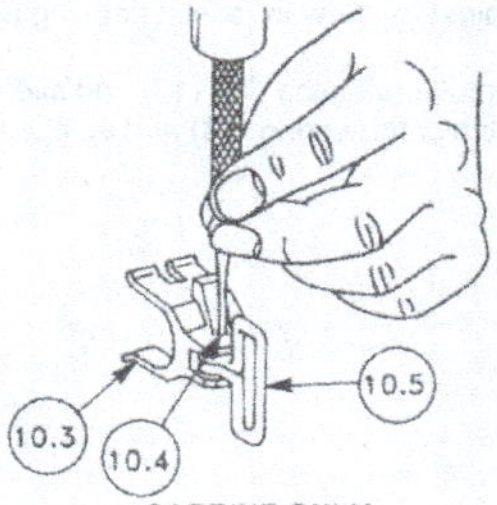

CARBINE ONLY

5D. Remove rivet (10 4) Separate front swivel (10 5) from swivel blank) mount (10 3)

2-16. UPPER RECEIVER AND BARREL ASSEMBLY, RIFLE BARREL ASSEMBLY, AND UPPER RECEIVER ASSEMBLY (CONT).

a. DISASSEMBLY (CONT)

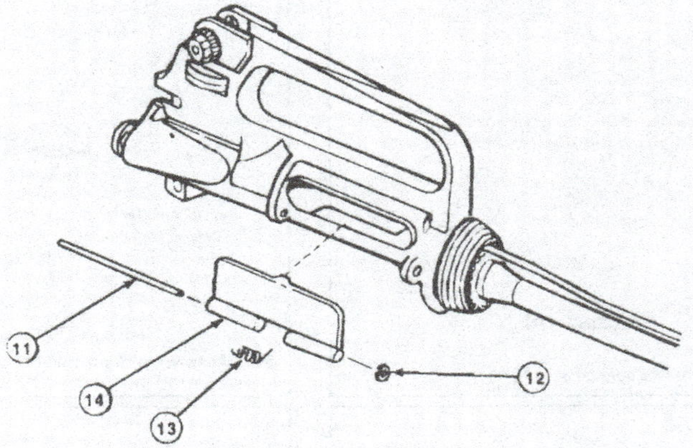

NOTE

Do not disassemble further unless repair is necessary. Headless grooved pin (11) may bind may bind against the forward assist housing and require some additional force to remove.

6. Remove retaining ring (12) and slide headles grooved pin (11) out to the rear.
7. Catch cover spring (13) and ejection port cover (14) to prevent loss as headless grooved pin is with-drawn.

M4/M4A1 ONLY

NOTE

Do not remove the carrying handle assembly nut (2 each) or clamp for normal maintenance. If carrying handle assembly nut or clamp are missing or the clamp is damaged to the point of being non-functional they may be replaced.

7.1. Disassemble as follows only if required to replace damaged or missing carrying handle assembly nut or clamp.

a. Observe the position of the clamp (15) in relation to the carrying handle (16).
b. Using care not to damage the surface finish, remove carrying handle assembly nut(s) (17) and clamp as required.
c. Inspect mating surfaces of the carrying handle (16) and the clamp (15) for damage. If damaged replace the damaged part.
d. Inspect for other damage, rust, etc. Repair or replace as required.
e. Apply a light coat of lubricant to all mating surfaces.
f. Install clamp (15) paying close attention to ensure that it is installed in its original position. The forward end of the clamp (15) must not extend past the forward end of the carrying handle (16).
g. Taking care not to cross thread, install handle assembly nut (17).

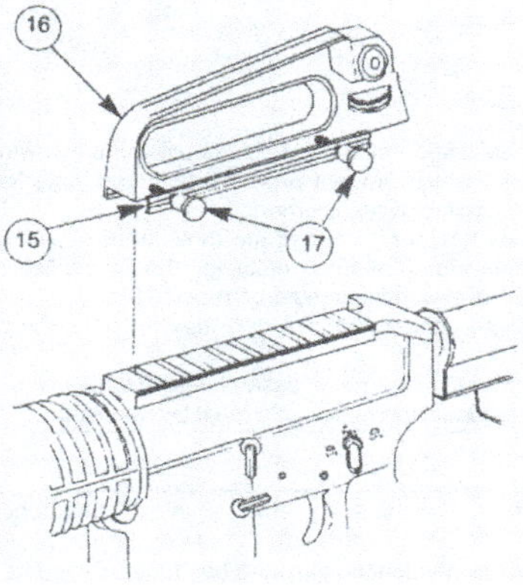

b. CLEANING

Clean all items (operator's manual).

c. INSPECTION

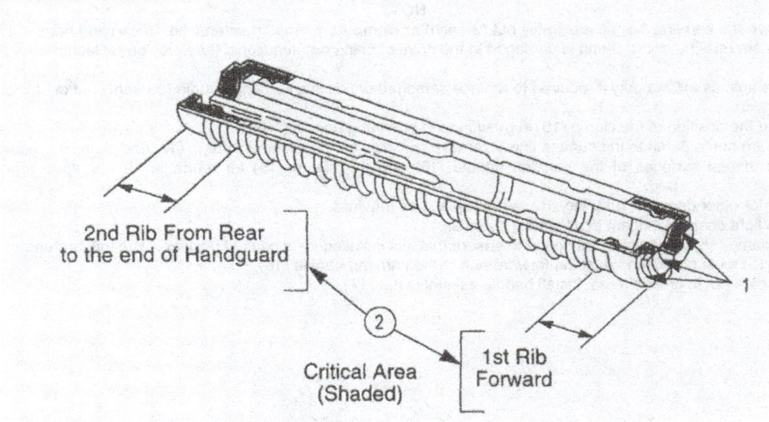

1. Inspect handguard assemblies for breaks separation, and cracks using the following guidelines:
   a. Breaks and separations of material which prevent proper retention or interfere with functioning of the rifle will be cause for handguard assembly rejection and replacement.
   b. M16A2 handguard assemblies may have up to two of the three front retaining tabs missing (1). M4/M4A1 may not have any front retaining tabs (1) missing. If all front retaining tabs for the M1 6A2 are missing, or any of the tabs for the M4/M4A1 are missing, handguard assemblies must be replaced.
   c. Cracks up to one inch in length are acceptable provided they do not extend into the retaining flange (CRITICAL AREA) (2).
   d. Replace severely cracked or damaged handguard assemblies. Handguard assemblies which have a heat shield which is loose enough to rattle when installed on the rifle must be replaced.

2. Inspect front sight assembly for chips, breaks, and cracks. Evacuate to support maintenance if broken, cracked, or bent.
3. Inspect front sight area for evidence of gas leakage around gas tube. Evacuate to support maintenance if short recoil results from gas leakage.
4. Inspect front sight post, front sight detent, and helical spring for damage. If damaged, replace.

2-16. UPPER RECEIVER AND BARREL ASSEMBLY, RIFLE BARREL ASSEMBLY, AND UPPER RECEIVER ASSEMBLY (CONT).

c. INSPECTION (CONT)

5. Inspect barrel for pits in bore, burrs, broken or worn locking lugs, and surface cracks and defects.
   a. Pits no wider than a land or groove and 3/8 inch (0.953 cm) or less in length are allowable in the bore.
   b. Uniformly fine pits in a densely pitted area of the bore are allowable.
   c. Lands that appear dark due to coating of gliding metal from projectiles are allowable.
   d. Striping of lands and grooves shall not be cause for rejection unless support maintenance determines by use of the barrel erosion gage.
   e. For pits other than mentioned above, broken or burred locking lugs, or surface cracks, evacuate to support maintenance.
6. Inspect bore for ringing. Definitely ringed bores or bores ringed sufficiently to bulge the outside surfaces of the barrel are causes for rejection. Evacuate to support maintenance.
7. Inspect chamber for pitting. Fine pits, or fine pits in a densely pitted area, are allowable. Pits 1/8 inch (0.318 cm) in length are cause for rejection. Evacuate to support maintenance.
8. Hand check compensator (3) for looseness on barrel. The third (middle) slot (4) must be straight up at top dead center (TDC). The alignment may vary as much as one half the width of the slot either direction. If loose or out of alignment, evacuate to support maintenance.
9. Inspect all items for serviceability in and tightness of latch assembly on ejection port cover. If items are damaged or nonfunctional, they are unserviceable.

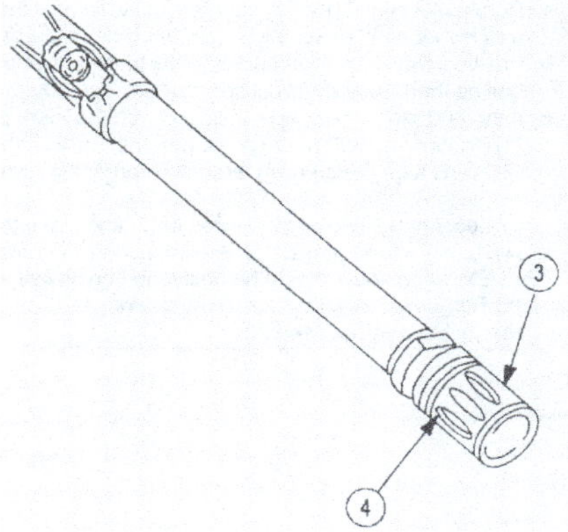

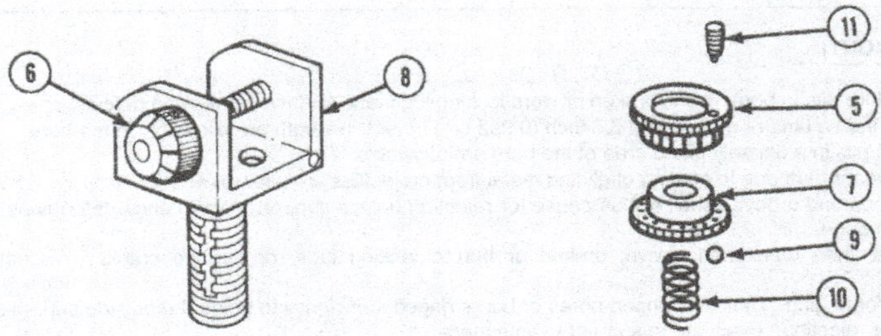

10. Rotate and test elevation index (5) and windage knob (6) for ease of functioning and legibility of markings.

11. Inspect elevation knob zero as follows:

a. Rotate elevation knob (7) counterclockwise until the rear sight assembly is all the way down. If a whole click is not felt as the rear sight assembly stops, the rear sight assembly has bottomed out and will not pivot freely.

b. Position elevation knob back slightly to Its last whole click so the rear sight assembly base (8) is under tension of the ball bearing (9) and helical spring (10). The 300 meter mark should align with the mark on the receiver.

c. If the 300 meter mark is not aligned with the mark on receiver, slip the range scale in the following manner:

(1) Position the 300 meter mark with the mark on the receiver.

(2) Insert a 1/16 inch allen wrench through the access hole of the rear sight assembly base and into the Index screw (11).

(3) Loosen the index screw three turns and leave the wrench in place.

(4) Rotate lower portion of elevation knob counterclockwise until it stops (range scale should not have moved). Elevation knob should be positioned on its last whole click.

(5) Tighten index screw and remove wrench.

(6) Check for proper setting.

2-16. UPPER RECEIVER AND BARREL ASSEMBLY, RIFLE BARREL ASSEMBLY, AND UPPER RECEIVER ASSEMBLY (CONT).

d. REPAIR

1. Replace all authorized unserviceable parts

2. Evacuate to support maintenance.

e. LUBRICATION

Lightly lubricate all Items (p 2 311 (oper ator's manual)

f. REASSEMBLY

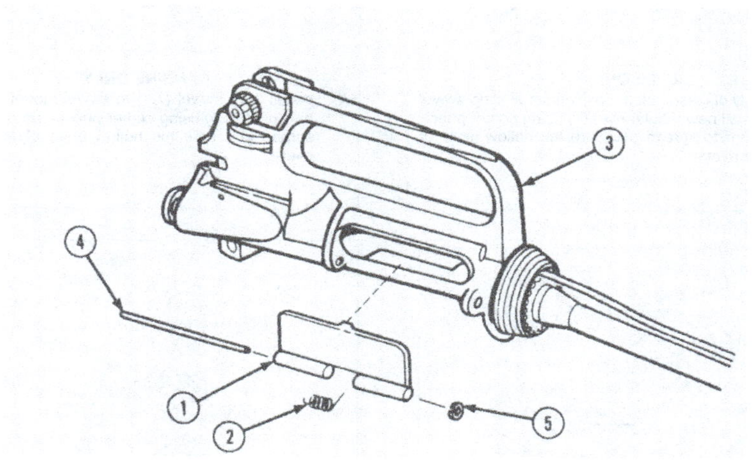

1. If previously disassembled, position election port cover 11) and helical spring (2) on Lip per receiver (3) with short leg of helical spring to the rear on Inside of ejection port cover

NOTE

Long legs of helical spring must be positioned and pretensioned before the headless grooved pin is installed

2. Hold helical spring (2) short leg In this position and turn long leg one half turn (180 degrees) with fingers of right hand

3. Position long leg of helical spring (21 against election port cover (1) Hold helical spring and election port cover in this position and Install headless grooved pin (4) Check for proper spring tension during installation of retaining ring (5)

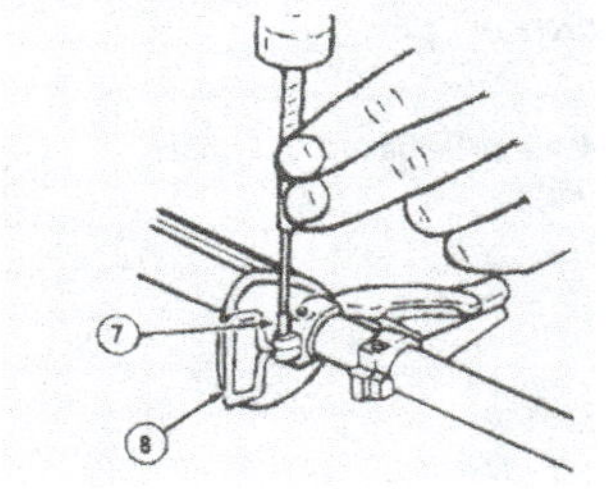

RIFLE ONLY

4. If previously disassembled, position small sling swivel (6) and install new tubular rivet (7) using center punch and hammer to spread and flare the hollow head of the tubular rivet.

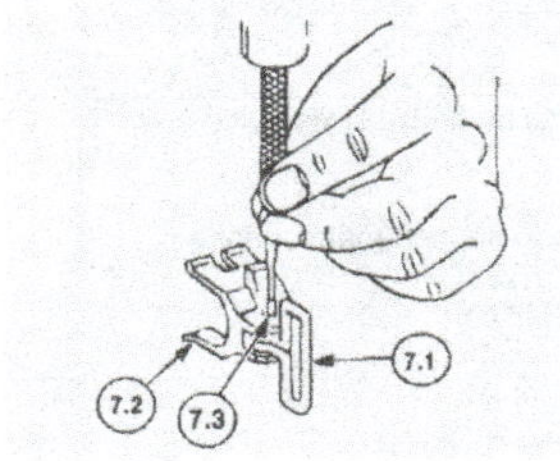

CARBINE ONLY

4A. Install front swivel (7.1) to swivel mount (7.2) with new rivet (7.3) using center punch and hammer to spread and flare the hollow head of the tubular rivet.

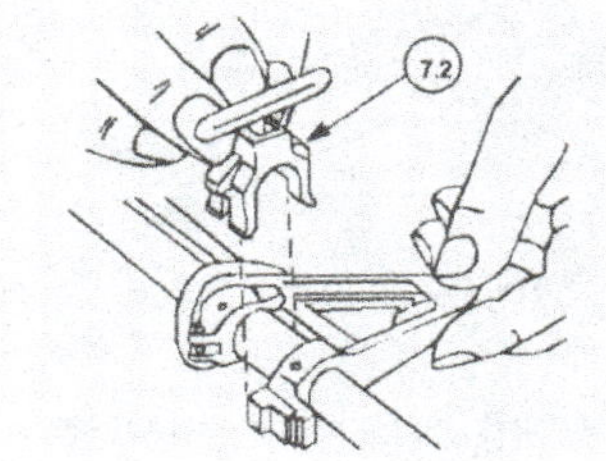

CARBINE ONLY

NOTE

It is recommended that the swivel mount be placed on the left side of the carbine for right handed operators and on the right side for left handed operators. This will keep the sling out of the way when the carbine is used.

4B. Install swivel mount (7.2).

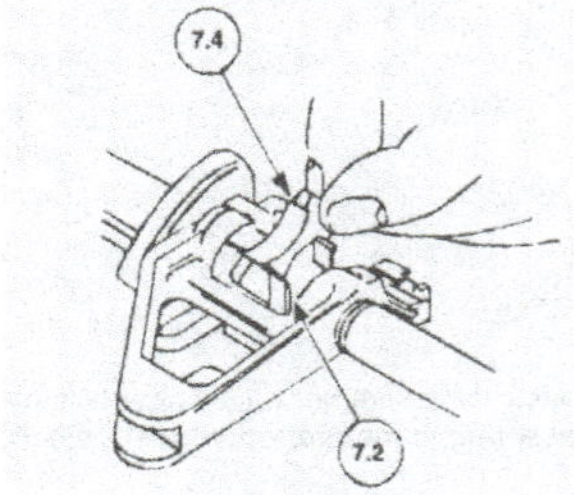

CARBINE ONLY

4C. Place swivel locking bar (7.4) in swivel mount (7.2).

2-16. UPPER RECEIVER AND BARREL ASSEMBLY, RIFLE BARREL ASSEMBLY, AND UPPER RECEIVER ASSEMBLY (CONT).

f. REASSEMBLY (CONT)

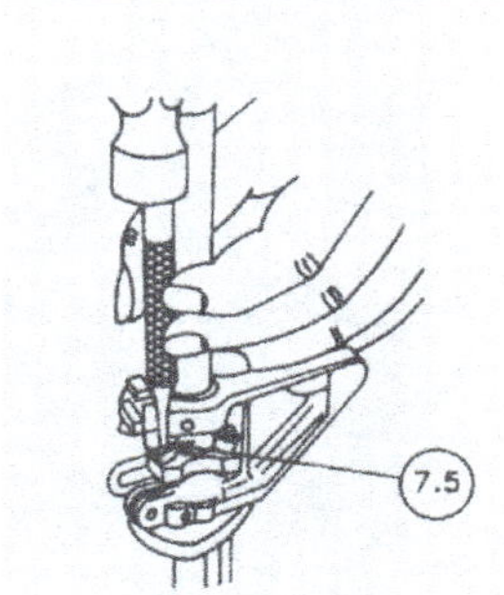

CARBINE ONLY

4D. Install two spring pins (7 5)

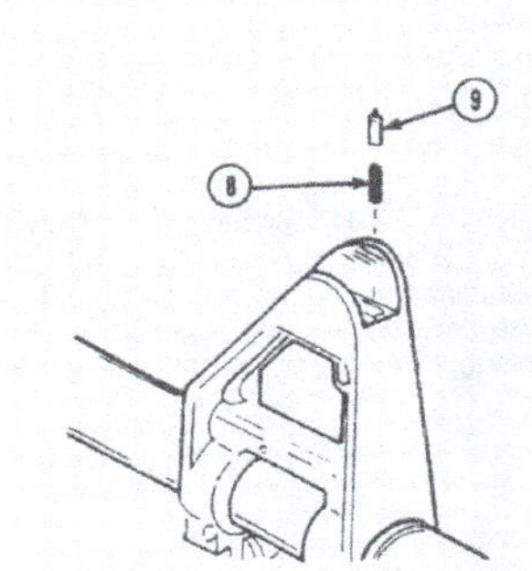

5. Position helical spring (8) and front sight detent (9).

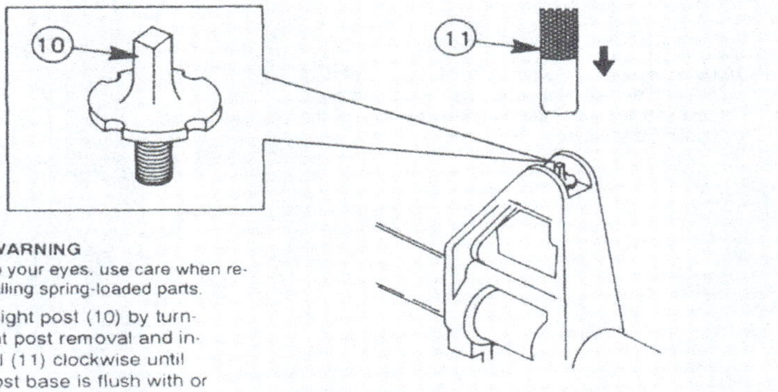

WARNING

To avoid injury to your eyes. use care when removing and installing spring-loaded parts.

6. Install front sight post (10) by turning front sight post removal and installation tool (11) clockwise until front sight post base is flush with or slightly below front sight frame.

2-16. UPPER RECEIVER AND BARREL ASSEMBLY, RIFLE BARREL ASSEMBLY, AND UPPER RECEIVER ASSEMBLY (CONT).

t. REASSEMBLY (CONT)

7. Mechanical Zero Procedures (A.F. Only).

a. Mark a piece of plastic card stock or rigid paper with lines from 1 to 5mm in 1 mm increments Set the card on the front sight frame and check the height of the top of the front sight post.

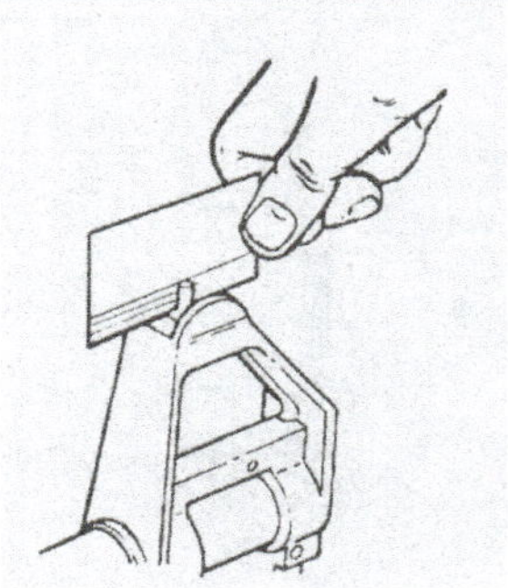

b. Using a dummy round or front sight post tool adjust the front sight so the top of the front sight post is 5mm above the machined surfaces of the front sight frame.

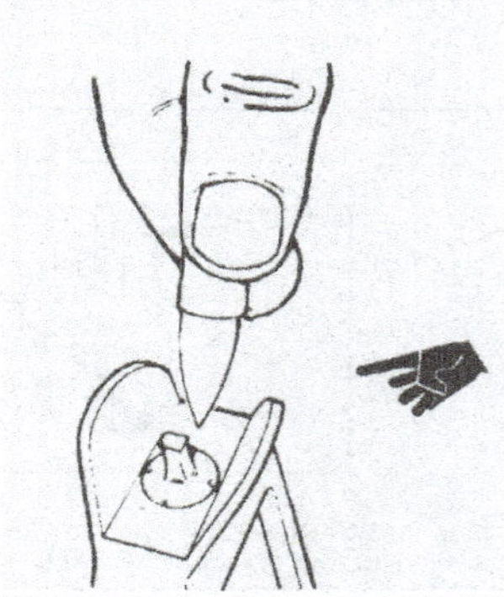

c. Visually check the front sight post top height by using the marked plastic or paper card. Card must set level on the machined surfaces of the front sight frame to obtain an accurate reading

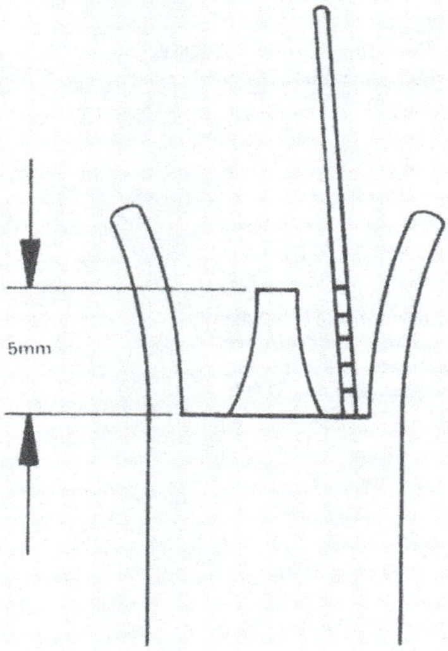

NOTE

This procedure, when used In conjunction with rear sight mechanical zero adjustment. will give an approximate battle sight zero to most M16A2 rifles The above steps can also be used before firing a new or newly assigned rifle Use the procedures to check rifles stored In preferred packaging during routine Inspections This will help ensure people armed with the rifles will stand a better chance of hitting an enemy If the rifles must be used before a live fire zero can be made Whenever possible, zeroing of the rifle should be accomplished using ball ammunition on a 25 meter zeroing target using the "L" aperture.

2-16. UPPER RECEIVER AND BARREL ASSEMBLY, RIFLE BARREL ASSEMBLY, AND UPPER RECEIVER ASSEMBLY (CONT).

f. REASSEMBLY (CONT)

NOTE
Refer to operator's manual for 'buddy system" procedure on installing hand guard assemblies

8. Install top of upper handguard assembly (12) In tube cap (13) while pushing down on handguard slip ring (14). Push bottom of upper handguard assembly (12) in place and release handguard slip ring (14) to lock handguard assembly In place

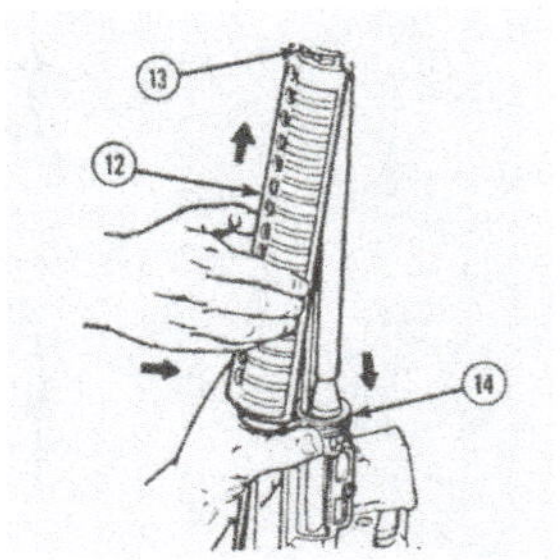

9. Install top of lower handguard assembly 1 5) In tube cap (13) while pushing down on handguard slip ring (14) Push bottom of lower handguard assembly (15) In place and release handguard slip ring {14) to lock both handguard assemblies In place

10. Reassemble rifle, refer to page 2-68

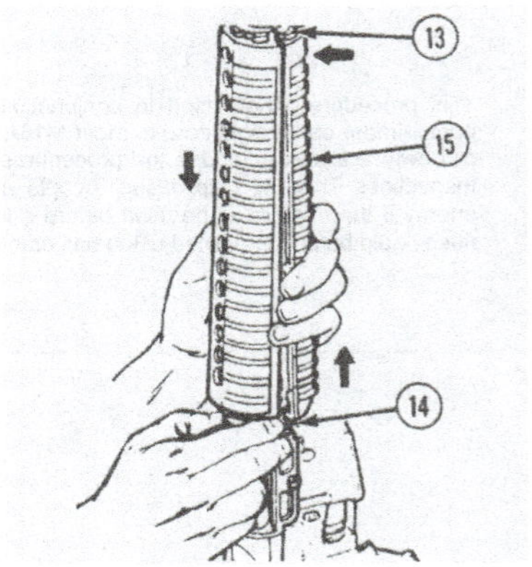

2-17. LOWER RECEIVER AND BUTTSTOCK ASSEMBLY.

This task covers:

a. Disassembly
b. Cleaning
c. Inspection
d. Repair
e. Lubrication
f. Reassembly

INITIAL SETUP

Tools
- (ARMY) Small Arms Repairman Tol Kit (item 3, app B)
- Pivot Pin Removal Tool (fig. E-3, app E)
- Pivot Pin Installation Tool (fig. E-4, app E)

Materials/Parts
- Lubricant, solid film (SFL) (item 21, app D)
- Screw, self-locking (item 6, p C-1 1)

References
- TM 9-1005-319-10 (operator's manual)

Equipment Conditions
- 2-34 Lower receiver and buttstock assembly removed

General Safety Instructions
- To avoid injury to your eyes, use care when removing and installing spring-loaded parts.
- When using solid film lubricant or dichloromethane, be sure the area is well ventilated.

a. DISASSEMBLY

1. Remove screw (1) and lockwasher (2).

WARNING
To avoid injury to you eyes, use care when removing and installing spring-loaded parts.

2. Carefully remove pistol grip (3) and catch helical spring (4) and safety detent (5) to prevent loss.

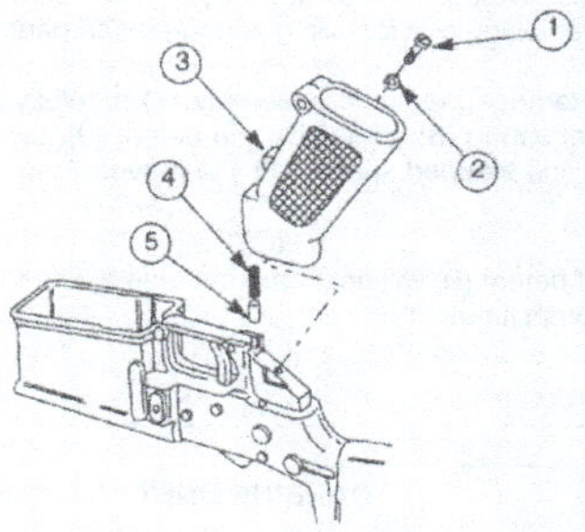

2-17. LOWER RECEIVER AND BUTTSTOCK ASSEMBLY (CONT).

a. DISASSEMBLY (CONT)

RIFLE ONLY
NOTE

If the self-locking screw is removed, it must be discarded and replaced with a new one

3. Remove self-locking screw (6).

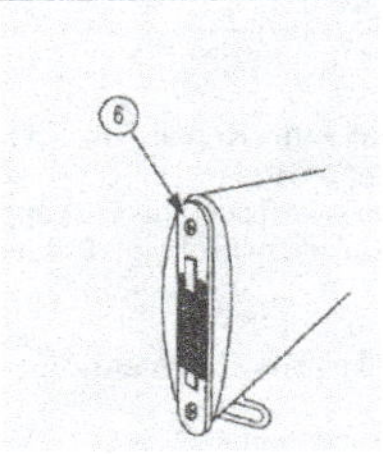

RIFLE ONLY
WARNING

To avoid injury to your eyes, use care when removing and installing spring-loaded parts.

4. Remove buttstock assembly (7) carefully and catch helical spring (8), takedown pin detent (9), takedown pin (10), and stepped spacer (11) to prevent loss.

NOTE
If detent (9) will not come out, use a wire to push it out.

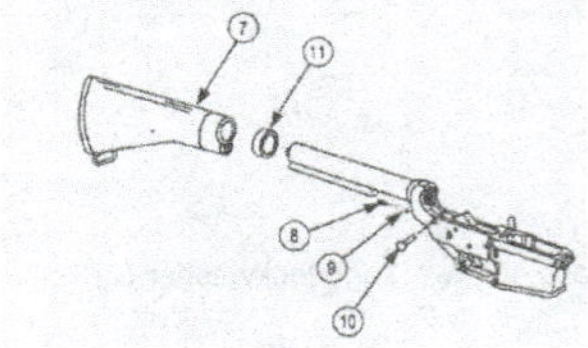

CARBINE ONLY

4A. Extend buttstock assembly (11.1).

4B Grasp the lock release lever (11.2) in the area of the retaining nut (11.3), pull down-ward, and slide buttstock to the rear to separate the buttstock assembly from the lower receiver extension

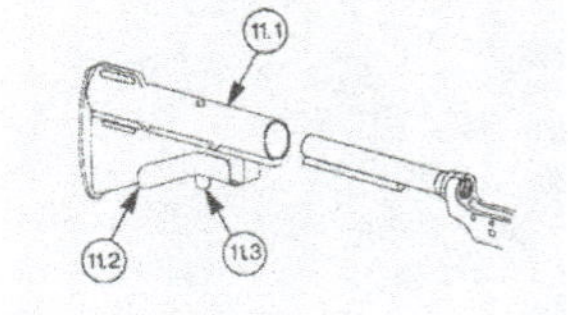

WARNING

To avoid injury to your eyes, use care when removing and installing spring-loaded parts

NOTE

Catch pivot pin detent and helical spring as pivot pin is removed (see step 6 on next page)

5. Depress pivot pin detent and spring and remove pin or Insert fabricated pivot pin removal tool (12) to compress pivot pin detent Turn pivot pin (13) a quarter turn Remove tool and pivot pin

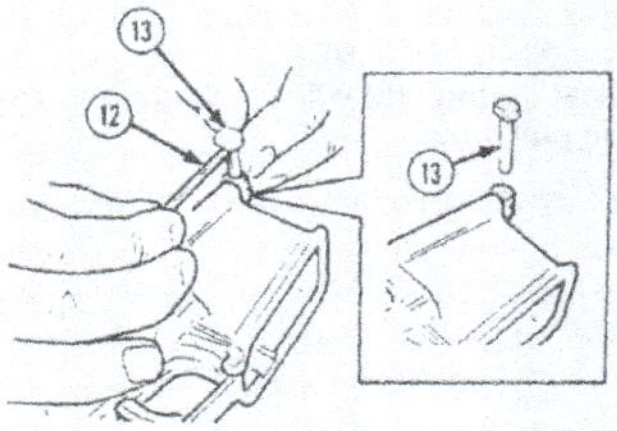

6. Be sure to hold cupped hand in front of pivot pin detent (14) and helical spring (15) to prevent loss of pivot pin detent and helical spring.

NOTE

If helical spring (15) will not come out, use a wire to pull it out.

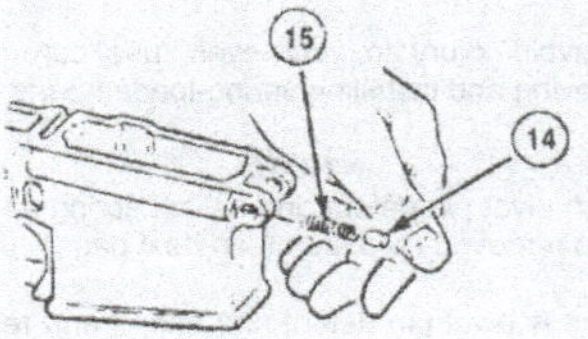

NOTE

Make sure the hammer is cocked and the selector lever is not set on BURST or AUTOMATIC before removing the buffer assembly.

7. Press buffer assembly (16) in about 1/4 inch (0.635 cm). Depress buffer retainer (17) and release buffer assembly (16) and action spring (18). Remove buffer assembly(16) and action spring (18) from receiver while depressing buffer retainer (17).

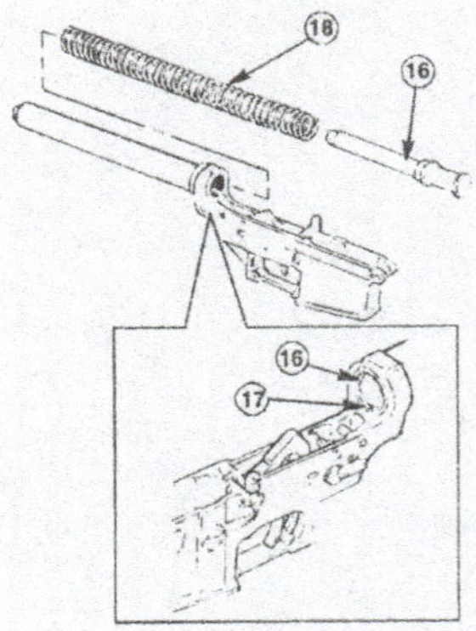

b. CLEANING

Clean all items (operator's manual). Remove carbon deposits

1 2-17. LOWER RECEIVER AND BUTTSTOCK ASSEMBLY (CONT).

c. INSPECTION I

1. Inspect buffer assembly (1, 2. 3, or 3 1) for cracks or damage

RIFLE ONLY

a. Some old buffers (1) have a hole with pin installed which protrudes equally on each side approximately 1/32 inch (O 08 cm)

b. Some buffers (2) have a hole In the housing but no pin

c. New buffers (3) do not have a hole In buffer body or a pin.

BOTH WEAPONS

2. If cracked or damaged, replace.

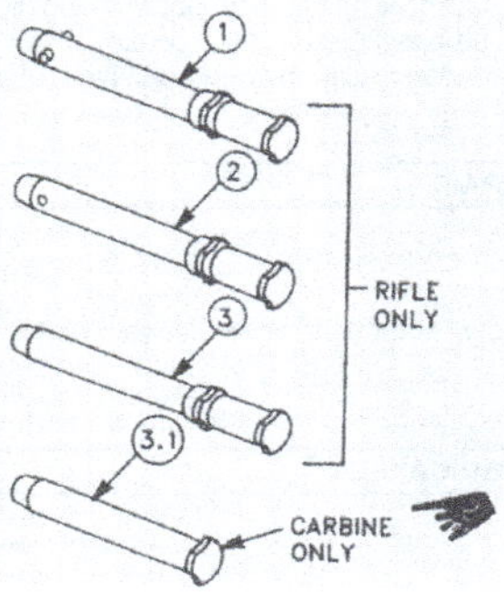

3. Check free length of action spring (4). The free length **FOR RIFLE ONLY** must be between 11 3/4 inches (29.85 cm) mini- mum and 13 1/2 inches (34 29 cm) maximum, **FOR CARBINE ONLY** must be between 10 1'/16 inches (25,56 cm) minimum and 11 114 Inches (28.58 cm) maximum, If not, replace Do not attempt to adjust the length by stretching the action spring

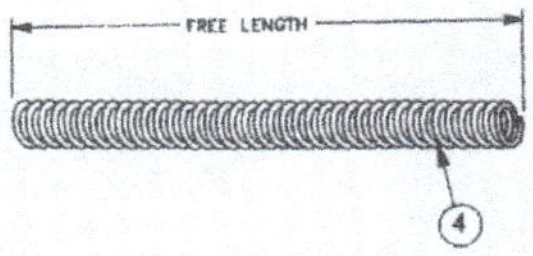

4. Inspect lower receiver (5) (without further disassembly) for legibility of serial number **ARMY ONLY**: If the serial number is hard to read, evacuate to direct support maintenance **AIR FORCE ONLY**: If the serial number is hard to read, evacuate to depot maintenance

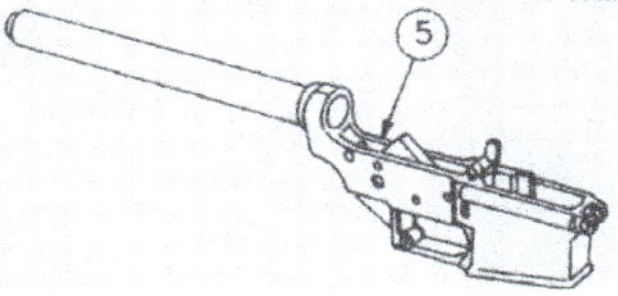

NOTE

AIR FORCE ONLY: Only depot maintenance is authorized to restamp the serial number.

5. Inspect for missing or damaged parts. Inspect finish of lower receiver for shiny spots. Touch up with solid film lubricant as required (p 2-34).

NOTE

If a M16A2 rifle or M4/M4A1 carbine lower receiver is missing one third or more of its exterior protective finish, resulting in an unprotected, light reflecting surface, it is candidate for overhaul. This missing finish will be considered a shortcoming. This shortcoming requires action to obtain a replacement weapon. Once a replacement has been received, evacuate the original weapon to depot maintenance for overhaul.

d. REPAIR

Replace all authorized unserviceable parts. If repair is not authorized at this level, evacuate to support maintenance.

e. LUBRICATION

Lightly lubricate all metal components (p 2-33) (operator's manual).

2-17. LOWER RECEIVER AND BUTTSTOCK ASSEMBLY (CONT).

f. REASSEMBLY

WARNING

To avoid Injury to your eyes, use care when removing and Installing spring loaded parts.

NOTE

Make sure the hammer Is cocked and the selector lever is not set on BURST before Installing the buffer assembly.

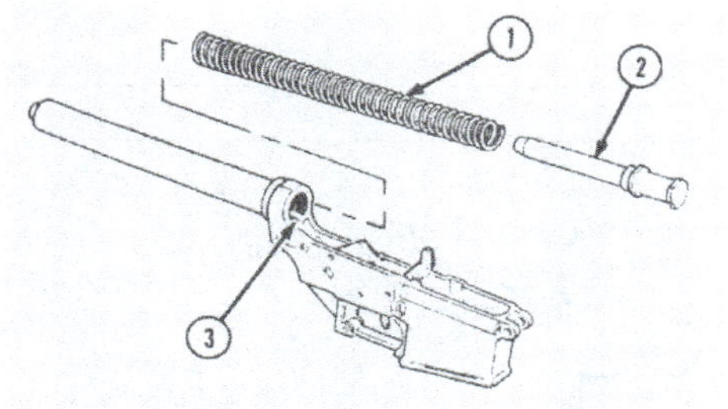

1. Press action spring (1) and buffer assembly (2) In until buffer retainer (3) snaps up and holds them In place.

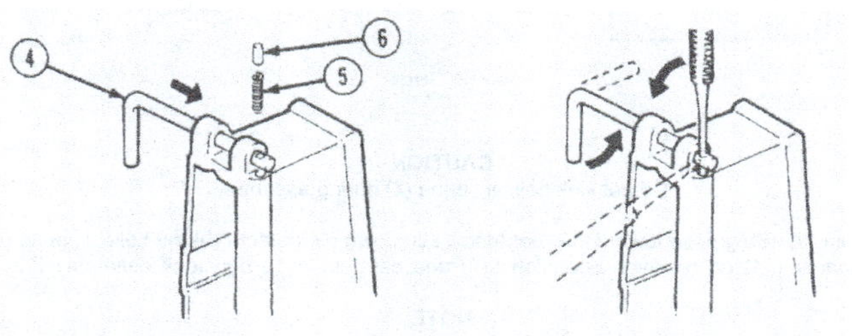

2. Install fabricated pivot pin installation tool (4) Insert helical spring 151 and pivot pin detent (6) Compress pivot pin detent in recess with 3 32 inch punch and rotate tool Remove 3'32 Inch punch

NOTE

Rounded end of pivot pin detent must be In the groove of the pivot pin (7) when assembly is complete.

3. Install pivot pin (7) while removing fabricated pivot pin Installation tool (4) Maintain pressure while sliding pivot pin (7) into hole Rotate pivot pin until pivot pin detent is inserted Into pivot pin groove

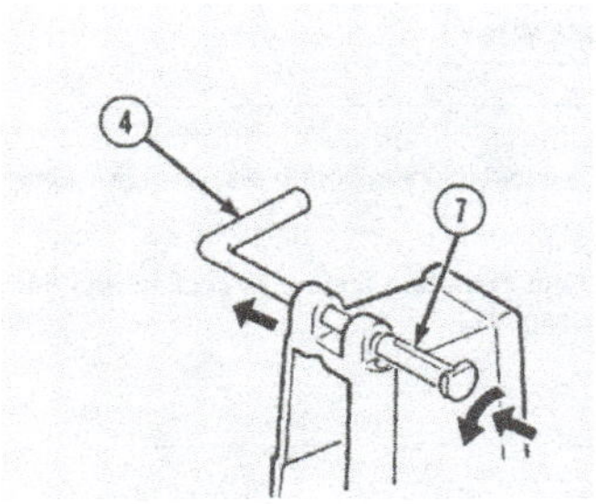

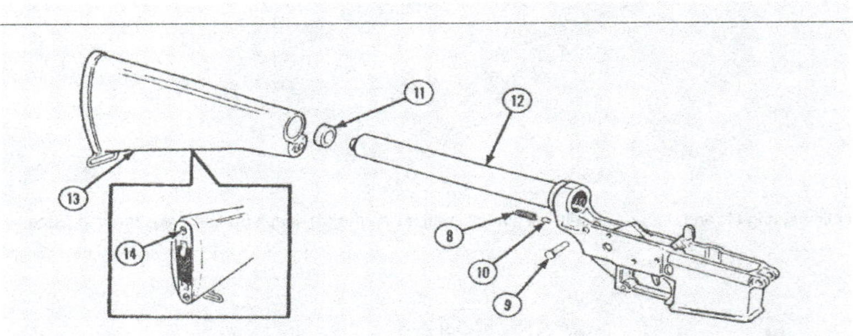

RIFLE ONLY

CAUTION

Do not kink helical spring (8) during assembly

4. Install takedown pin (9) with groove toward the rear Install takedown pin detent (10)and helical spring (8) from the rear
5. Install stepped spacer (11) on receiver extension (12) and carefully slide buttstock assembly (13) on to compress helical spring (8)

NOTE

Self-locking screw (14), If removed must be discarded and replaced with a new one 6. Install self-locking screw (14) to secure buttstock assembly (13)

6. Install self-locking screw (14) to secure buttstock assembly (13)

1 2-17. LOWER RECEIVER AND BUTTSTOCK ASSEMBLY (CONT).

f. REASSEMBLY (CONT)

CARBINE ONLY

6A. Grasp the lock release lever (14 1) In the area of the retaining nut (14 2) and pull to reinstall the buttstock assembly (14.3) onto the lower receiver extension (14 3) onto the lower receiver extension (14 4)

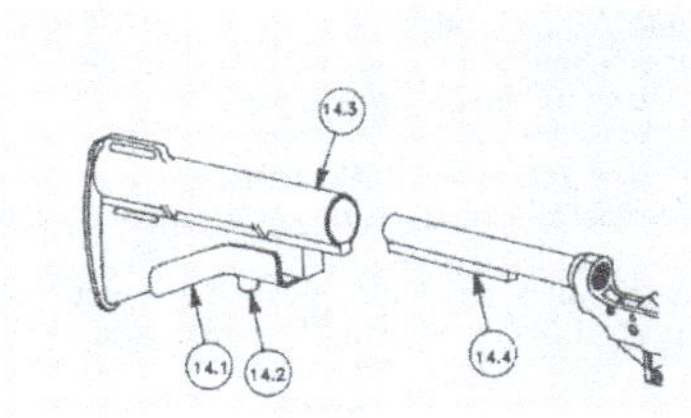

2-17. LOWER RECEIVER AND BUTTSTOCK ASSEMBLY (CONT).

f. REASSEMBLY (CONT)

WARNING

When utilizing the enhanced rifle grip (it has a bump between the second third finger for a better grip) PN 9349127, rifle grip screw. PN AN 501D416 18 (1-1/8 in.) or AN 501 D416 16 (1 in), is authorized to be(used with the enhanced grip Any screw longer than 1-1 8 In used with 9349127 could cause a hazardous situation Also, ensure the washer Is in 11place

CAUTION

Do not kink helical spring (15) during assembly

7. Install safety detent (16), pointed end first, and helical spring (1 5) into bottom of lower receiver (17)

NOTE

A portion of the helical spring will fit In a hole in the pistol grip

8. Carefully install pistol grip (18) to compress helical spring 115). Secure pistol grip 118) in place with lockwasher (19) and screw 120)

9. Reassemble rifle, refer to page 2 68

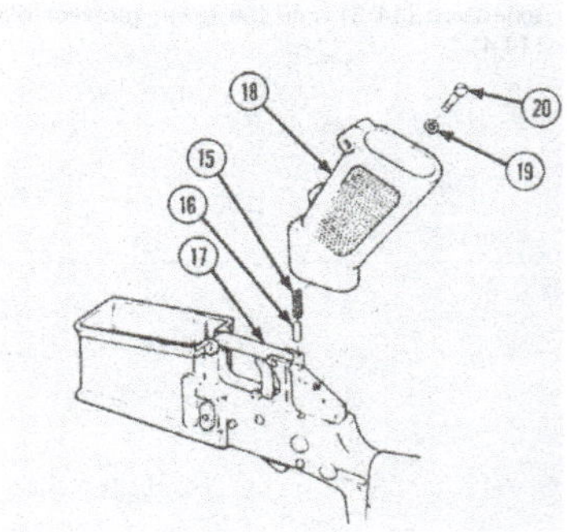

2-18. BUTTSTOCK ASSEMBLY.

This task covers:

a. Disassembly
b. Cleaning
c. Inspection
d. Repair
e. Lubrication
f. Reassembly

INITIAL SETUP

Tools

(ARMY) Small Arms Repairman Tool Kit (item 3, app B)

Equipment Conditions

2-57 Buttstock assembly removed from lower receiver and buttstock assembly

a. DISASSEMBLY

RIFLE ONLY

1. Remove self-locking screw (1), small sling swivel (2), and buttplate group (3) from buttstock (4)

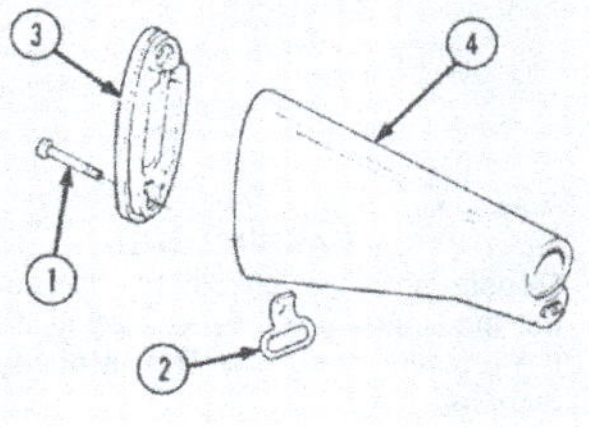

RIFLE ONLY

2. Push down on plunger (5) and lift door assembly (6) out of buttplate (7).

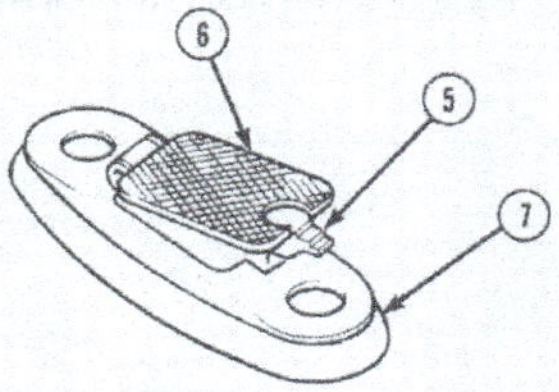

RIFLE ONLY

3. Remove straight pin 181 and separate hinge (9) and door assembly (61

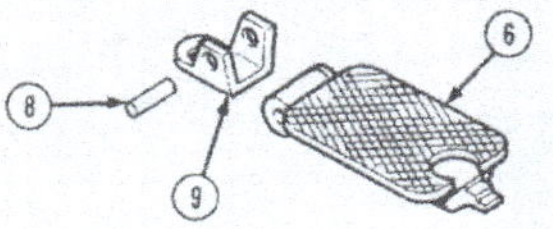

2-18. BUTTSTOCK ASSEMBLY (CONT).

a. DISASSEMBLY (CONT)

CARBINE ONLY

4. Disassemble the buttstock assembly by tapping out the spring pin (10) located in ,he oval slot of the retaining nut (1 1) This is done using a 1 16 Inch punch

5. Insert your Index finger into the forward end of the buttstock 112) and push down on the locking pin (13 Unscrew the retaining nut 11 and remove the release lever (14) locking pin (13) and spring (15)

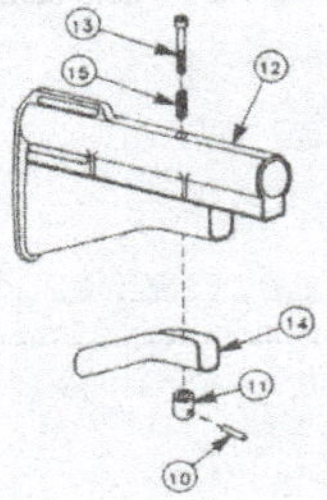

b. CLEANING (RIFLE ONLY)

Clean all parts with CLP (operator s manual) Use brush to clean knurled surface of door assembly

2-18. BUTTSTOCK ASSEMBLY (CONT).

c. INSPECTION

NOTE

M16A2 buttstocks, PN 9349121, with unauthorized markings may be used under the following conditions:

a. The only authorized markings are those which are temporary In nature, i e , paint, tape, etc

b. When marking a buttstock, only use temporary markings

c. Buttstocks with unauthorized markings that have been stamped into the surface of the buttstock will not be used

d. Unauthorized markings that have previously been scratched, etched, carved, etc may continue In use if the marks do not extend Into the fiber of the buttstock Cutting Into the fiber of the buttstock may weaken it

e. These marks may be at any location on the buttstock Unauthorized markings are not desirable. However, if previously applied, they will be allowed to continue In use due to the cost of the buttstock.

1. Inspect buttstock for cracks using the following guidelines:

   a. Under the following conditions, hairline cracks (no chipped away material allowed) originating from buttplate end of buttstock are acceptable.
      (1) One hairline crack, not to exceed 1 in. 12.54 cm) In length, per side of buttstock.
      (2) Two additional hairline cracks up to 0 25 in (0 64 cm) In length, per side of buttstock
      (3) A total of three cracks per side of buttstock, originating from buttplate end, are allowable.

   b. Cracks In the critical area at the front end of the buttstock are not acceptable and these buttstocks must be replaced.

2. While buttplate Is Installed on rifle, Inspect for cracks around the mounting holes Check for cracks In excess of 0 25 In. (O 64 cm) In length which extend through the buttplate Replace If cracked.

3. Inspect door assembly for cracks, corrosion, stuck plunger, separations on outer face, or other damage Replace If defective.

d. REPAIR

Replace all authorized unserviceable Items Unserviceable Items are those Items which are damaged.

e. LUBRICATION

Lubricate all metal components (p 2-33) (operator's manual)

f. REASSEMBLY

RIFLE ONLY

1. Position hinge (1) on door assembly (2) and Install straight pin (3)

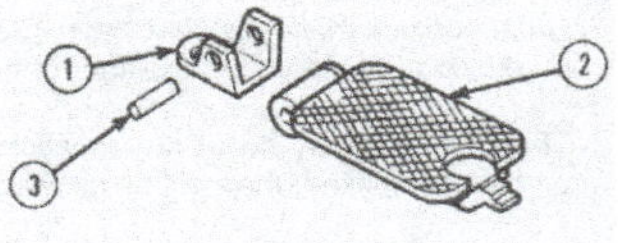

RIFLE ONLY

2. Install door assembly (2) into buttplate (4) and press plunger (5) to lock

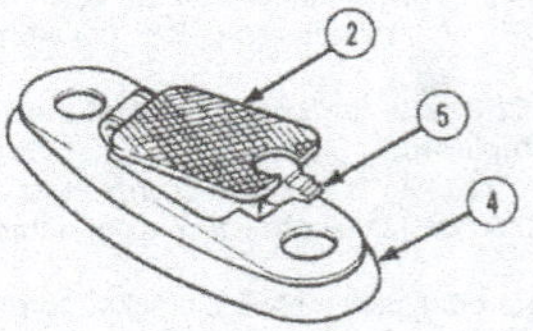

RIFLE ONLY

3. Position buttplate group (6) and small sling swivel (7) to the buttstock (8) and secure with self-locking screw (9).

NOTE

See page 2-62, reassembly, for reassembly of buttstock assembly to lower receiver

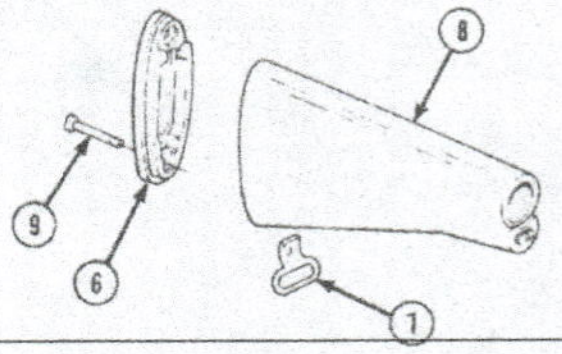

2-18. BUTTSTOCK ASSEMBLY (CONT).

f. REASSEMBLY (CONT)

CARBINE ONLY

4. Insert locking pin spring (10) onto locking pin (11)

5. Insert locking pin (11) and spring (10) Into the hole on top of the buttstock (12), threaded end first

6. Insert your index finger into the forward end of the buttstock (12) and push down on locking pin (11)

7. Install the release lever (13) onto the threaded portion of the locking pin (11) protruding through the bottom of the buttstock (12)

8. Screw on the locking nut (14) until flush with locking pin (11) Align the slot In locking nut (14) with the spring pin hole in the locking pin (11)

9. Lightly tap in spring pin (15) until flush on both sides of the retaining nut (14)

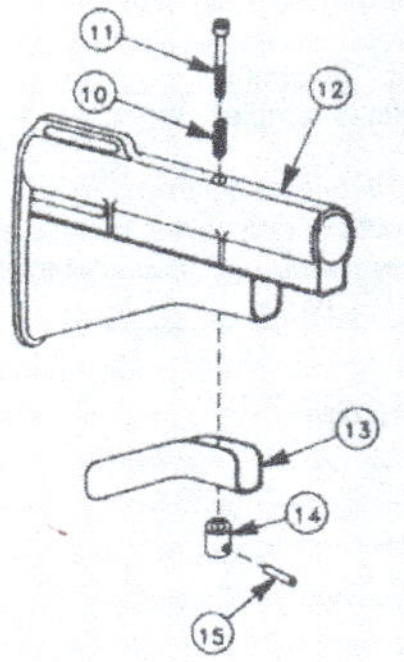

2-19. MAJOR COMPONENTS OF M16A2 RIFLE.

This task covers

a. Reassembly
b. Inspection
c. Stowage

INITIAL SETUP

References
TM 9 1005-319 10

Equipment Conditions
Rifle disassembled into major components

General Safety Instructions
To avoid injury to eyes, use care when removing and installing spring-loaded parts

Do not Interchange bolt assemblies from one rifle to another Doing so may result In Injury to, or death of. personnel

Do not keep live ammunition near the work area

a. REASSEMBLY

Refer to operator's manual and assemble lower receiver and buttstock assembly (1) upper receiver and barrel assembly (2). charging handle assembly (3), bolt carrier assembly (4), sling (5), and magazine(6).

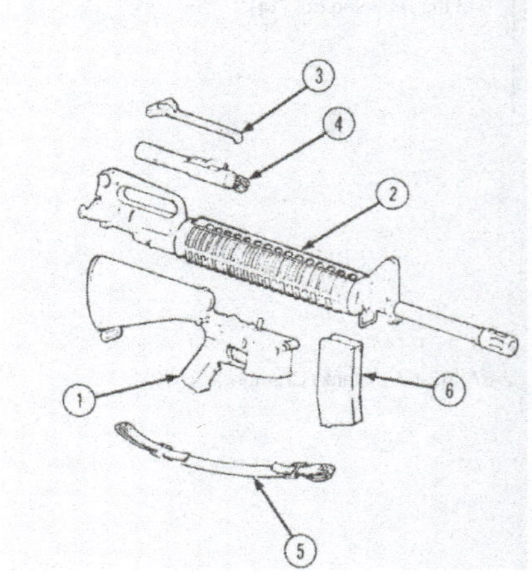

b. INSPECTION

Perform the following function checks on assembled weapon:

1. Remove magazine if installed. Pull charging handle assembly to rear. Check that chamber is clear. Let bolt and bolt carrier close. Do not pull trigger. Leave hammer in cocked position.

WARNING

If rifle fails any of the following tests, continued use of the rifle could result in injury to, or death of, personnel.

2. Place selector lever in SAFE position and pull trigger. Hammer should not fall.

3. Place selector lever in SEMI position. Pull trigger. Hammer should fall.

M16A2 and M4 ONLY
NOTE

For the purpose of the following check "SLOW" is defined as 1/4 to 1/2 the normal rate of trigger release.

4. Hold trigger to the rear, charge weapon, and release the trigger with a slow, smooth motion, without hesitations or stops, until the trigger is fully forward; an audible click should be heard. Hammer should not fall.

5. Repeat the SEMI position test five times. The weapon must not malfunction during any of these five tests. If the weapon malfunctions during any of these five tests, evacuate rifle to support maintenance for repair.

6. Place selector lever in BURST position. Charge weapon and pull trigger. Hammer should fall.

7. Hold trigger to the rear, pull charging handle assembly to rear and release three times. Release trigger. Hammer should not fall. The burst disconnector should have held the hammer to the rear while the trigger was in the pulled position.

8. Pull trigger. Hammer should fall. This should be the first round of a three round burst.

9. With hammer in forward position, attempt to place the selector lever in the SAFE position. If selector lever can be placed on SAFE, evacuate the weapon to support maintenance.

M4A1 ONLY

10. Place selector lever in AUTO position. Charge weapon and pull trigger. Hammer should fall.

11. Hold trigger to the rear, charge weapon and release trigger. Pull trigger. Hammer should not fall. AUTO sear should have released hammer while holding trigger in the squeezed position before releasing and resqueezing the trigger.

12. With hammer in forward position, attempt to place the selector lever in the SAFE position. If selector lever can be placed on SAFE, evacuate the weapon to support maintenance.

1 2-19. MAJOR COMPONENTS OF M16A2 RIFLE (CONT).

c. STOWAGE

Prior to stowing the weapon in arms room, perform the following procedures:

NOTE

Weapon passed into arms room issue window should be passed butt first with the bolt locked to the rear.

1. Clear. Refer to operator's manual.
2. Place selector lever in SEMI position.
3. Pull trigger. Hammer should fall.
4. Close ejection port (dust) cover.
5. Place weapon in rack.

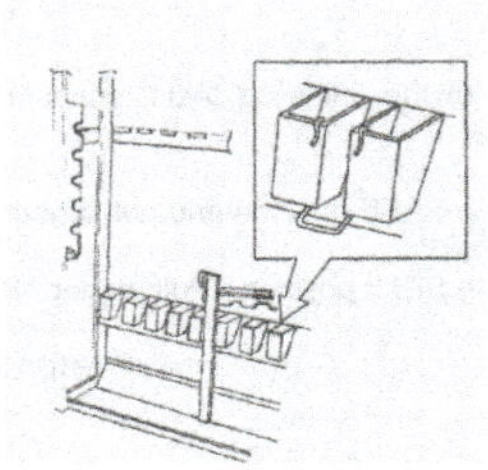

6. M4/M4A1 Carbine Only, use M12 Arms Rack.

a. The M12 arms rack is the correct arms rack in which to store the M4/M4A1 Carbine. The carbine must be stored with buttstock extended. When storing the M4/M4A1 Carbine in the M12 arms rack a mounting bracket, NSN 5340-01-230-3181, (app D, item 23.1) may be used for each M4/M4A1 Carbine being stored. This option is for the convenience of the person who opens and closes the arms rack to store the carbines.

b. To install the mounting bracket on the M12 arms rack, for use with the M4/M4A1 Carbine, install the bracket with the hooks of the bracket facing toward the carbine, so that the lower receiver extension will contact the bent end of the bracket. The bent end of the bracket will hold the carbine upright when the arms rack is opened. The bracket can be turned around when not in use for the carbine to allow storage of the M16A2 rifle.

c. When storing the M4/M4A1 Carbine in the M12 arms rack, an adapter bar (fig E-6, app E), MUST be used for security reasons. To install the adapter bar to the M12 arms rack:

(1) Remove all weapons from the rack and position the rack to gain access to the back.

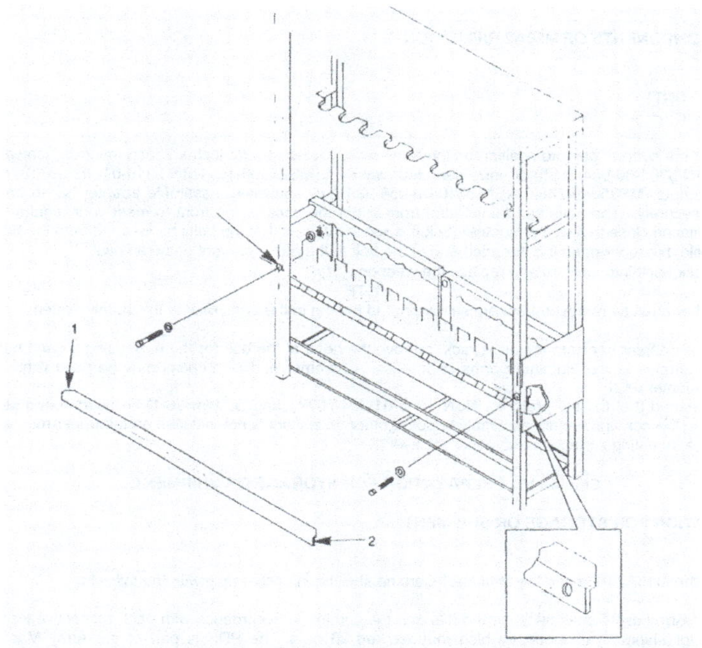

(2) The side of the adapter bar with the corners (1) cut off is the top and the side with square corners (2) is the back. The portion of the bar must be placed so the cut off corners (1) face the front of the rack.

(3) Holding the adapter bar at an angle, place one end into position inside the rear leg of the arms rack. Lower the other end of the bar into position. Allow the adapter bar to rest on the rack.

(4) Clamp both ends of the adapter bar into position. Mark the center line of the leg and adapter bar where they meet (see illustration) Using a center punch, mark the location of the holes to be drilled where the center lines cross. The holes must be centered on both the leg and the adapter bar. Drill a 1/8 inch pilot hole through both arms rack legs and the ends of the adapter bar. Drill a 3/8 inch hole through both arms rack legs and the ends of the adapter bar using the pilot hole as a guide. Remove the adapter bar. File the edges of all holes smooth Paint all bare metal surfaces with olive drab enamel paint, NSN 8010-01-350-5249, or equivalent.

2-19. MAJOR COMPONENTS OF M16A2 RIFLE (CONT).

c. STOWAGE (CONT)

(5) Reinsert the adapter bar into position on the arms rack. Using two 3/8 inch x 2 inch machine screws (MS35206-315, NSN 5305-00-984-5695) or equivalent, four washers (MS21306-1C, NSN 5310-01- 327-9713) or equivalent and two nuts (MS35649-2382, NSN 5310-00-056-3395) or equivalent assemble adapter bar to arms rack and tighten securely. The bolts can be inserted from either the back or the front to meet your requirements. If the rack is placed close to a wall or another rack It is recommended that the bolts be inserted from the back

(6) Tack weld, braze, or peen the threaded end of the bolt to the nut to prevent easy removal

(7) Place rack back into position and replace the weapons

NOTE

Adapter bar must be removed from arms rack prior to turning in the arms rack to the supply system.

d. To remove the adapter bar from the arms rack, remove the head of the bolt (or the nut), using a hand held grinder or file to avoid damage to the aluminum arms rack legs. A hammer and cold chisel may be used If no other way to remove the adapter exist

e. It is recommended that the rail protector NSN 1005-01-394-7677, (app D, item 26.1) be used during storage of the carbines when the carrying handle assembly or some other accessory is not installed on the upper receiver to prevent damage to the mounting surface on the upper receiver.

Section VI. PREPARATION FOR STORAGE OR SHIPMENT

2-20. PREPARATION FOR STORAGE OR SHIPMENT.

a. Packaging of the M16A2 Rifle and the M4/M4A1 Carbine shall be in accordance with the following

ARMY ONLY: Army users shall package the rifle and the carbine in accordance with each respective Packaging Data Sheet (PDS) for shipment or storage which may exceed 90 days The PDS is part of the Army Master Data File Retrieval Microform System (ARMS) Packaging
File

AIR FORCE ONLY: Air Force users shall package the rifle in accordance with each respective Special Packaging Instruction (SPI) 00-856-6885 for shipment or storage which may exceed 90 days The SPIs are part of the Army Master Data File Retrieval Microform System (ARMS) Packaging File

b. Packaging, if required, for shipping/storage which will not exceed 90 days shall be as follows

(1) Clean in accordance with operator's manual

(2) Wrap with MIL-B-121 waterproof material.

(3) Place in barrier bag MIL-B-117, Type I, Class C, or wrap with MIL-B-121, Type I, Grade A, and seal with tape, PPP-T-76.

(4) Place one or more of item in minimum size container. Block and brace in accordance with MIL-STD-1186. Cushion the M16 and similar weight items with PPP-C-843, and use PPP-F-320 as filler, to create a tight pack.

(a) Fiber board containers shall be in accordance with PPP-B-636 and may be Class Domestic. Gross weight and size of material shall determine grade of fiberboard container. PPP-B-640 may also be used.

(b) Wood containers shall be in accordance with PPP-B-601 or PPP-B-621.

(5) Equivalent materials may be used.

c. NSNs are not assigned to all the specified material. If it is necessary to specify an NSN in the TMs, the packing materials will have to be spared and part numbers and NSNs assigned.

d. The specifications used are:

(1)	MIL-B-121	Barrier Material, Greaseproofed, Waterproofed, Flexible (NSN 8135-00-753-4661)
(2)	MIL-B-117	Bag, Sleeve and Tubing -- Interior Packaging (NSN 8135-00-543-6574)
(3)	PPP-B-636	Boxes, Shipping, Fiberboard
(4)	PPP-B-601	Boxes, Wood, Cleated Plywood
(5)	MIL-STD-129	Marking for Shipment and Storage
(6)	PPP-T-76	Tape, Packaging, Paper
(7)	MIL-STD-1186	Cushioning, Anchoring, Bracing, Blocking and Waterproofing; with Appropriated Test Methods
(8)	PPP-C-843	Cushioning Material, Cellulosic
(9)	PPP-F-320	Fiberboard, Corrugated and Solid Sheet Stock (Container Grade), and Cut Shapes
(10)	PPP-B-640	Boxes, Fiberboard, Corrugated, Triple-Wall
(11)	PPP-B-621	Boxes, Wood, Nailed and Locked -- Corner

CHAPTER 3

DIRECT SUPPORT MAINTENANCE INSTRUCTIONS

CHAPTER OVERVIEW

This chapter provides information and instructions to keep the weapon in good repair and contains the following sections:

a. Repair Parts and Special Tools
b. Direct Support Troubleshooting
c. Direct Support Maintenance Procedures for the M1 6A2 Rifle and M4/M4A1 Carbine
d. Preparation for Storage or Shipment
e. Preembarkation Inspection of Material in Units Slated for Overseas Movement

Section I. REPAIR PARTS, SPECIAL TOOLS, TMDE, AND SUPPORT EQUIPMENT

3-1. COMMON TOOLS AND EQUIPMENT. For authorized common tools and equipment, refer to the Modified Table of Organization and Equipment (MTOE) applicable to your unit. Air Force users must maintain the following common tools:

Flat tip screwdriver
Socket wrench handle and socket head screw socket wrench
Vise jaw caps
Machinist's vise
Solid center punch
Hammer
Combination wrench
Torque wrench
Retaining ring pliers
(Two) 8-inch adjustable wrenches
Flat file
Ball-peen hammer
Trigger pull test fixture rod and weights
5/64-inch drive pin punch
1/8-inch drive pin punch
1/16-inch drive pin punch
3/32-inch drive pin punch

3-2. SPECIAL TOOLS, TMDE, AD SUPPORT EQUIPMENT. Special tools required for direct support maintenance are listed in appendixes B and C. Fabricated tools are listed and illustrated in appendix E.

3-3. REPAIR PARTS. Repair parts are listed and illustrated in appendix C of this manual.

NOTE

Bolt assemblies, and/or barrel assemblies may be interchanged , at the Direct Support Maintenance level, from one rifle to another, under the provisions of the note on page C-3: If these parts are interchanged the rifle must be checked/inspected as depicted on pages 3-17, 3-21, and 3-33. While performing these checks and inspections, pay special attention to the headspace requirements on page 3-47.

Section II. TROUBLESHOOTING

3-4. GENERAL.

a. This section contains direct support level troubleshooting information for locating and correcting most of the operating troubles which may develop in the M16A2 rifle and M4/M4A1 carbine. Each malfunction for the individual part or assembly is followed by a list of tests or inspections which will help you to determine the corrective action to take. You should perform the tests/inspections and corrective actions in the order listed.

b. This manual cannot list all malfunctions that may occur, nor all tests or inspections and corrective actions. If a malfunction is not listed or is not corrected by listed corrective actions, see individual repair sections in the maintenance procedures on each major assembly.

3-5. TROUBLESHOOTING PROCEDURES. Refer to troubleshooting table for malfunctions, tests, and corrective actions. The symptom index is provided for a quick reference of the malfunctions covered in the table.

SYMPTOM INDEX

TROUBLESHOOTING

MALFUNCTION
TEST OR INSPECTION
CORRECTIVE ACTION

1. FAILURE OF MAGAZINE TO LOCK IN RIFLE.
   Step 1. Dirty or corroded magazine catch (1).
   Disassemble and clean.
   Step 2. Defective magazine catch spring (2).
   Replace magazine catch spring (2) (p 3-62).
   Step 3. Worn or broken magazine catch (1).
   Replace magazine catch (1) (p 3-62).

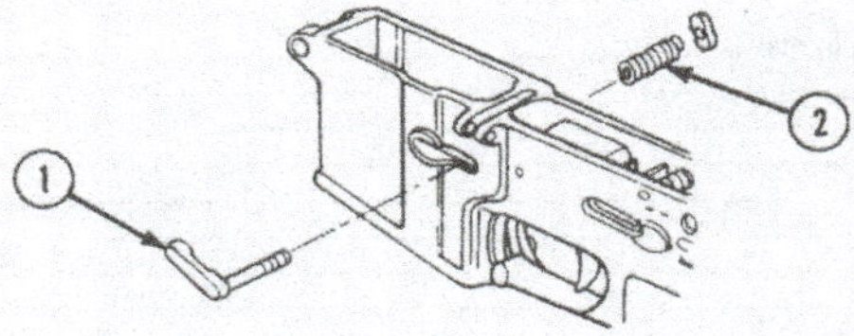

2 FAILURE TO FEED
   Step 1 Magazine catch spring (1) weak or broken.
   Replace magazine catch spring I1) (p 3-62)

   Step 2 Short recoil
   Refer to page 3-9

3 FAILURE TO CHAMBER
   Short recoil.
   Refer to page 3-9.

3-5. TROUBLESHOOTING PROCEDURES (CONT).

TROUBLESHOOTING (CONT)

MALFUNCTION
TEST OR INSPECTION
CORRECTIVE ACTION

4. FAILURE TO LOCK.

Step 1. Damaged bolt carrier key (1).
Repair or replace bolt carrier key (1) (p 3-25)

Step 2 Loose screws (2) on bolt carrier key (1).
a. Disassemble and repair (p 3-25)
b. Reassemble using new screws.

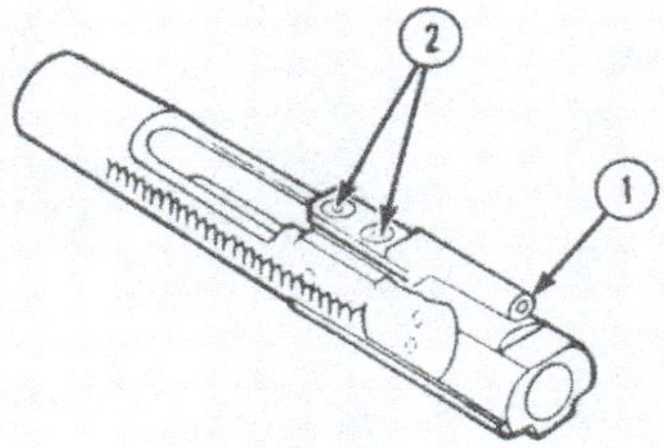

Step 3 Bent gas tube (3).
a Adjust by bending gas tube 13) In area of handguards
b Replace gas tube (3) and check alignment (p 3-29)

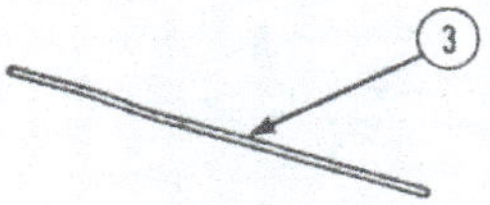

Step 4. Short recoil
Refer to page 3 9

TROUBLESHOOTING (CONT)

MALFUNCTION
TEST OR INSPECTION
CORRECTIVE ACTION

5. FAILURE TO FIRE.

Step 1. Broken hammer (1).
Replace hammer (1) (p 3-73).

Step 2. Weak or broken hammer spring (2).
Replace hammer spring (2) (p 3-73).

Step 3. Hammer spring (2) improperly assembled.
Assemble properly.

Step 4. **(M16A2 and M4 ONLY)** Burst cam (3) and/or cam spring (4) frozen or improperly assembled
Disassemble, clean, lubricate, and reassemble correctly (p 3-73).

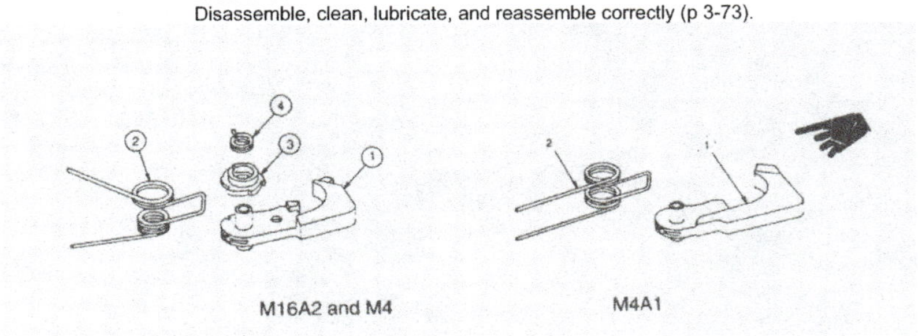

Step 5. Selector lever (5) frozen on SAFE position.
Disassemble and clean (p 3-62)

Step 6. Broken firing pin (6) or firing pin does not meet gage protrusion requirement.
Replace firing pin (6) (p 3-16).

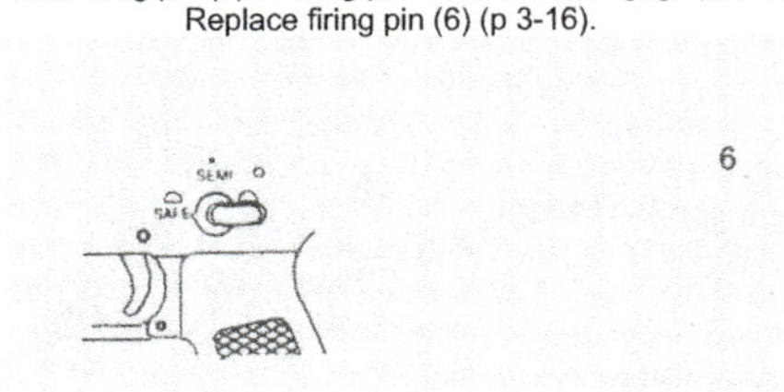

3-5. TROUBLESHOOTING PROCEDURES (CONT).

TROUBLESHOOTING (CONT)

MALFUNCTION
TEST OR INSPECTION
CORRECTIVE ACTION

6. FAILURE TO UNLOCK.

Step 1 Burred locking lugs (1) on bolt assembly.
Remove burrs.

Step 2 Burred lugs (2) on barrel extension.
Remove burrs.

Step 3. Short recoil.
Refer to page 3-9.

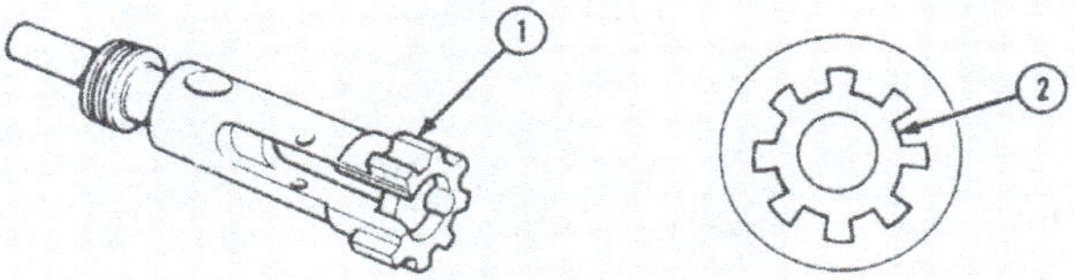

7. FAILURE TO EXTRACT

Step 1. Inspect cartridge extractor and extractor spring assembly.
Replace if cracked or broken (p 2-35).

Step 2. Inspect badly pitted chamber with reflector tool (Item 2, fig. C-1 7).
Replace barrel assembly If chamber Is badly pitted (p 3-29).

8 FAILURE TO EJECT.

Short recoil.
Refer to page 3-9.

TROUBLESHOOTING (CONT)

MALFUNCTION
TEST OR INSPECTION
CORRECTIVE ACTION

9. FAILURE TO COCK.

Step 1. Worn or broken trigger nose (1) or trigger spring (2).
Replace trigger (3) or defective trigger spring (2) (p 3-62).

Step 2. Worn or broken hammer trigger notch (4).
Replace hammer (5) (p 3-62).

Step 3. Worn or broken hammer disconnector hook (6).
Replace hammer (5) (p 3-62).

Step 4. Worn or broken hammer automatic sear hook (7).
Replace hammer (5) (p 3-62).

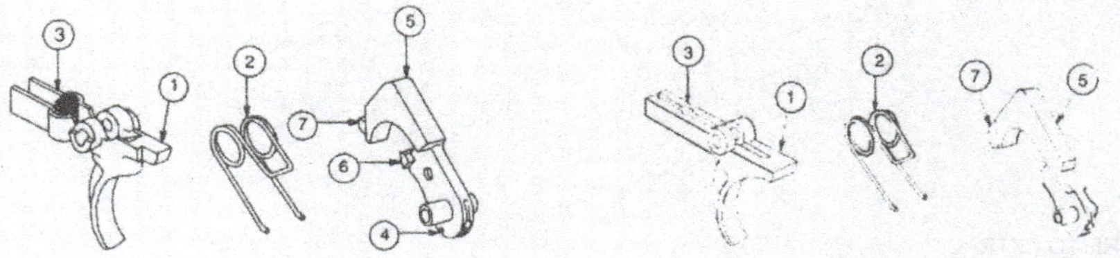

M16A2 and M4 ONLY

M4A1 ONLY

3-5. TROUBLESHOOTING PROCEDURES (CONT).

TROUBLESHOOTING (CONT)

MALFUNCTION
TEST OR INSPECTION
CORRECTIVE ACTION

Step 5 **(M4A1 has only 1 disconnector)** Worn or broken disconnector hooks (8).
Replace defective disconnectors (9) (p3-62).

Step 6 **(M4A1 has only 1 spring)** Weak, broken, or missing disconnector springs (10).
Replace disconnector springs (10) (p 3-62).

Step 7. Worn, broken, or missing automatic sear (11).
Replace automatic sear (11) (p 3-62).

Step 8 Weak or broken automatic sear spring (12).
Replace automatic sear (11) (p 3-62).

Step 9. Long leg (13) of automatic sear spring incorrectly assembled in receiver.
Remove automatic sear assembly (11) and install correctly (p 3-62).

Step 10. **(M16A2 and M4 ONLY)** Burst cam (14) or clutch spring (15) frozen or improperly assembled.
Disassemble, inspect, clean, lubricate, or replace as required (p 3-73).

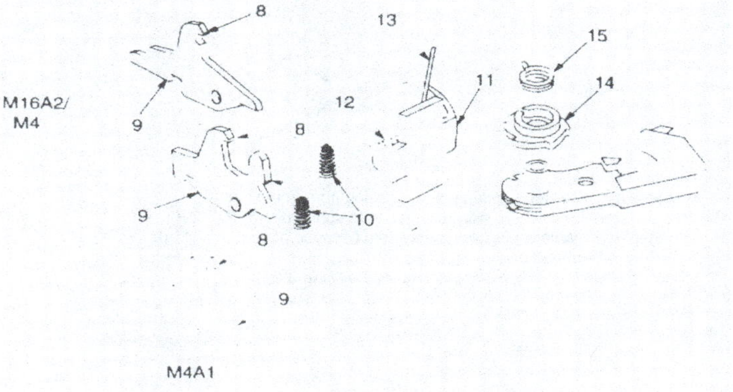

TROUBLESHOOTING (CONT)

MALFUNCTION
TEST OR INSPECTION
CORRECTIVE ACTION

10. SHORT RECOIL.

Step 1. Improper gap space or worn, missing, or broken bolt rings (1).

a. Stagger bolt ring gaps (p 3-21).

b. Replace bolt rings (1) and stagger gaps (p 3-21).

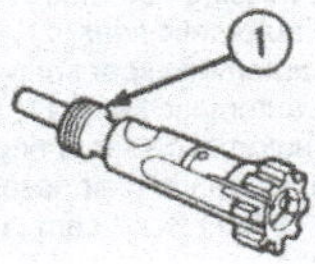

Step 2. Broken or bent gas tube (2).

Adjust by bending in area of handguards or replace gas tube (2) (p 3-29).

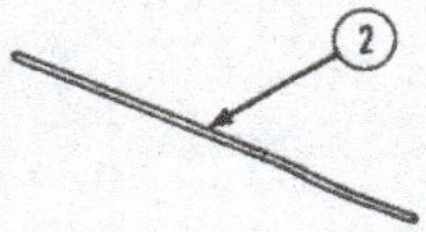

Step 3. Gas tube spring pin (3) missing from front sight (4).

Replace gas tube spring pin (3) (p 3-29).

Step 4. Partially plugged gas system because of carbon build-up in the gas tube (2).

Replace gas tube (2) (p 3-29).

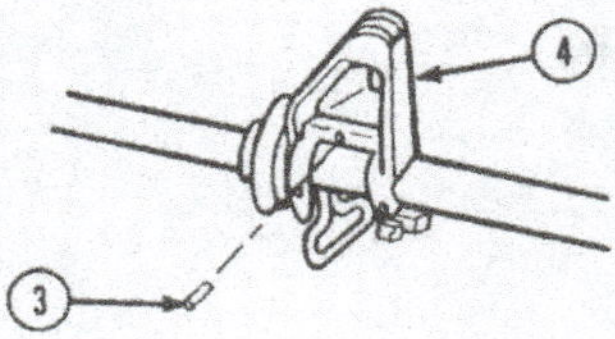

3-5. TROUBLESHOOTING PROCEDURES (CONT).

TROUBLESHOOTING (CONT)

MALFUNCTION
TEST OR INSPECTION
CORRECTIVE ACTION

10 SHORT RECOIL (CONT).

WARNING

When using carbon removing compound (Item 8, app D), avoid skin contact. If it comes in contact with the skin, wash off thoroughly with running water. The use of a good lanolin base cream after exposure to compound Is helpful. The use of gloves and protective equipment Is required. Step 5. Carbon build-up in barrel gas port (5). Remove carbon build-up by soaking barrel In carbon removing compound (item 8, app D). Use rubber gloves (Item 18, app D) with carbon removing compound. Use a bore small arms cleaning brush (Item 4, app D).

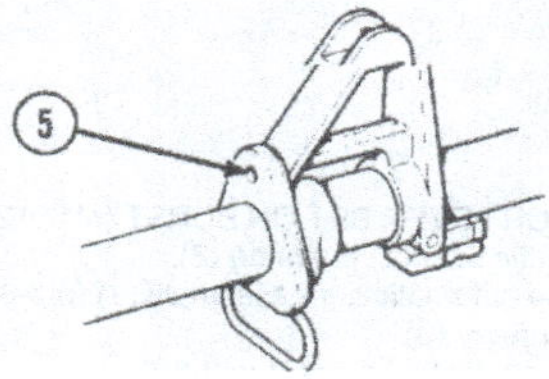

11. RIFLE CANNOT BE ZEROED.

Step 1 Inspect for defective or bent barrel assembly (1) (p 3-29).
Replace barrel assembly (1) (p 3-29).

Step 2 (For windage) barrel assembly (1) out of alignment with rear sight on up per receiver.
Align barrel assembly (1) and upper receiver (p 3-291

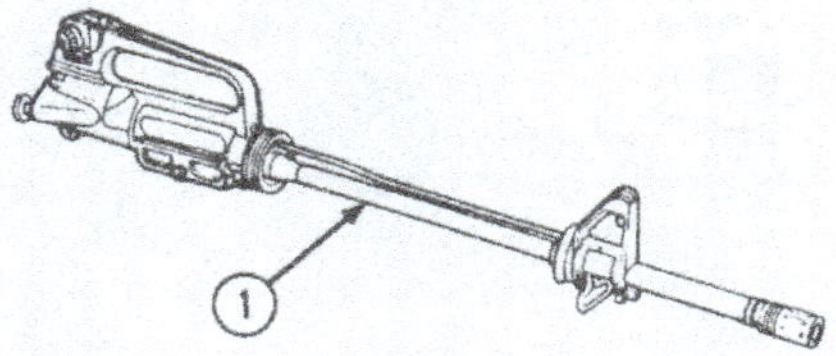

TROUBLESHOOTING (CONT)

MALFUNCTION
TEST OR INSPECTION
CORRECTIVE ACTION

Step 3. (For elevation) defective front sight (2) or rear sight (3).
Repair as required (p 3-48).

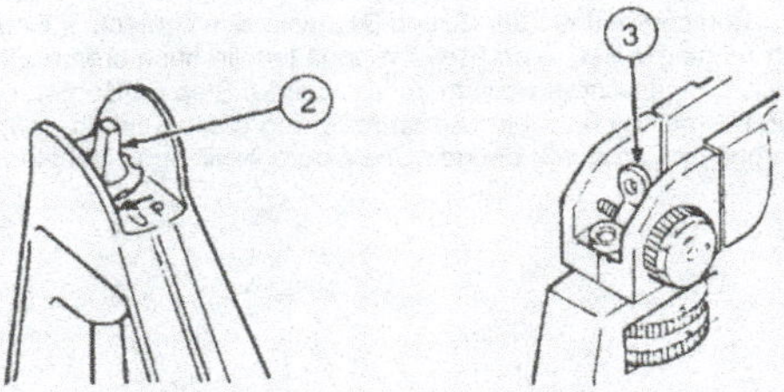

12. FAILURE TO CYCLE WITH SELECTOR LEVER SET ON BURST (M16A2 and M4 ONLY).

Step 1. Broken automatic sear (1) or spring (2).
Replace automatic sear assembly (1) (p 3-62).

Step 2. Faulty selector lever (3).
Replace selector lever (3) (p 3-62).

Step 3. Broken tooth on burst cam (4).
Replace burst cam (4) (p 3-73).

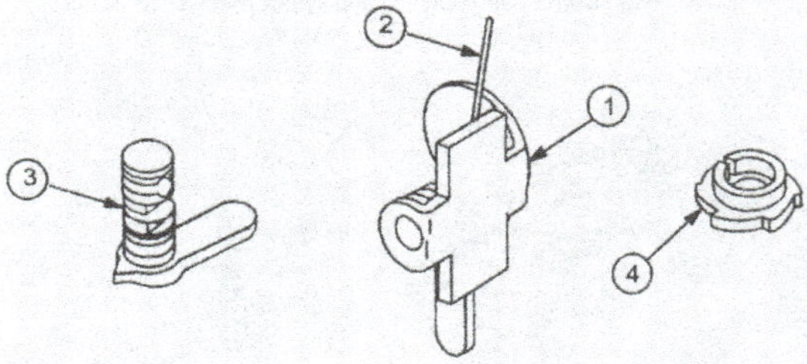

M16A2 and M4

3-5. TROUBLESHOOTING PROCEDURES (CONT).

TROUBLESHOOTING (CONT)

MALFUNCTION
TEST OR INSPECTION
CORRECTIVE ACTION

12 FAILURE TO CYCLE WITH SELECTOR LEVER SET ON BURST (M16A2 and M4 ONLY) (CONT).

Step 4 Broken cam clutch spring (5) Cam clutch spring should be bent and properly formed without any sharp edges or corners

Inspect and replace If required.

Step 5 The bend In the cam clutch spring (5) installed backwards (toward outside).

Install cam clutch spring (5) properly with the bend to the inside (p 3-73).

NOTE

When hammer Is rotated back to cocked position, cam should rotate to allow the burst disconnector to latch in the next notch.

Step 6. Cam clutch spring (5) fails to "clutch" and burst cam (4) fails to rotate back with hammer (6).

Replace cam clutch spring (5) (p 3-73). If problem continues, replace hammer (6) and cam (4) (p 3-73).

Step 7. Broken front hook (7) on burst disconnector (8).

Replace burst disconnector (8) (p 3-62).

Step 8. Short recoil.

Refer to page 3-9

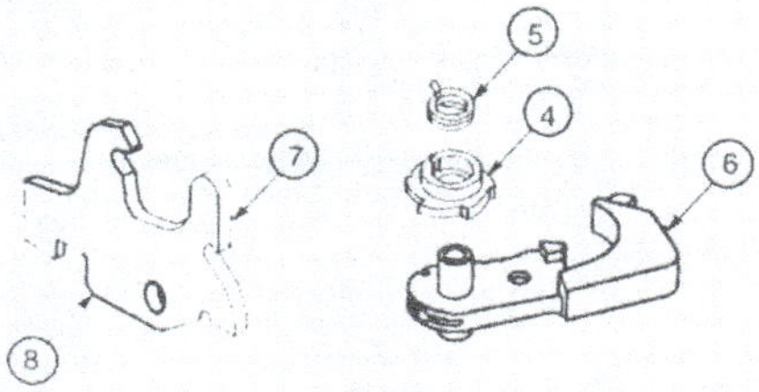

M16A2 and M4

TROUBLESHOOTING (CONT)

MALFUNCTION
TEST OR INSPECTION
CORRECTIVE ACTION

12.1. FAILURE TO CYCLE WITH SELECTOR LEVER SET ON AUTO (M4A1 ONLY)

Step 1. Broken automatic sear (1) or spring (2).
Replace automatic sear assembly (1).

Step 2. Faulty selector lever (3).
Replace selector lever (3).

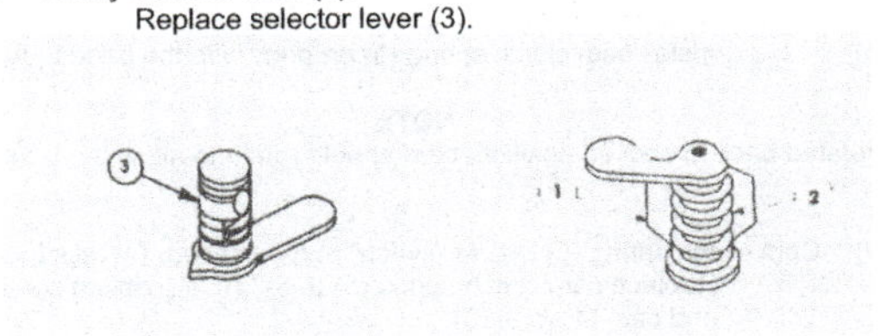

Step 3. Short recoil.
Refer to page 3-9.

13. FIRES TWO ROUNDS WITH ONE PULL OF TRIGGER WITH SELECTOR LEVER SET ON SEMI (DOUBLE FIRING).

Step 1. Defective semiautomatic disconnector (1). Replace semiautomatic disconnector (1) (p 3-62).

Step 2 Worn or broken trigger notch (2) of hammer (3) (searing portion). Replace hammer (3) (p 3-62).

Step 3. Worn or broken disconnector notch (4) of hammer (3). Replace hammer (3) (p 3-62).

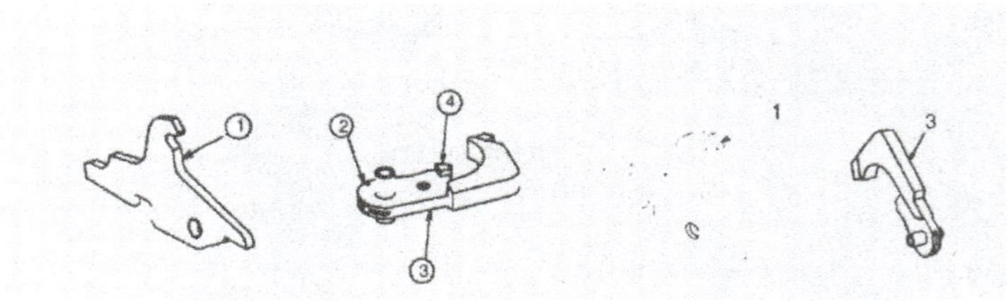

M16A2 and M4 M4A1

3-5. TROUBLESHOOTING PROCEDURES (CONT).

TROUBLESHOOTING (CONT)

MALFUNCTION
TEST OR INSPECTION
CORRECTIVE ACTION

13 FIRES TWO ROUNDS WITH ONE PULL OF TRIGGER WITH SELECTOR LEVER SET ON SEMI (DOUBLE FIRING) (CONT).

Step 4 Worn or broken trigger (5) (searing portion).
Replace trigger (5) (p 3-62).

Step 5 Worn trigger or hammer pin hole (6).
Gage trigger pin hole (7) and hammer pin hole (6) (p 3-62). If test falls, replace weapon.

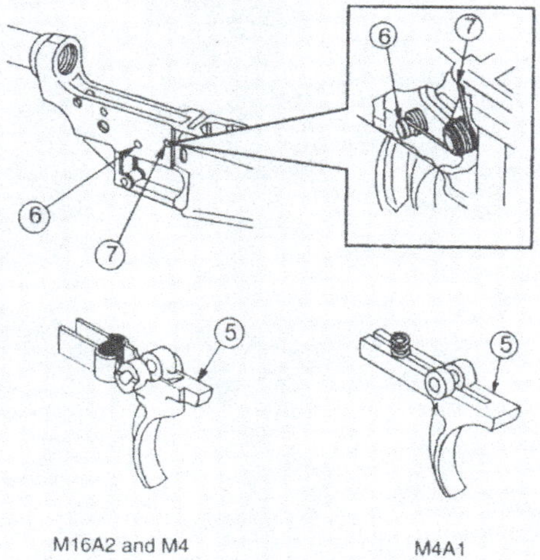

TROUBLESHOOTING (CONT.)

MALFUNCTION

TEST OR INSPECTION

CORRECTIVE ACTION

14. FIRES WITH SELECTOR LEVER ON SAFE OR WHEN TRIGGER IS RELEASED WITH SELECTOR LEVER ON SEMI.

Step 1. Defective selector lever (1).

Replace selector lever (1) (p 3-62).

Step 2. Worn or broken trigger (rear portion) (2).

Replace trigger (3) (p 3-62).

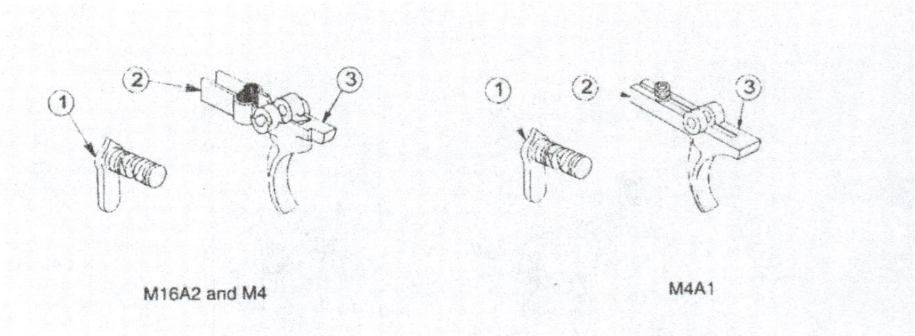

3-5. TROUBLESHOOTING PROCEDURES (CONT.).

TROUBLESHOOTING (CONT.)

MALFUNCTION

TEST OR INSPECTION

CORRECTIVE ACTION

15. HAMMER PIN "WALKS".

Hammer pin (1) "walks" or works loose during firing or hammer pin Is very easy to push out of receiver when hammer Is Installed.

Replace hammer assembly (p 3-62).

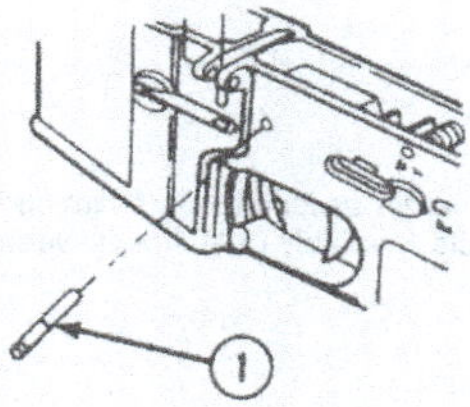

16. BOLT ASSEMBLY FAILS TO LOCK TO REAR AFTER FIRING LAST ROUND.

Step 1. Broken bolt catch (1).

Replace bolt catch (1) (p 3 62,.

Step 2. Weak or broken bolt catch spring (2).

Replace bolt catch spring (2) (p 3-62)

Step 3 Restricted movement of bolt catch (1).

Disassemble and clean.

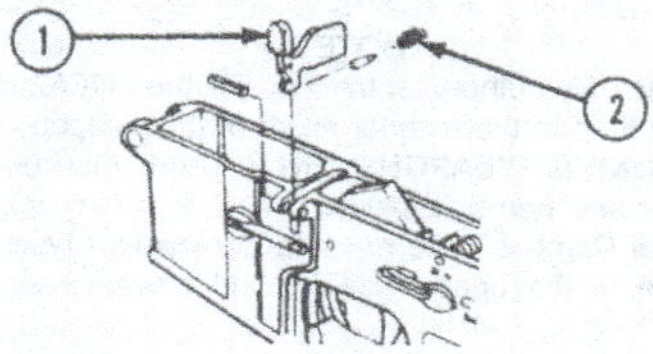

Section III. MAINTENANCE PROCEDURES FOR THE M16A2 RIFLE AND M4/M4A1 CARBINES

3-6. MAJOR COMPONENTS OF M16A2 RIFLE AND M4/M4A1 CARBINES.

This task covers disassembly.

INITIAL SETUP

Tools
(ARMY) Small Arms Repairman Tool Kit (item 3, app B)

References
TM 9-1005-319-10 (operator's manual)

Equipment Conditions
Weapon assembled

General Safety Instructions
Before starting an inspection, be sure to clear the weapon Do not pull the trigger until the weapon has been cleared.
Inspect the chamber and receiver to ensure no ammunition is present Do not keep live ammunition near work area.

DISASSEMBLY

a. Refer to operator's manual.

b. Remove magazine (1), sling (2), bolt carrier assembly (3), charging handle assembly (4), and upper receiver and barrel assembly (5), carrying handle assembly (5.1), from lower receiver and buttstock assembly (6).

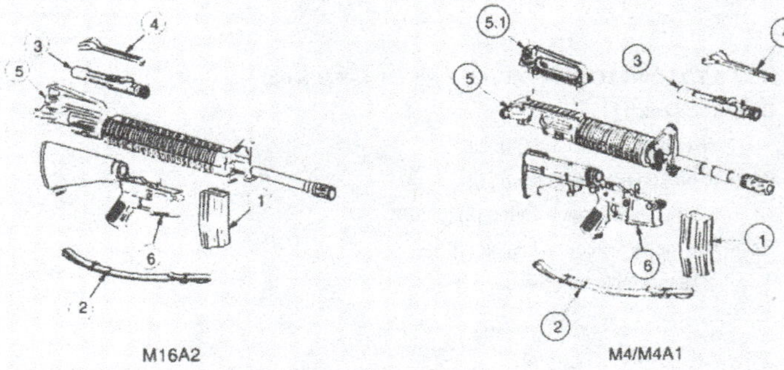

NOTE

Solid Film Lubricant (SFL) is the only authorized touchup for the M16A2 rifle and M4/M4A1 carbines and may be used on up to one third of the exterior finish of the weapon. **FOR ARMY CONUS USE ONLY AND AIR FORCE TRAINING WEAPONS ONLY**: SFL may be used as touchup without limitation on the upper receiver and barrel assembly. This is to say that units which DO NOT fall under the category of Divisional Combat Units or rapid deployment type units may have up to 100 percent of the exterior surface of the upper receiver and barrel assembly protected with SFL if necessary.

3-7. BOLT CARRIER ASSEMBLY.

This task covers:

a	Disassembly	**d**	Test
b	Cleaning	**e**	Repair
c	Inspection	**f**	Reassembly

INITIAL SETUP

Test Equipment
Tool and Gage Set (item 2, app B)

Tools
(ARMY) Small Arms Repairman Tool Kit
(item 3, app B)

Materials/Parts
Cleaner, lubricant, and preservative (CLP)
(item 9, app D)
Cleaning compound, rifle bore (RBC)
(item 12, app D)
Pipe cleaner (item 11, app D)

Equipment Conditions
3-1 5 Bolt carrier assembly removed

a. DISASSEMBLY

1. Remove firing pin retaining pin (1).
2. Tip key and bolt carrier assembly (2) allowing firing pin (3) to drop out. Catch the firing pin.
3. Rotate bolt cam pin (4) one quarter turn and lift straight up to remove.
4. Remove bolt assembly (5) from key and bolt carrier assembly (2)

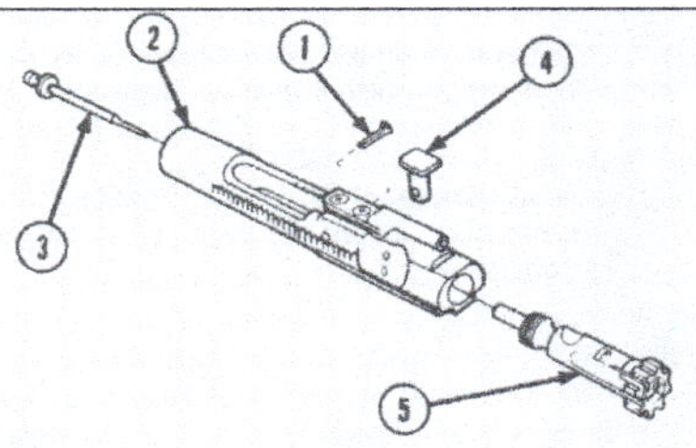

b. CLEANING

Clean carrier key and gas relief ports using CLP or RBC and pipe cleaner

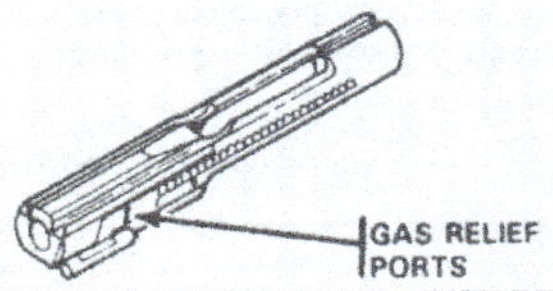

c. INSPECTION

1. Inspect bolt carrier assembly using the following guidelines.

(a) Inspect bolt carrier assembly (1) for burrs, cracks, wear, and evidence of gas loss.

(b) Visually inspect the carrier and key screws (2) for looseness and proper staking as shown below.

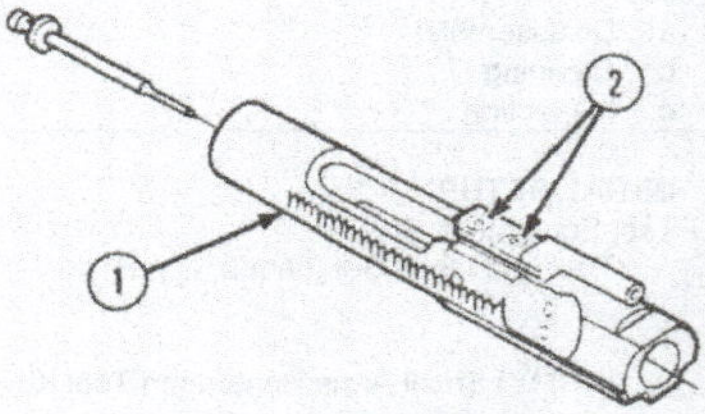

NOTE

Do not attempt to retorque if there is no loosening of the screws indicated by the staking marks.

Surface "A" must not Indicate distortion or damage which Impairs parallelism.

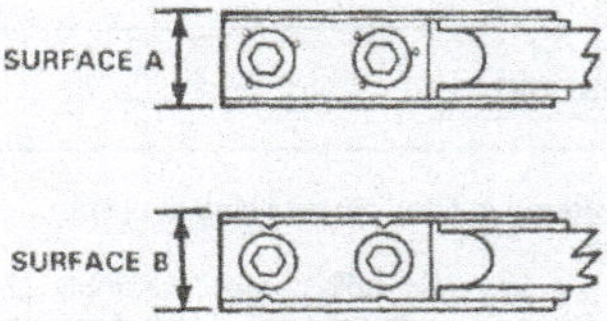

NOTE

A maximum of 0.025 in. (0.064 cm) protrusion In an upward direction is permissible.

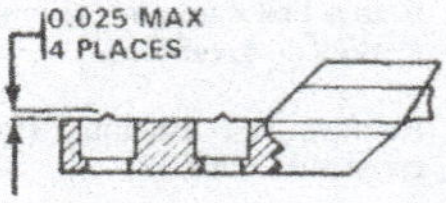

NOTE

There are bolts and bolt carriers on fielded rifles, some with chrome-plated exterior surface finishes and some with phosphate coating Both finishes are acceptable under certain operational requirements and or restrictions Phosphate-coated bolt carriers are required for divisional combat units Chrome plated bolt carriers are acceptable for divisional noncombat units and training center units. Chrome-plated and phosphate-coated bolt assemblies, bolt carrier assemblies, and repair parts for these assemblies may be intermixed In any combination, with the following exception:

Phosphate-coated bolt carriers are required for all deployable and deploying units Chrome-plated bolt carriers are acceptable for nondeployable and training center units.

3-7. BOLT CARRIER ASSEMBLY (CONT.).

c. INSPECTION (CONT.)

2. Inspect firing pin 131 tip for proper con tour Inspect for pitting, wear, and burrs Pits or wear in area (4) Is permissible Replace firing pin If defective

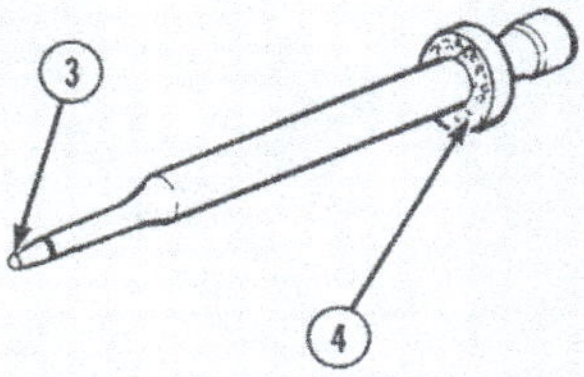

3. Prior to reassembly, insert bolt assembly (5) Into key and bolt carrier assembly (6) (do not insert bolt cam pin) and exercise bolt assembly in and out of key and bolt carrier assembly Check for binding

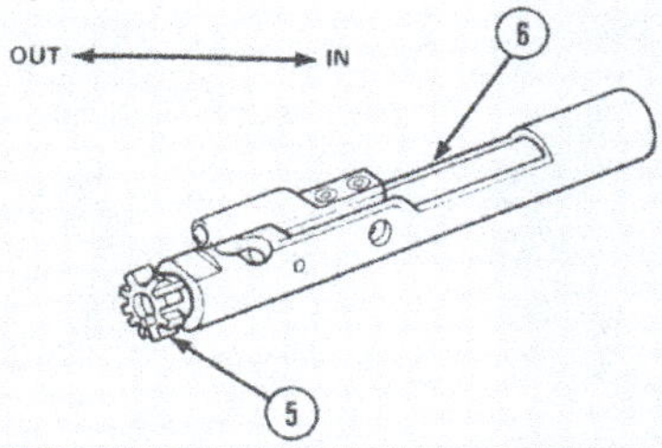

4. Check bolt assembly (5) for proper fit with bolt cam pin removed Turn key and bolt carrier assembly (61 and suspend so the bolt assembly is pointed down

NOTE

The bolt assembly must not drop out If weight of bolt assembly allows It to drop out of key and bolt carrier assembly, replace bolt rings (p 3-21)

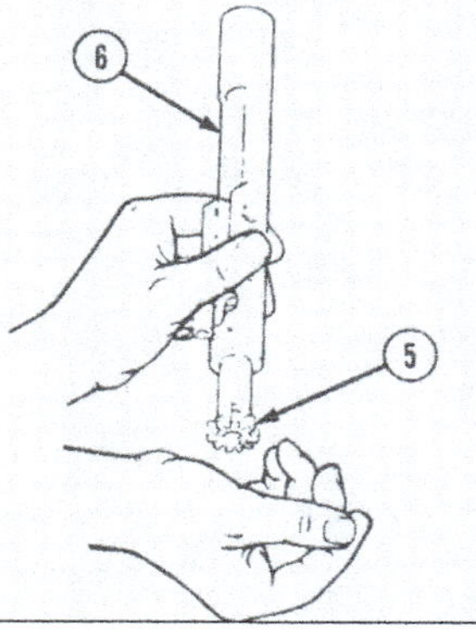

This page is blank. Procedures and figure were moved to page 2-36.

3-7. BOLT CARRIER ASSEMBLY (CONT.).

d. TEST

1. Insert firing pin (1) through bolt (2).
2. Position firing pin protrusion gage (3) PN 7799735 to check for proper firing pin (1) protrusion (minimum 0.028 in. (0 07 cm)- maximum 0 036 In. (0.09 cm))

NOTE
Firing pin should touch the gage on minimum but should not touch on maximum.

3. Replace a defective firing pin.

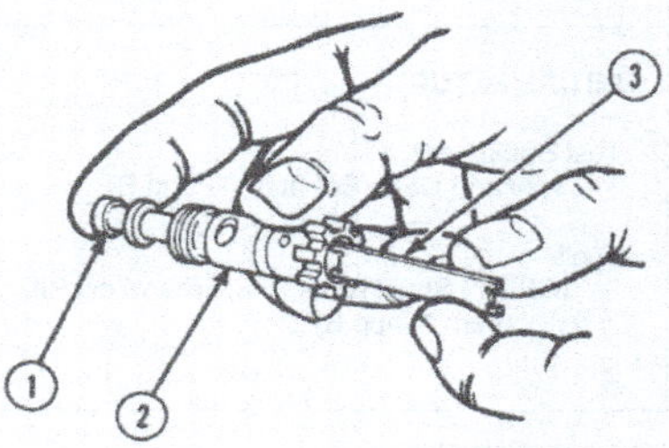

e. REPAIR

Replace all authorized unserviceable items. Retest all replaced parts.

f. REASSEMBLY

1. Install bolt assembly (1) into key and bolt carrier assembly (2).
2. Install bolt cam pin (3) and rotate one quarter turn to secure bolt assembly (1)
3. Hold key and bolt carrier assembly (2) with bolt assembly (1) down and drop In firing(pin (4)
4. Install firing pin retaining pin (5) from left side only to ensure proper Installation Check Installation by attempting to shake out firing pin
5. Reassemble rifle, refer to page 3-77

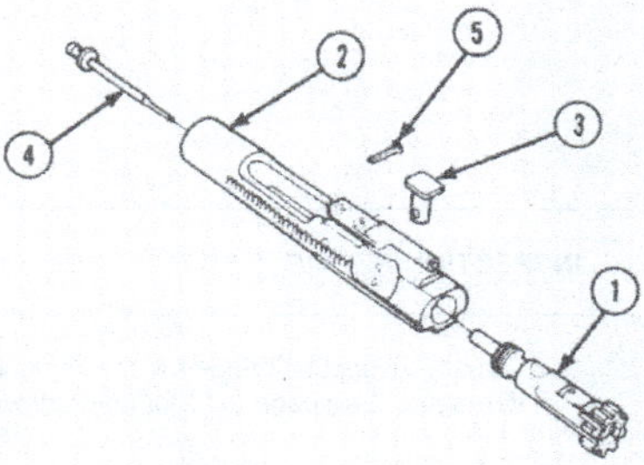

3-8. BOLT ASSEMBLY.

This task covers:

a. Disassembly
b. Inspection/Repair
c. Test
d. Reassembly

INITIAL SETUP

Test Equipment
Tool and Gage Set (Item 2, app B)

Tools
(ARMY) Small Arms Repairman Tool Kit (Item 3, app B)

Materials/Parts
Penetrant kit (item 25, app D)
Rag, wiping (item 26, app D)

Equipment Conditions
3-16 Bolt assembly removed

a. DISASSEMBLY

NOTE
Do not remove bolt rings unless they require replacement and three new replacement bolt rings are on hand.

Using small flat tip jeweler's screwdriver, remove the three bolt rings (1) from the bolt (2).

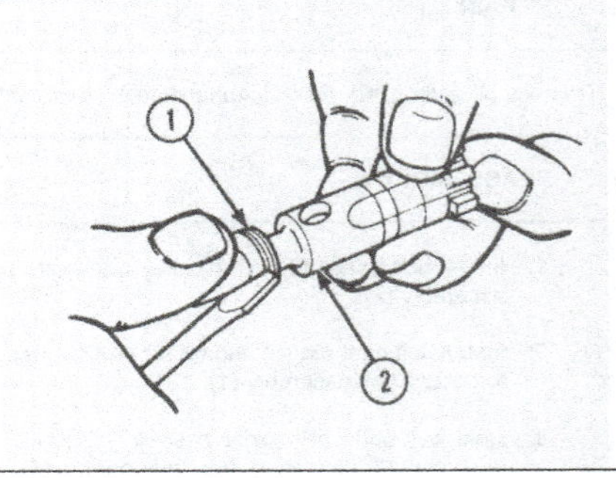

b. INSPECTION/REPAIR

1. Visually Inspect bolt rings for cracks, kinks, and bends. Replace all three bolt rings if one or more bolt rings is damaged. See page 3-16 for bolt ring wear check.

3-8. BOLT ASSEMBLY (CONT.).

b. INSPECTION/REPAIR (CONT.)

2. Inspect bolt for pits, burrs, and wear as follows.

(a) Bolt faces with a cluster of pits which are touching or tightly grouped, covering an area measuring approximately 1,8 Inch across, will be rejected and replaced.

(b) Bolts which contain individual pits or a scattered pattern will not be cause for rejection.

(c) Bolts that contain pits extending Into the firing pin hole will not be rejected unless firing pin hole gaging check determines excess wear.

(d) Rings on the bolt face (machine tool marks), grooves, or ridges less than approximately 0.010 inch will not be cause for rejection.

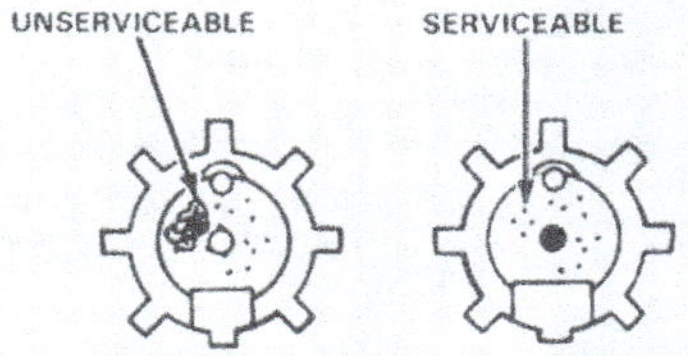

3. Inspect bolt for cracks In the locking lugs and the bolt cam pin hole area. Use black light If available; otherwise, use a glass of no more than 3X magnification or use a penetrant kit (item 25, app D). Pay close attention to the area where the locking lugs meet the body. Replace bolt assembly if bolt is defective.

WARNING

Dry cleaning solvent is flammable and toxic and should be used in a well-ventilated area. The use of rubber gloves is necessary to protect the skin when washing rifle parts.

4. Use penetrant kit (Item 25, app D) to check for cracks in bolt as follows:

(a) The area to be inspected must be clean, free of oil, etc. Spray a small amount of remover on the area to be inspected, let dry and wipe off with a wiping rag.

(b) Spray penetrant (only enough to wet the area) on the area of the bolt (1) to be inspected.

(c) Spray developer over the penetrant and let the developer work. Cracks will be indicated by a change in color where there is a crack. If there are cracks, the component 1' unserviceable.

(d) Pay close attention to the area where the locking lugs (2) meet the body.

(e) If there are no cracks, spray remover on the area; let dry and wipe off with a wiping rag. Oil the area to prevent corrosion.

(f) Replace bolt assembly If bolt (1) is defective.

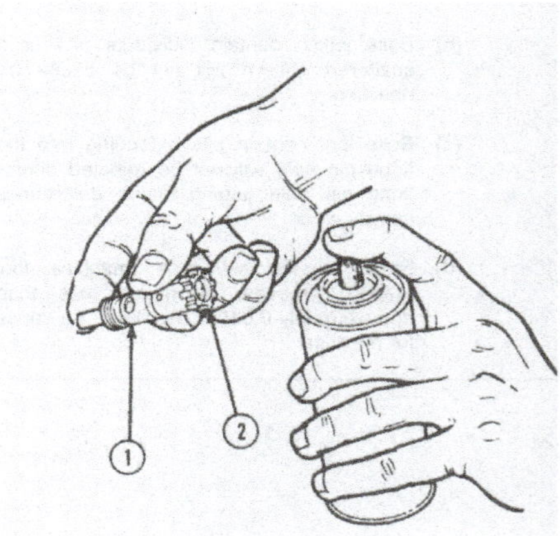

NOTE

Replacement of the bolt assembly will require that the headspace be tested (p 3-45, TEST).

3-8. BOLT ASSEMBLY (CONT.).

c. TEST

Test bolt (1) for elongated or oversized firing pin hole using special no-go plug gage (2) PN 12620101.

NOTE

Bolts with firing pin holes which permit the special no-go plug gage to fully penetrate at any position on the circumference will be rejected and replaced.

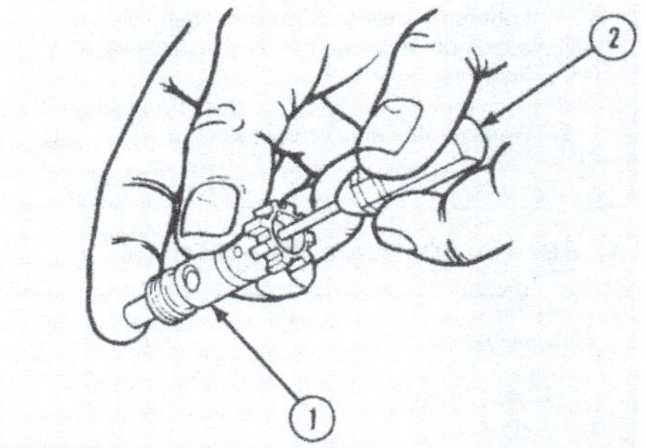

d. REASSEMBLY

NOTE

To install a bolt ring, carefully place one end in the bolt ring groove and hold In place with the thumb of one hand With the index finger of the other hand, gently guide and push the rest of the bolt ring into the groove a little bit at a time until the entire bolt ring is in place.

1. Install the three bolt rings (1) one at a time onto the bolt (2) using care not to bend or "spring" new bolt rings Stagger the bolt ring gaps (approximately ' turn apart).

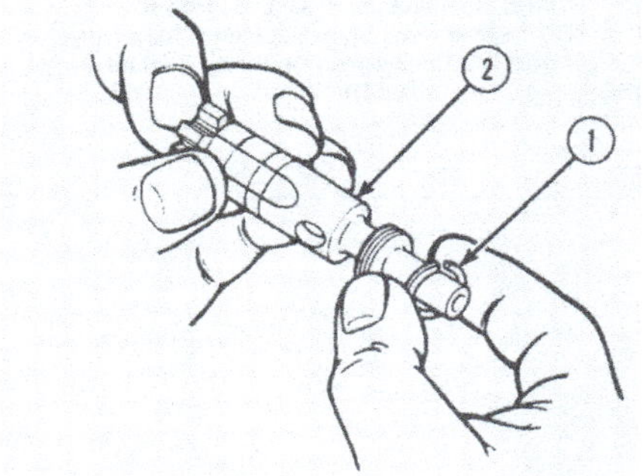

NOTE

Make certain bolt ring gaps are staggered to prevent loss of gas pressure. New bolt rings will make installing the bolt assembly difficult.

Lubricate inside key and bolt carrier assembly and use gentle pressure when installing.

2. Reassemble rifle, refer to page 3-77.

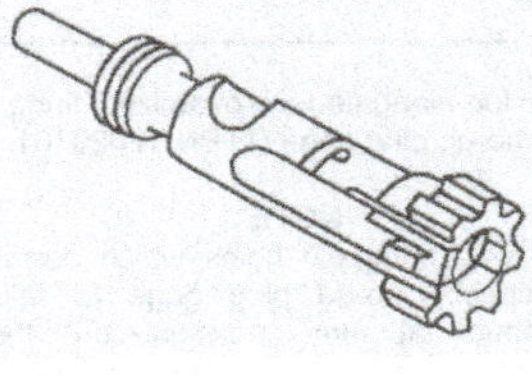

3-9. KEY AND BOLT CARRIER ASSEMBLY.

This task covers:

a Disassembly
b Repair
c Reassembly

INITIAL SETUP

Tools
Field Maintenance Basic Less Power Small Arms Shop Set (item 1, app B)
(ARMY) Small Arms Repairman Tool Kit (item 3, app B)

Materials/Parts
Carrier and key screws (2) (8448508)

Equipment Conditions
3-16 Key and bolt carrier assembly removed

3-9. KEY AND BOLT CARRIER ASSEMBLY (CONT.).

a. DISASSEMBLY

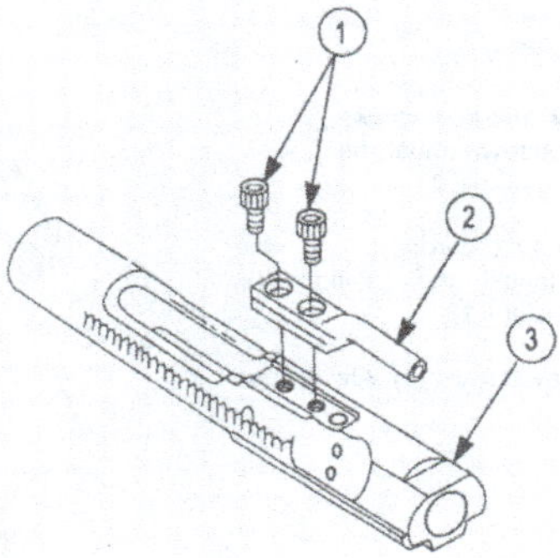

NOTE

Do not disassemble the key and bolt carrier assembly unless the bolt carrier key is defective as determined by inspection procedures described in section III, preceding.

1. Using socket wrench handle and tight fitting 1/8 inch socket head screw socket wrench attachment, remove two carrier and key screws (1).

NOTE

The heads and part of the bolt carrier key may be ground off in order to remove bolt carrier key from bolt carrier if carrier and key screws cannot otherwise be removed.

2. Remove bolt carrier key (2) from bolt carrier (3).

b. REPAIR

NOTE

Do not retorque carrier and key screws if staking marks do not indicate loosening screws.

Repair by replacing, torquing, and restaking carrier and key screws. Refer to the following reassembly procedures.

c. REASSEMBLY

NOTE
Do not reuse old carrier and key screws New carrier and key screws must be used at assembly.

1. If disassembled, place bolt carrier (1) In vise using vise law caps Install and position bolt carrier key (2) on bolt carrier (1).

2. Install two carrier and key screws (3). Always use new screws.

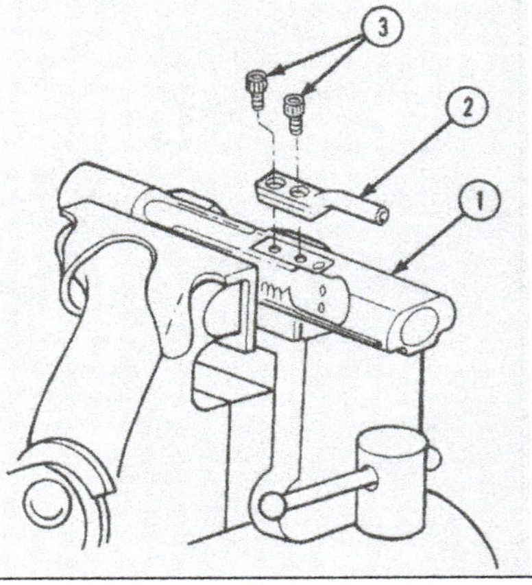

3. Use a tight-fitting 1 8 inch socket head screw wrench attachment and an Inch-pound torque wrench to torque the carrier and key screws (3) to 35 to 40 Inch-pounds (3 95 to 4 52 N m)

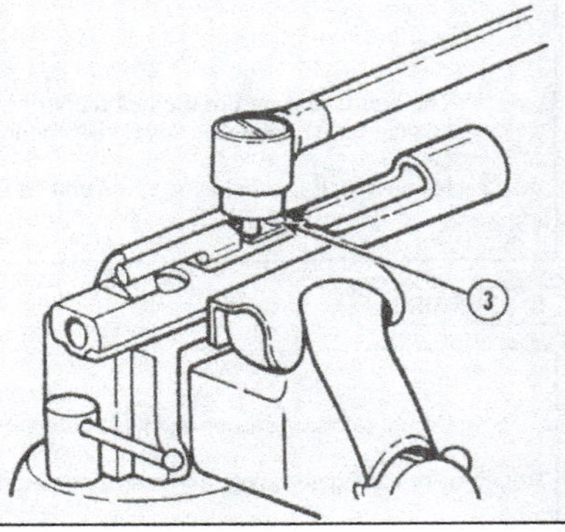

3-9. KEY AND BOLT CARRIER ASSEMBLY (CONT.).

c. REASSEMBLY (CONT.)

NOTE

Field staking method will be used by field units.

4. Use solid center punch and hand hammer to stake the two carrier and key screws (3) In three places.

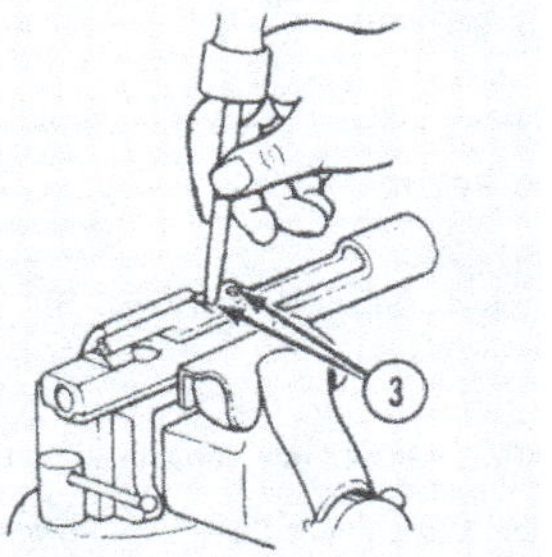

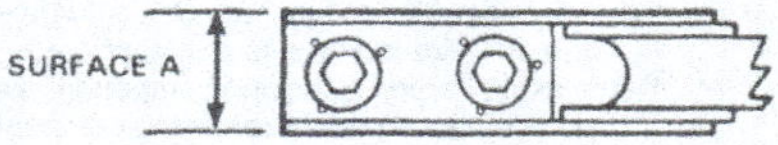

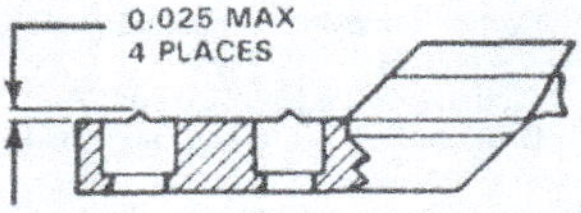

FIELD REPLACEMENT STAKING

5. Reassemble rifle, refer to page 3-77

CAUTION

If blanks are used, blank firing attachment (BFA) must be attached

NOTE

If the bolt carrier key is replaced, three to eight rounds of blank or ball ammunition must be fired to ensure a seal is created. Manual operation of the rifle may be required. If blank ammunition is utilized, M16A2 Blank Firing Attachment must be adapted

3-10. UPPER RECEIVER AND BARREL ASSEMBLY.

This task covers:

a Disassembly
b Inspection/Cleaning
c Repair
d Reassembly
e Test

INITIAL SETUP

Test Equipment
Tool and Gage Set (item 2, app B)

Tools
(ARMY) Small Arms Repairman Tool Kit (item 3, app B)
Field Maintenance Basic Less Power Small Arms Shop Set (item 1, app B)
Modified Needle Nose Pliers (fig E-7, app E)

Materials/Parts
Brush, cleaning, small (item 3, app D)
Carbon removing compound (item 8, app D)
Cloth, abrasive (item 13, app D)
Dichloromethane, technical (item 15, app D)
Dry cleaning solvent (item 16, app D)
Gloves, chemical and oil protective (item 18, app D)
Grease, molybdenum disulfide (item 19, app D)
Lubricant, solid film (item 21, app D)
Pan, wash (item 24, app D)
Polyethylene (item 32, app D)
Sealing compound (item 28, app D)
Target (item 31, app D)

References
FM 23-9
TM 9-1005-319-10

Equipment Conditions
3-15 Upper receiver and barrel assembly removed
2-43 Handguard assemblies removed

General Safety Instructions
To avoid injury to your eyes, use care when removing and installing spring-loaded parts.
When using solid film lubricant or dichloromethane, be sure the area is well ventilated.
When using carbon removing compound, avoid skin contact If carbon removing compound comes in contact with the skin, wash thoroughly with running water Using a good lanolin base cream after exposure to the compound is helpful Using gloves and protective equipment is required.

3-10. UPPER RECEIVER AND BARREL ASSEMBLY (CONT.).

a. DISASSEMBLY

1. Using ball-peen hammer and 5/64 inch diameter drive pin punch, drive spring pin (1) (which retains the gas tube) out of front sight assembly (2).

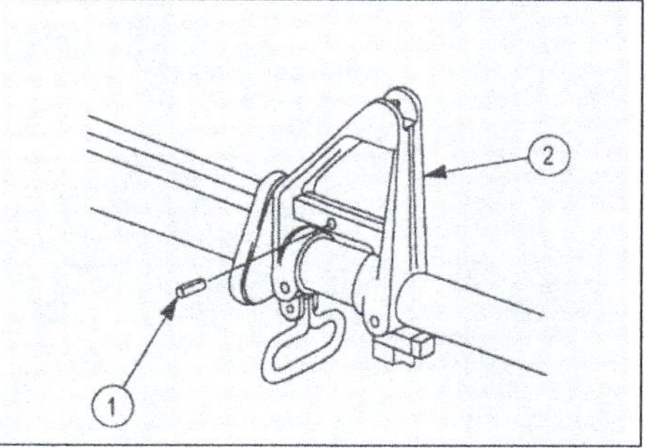

2. Slide gas tube (3) back into upper receiver assembly (4) to clear front sight assembly (2). Then lift slightly, pull forward, and remove gas tube (3).

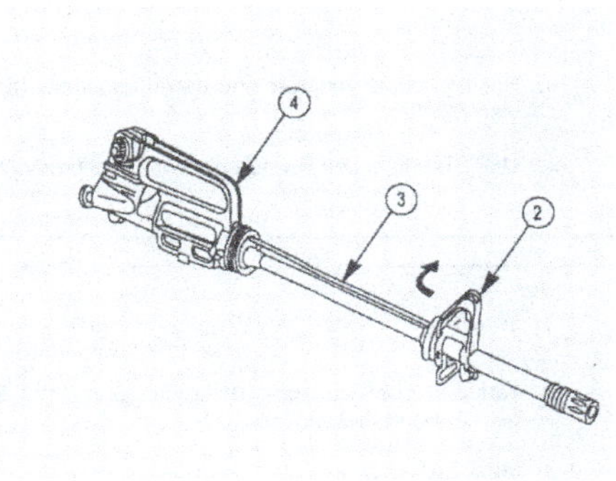

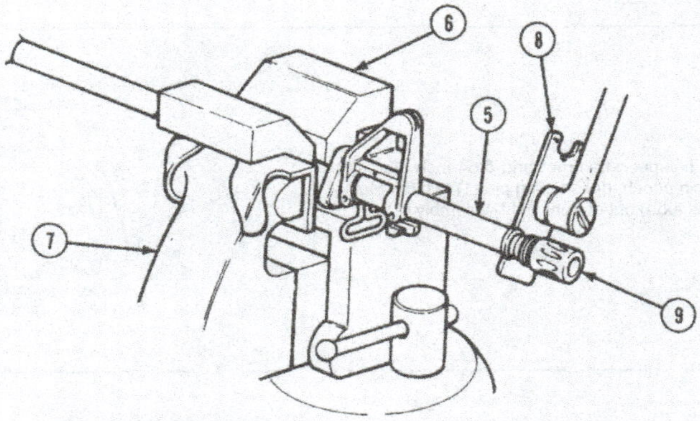

3. Position upper receiver and barrel assembly (5) in barrel removal fixture (6) and secure both In machinist's vise (7).

4. Using combination wrench (8) and 12 inch drive handle, remove compensator (9).

5. Remove peel washer (10) being careful not to lose or bend thin sections.

6. Remove upper receiver and barrel assembly (5) from barrel removal fixture and machinist's vise.

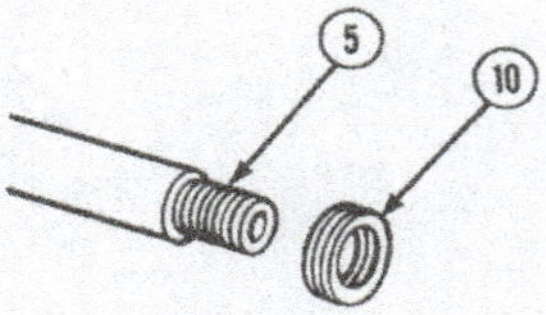

3-10. UPPER RECEIVER AND BARREL ASSEMBLY (CONT.).

a. DISASSEMBLY (CONT.)

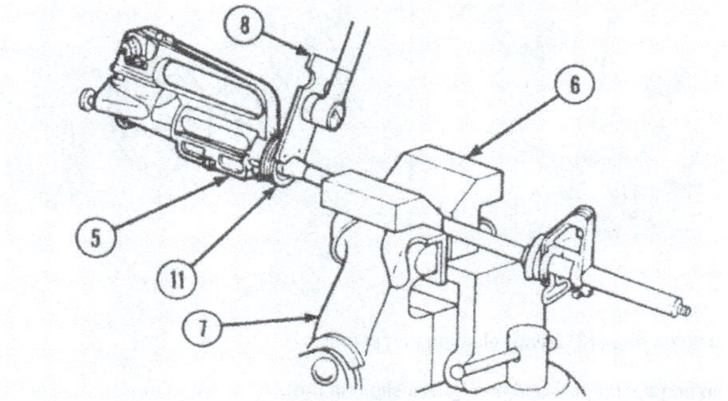

7. Place upper receiver and barrel assembly (5) Into barrel removal fixture (6) and clamp Into machinist's vise (7).

NOTE

Be sure all three drive pins on combination wrench are fully engaged with barrel nut assembly. Wrench must be pushed toward upper receiver assembly to compress the slip ring spring In barrel nut assembly. Do not use a torque wrench to loosen the barrel nut assembly.

8. Using '2 Inch drive handle and combination wrench (8), loosen barrel nut assembly (11)

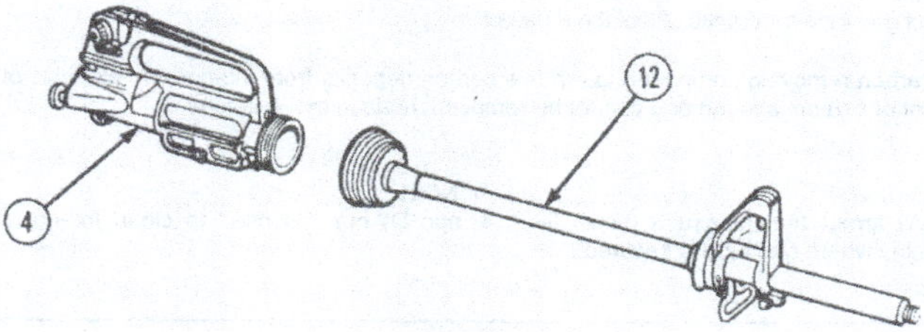

9. Separate upper receiver assembly (4) from barrel assembly (12)
10. Remove barrel assembly (12) from machinist's vise and barrel removal fixture

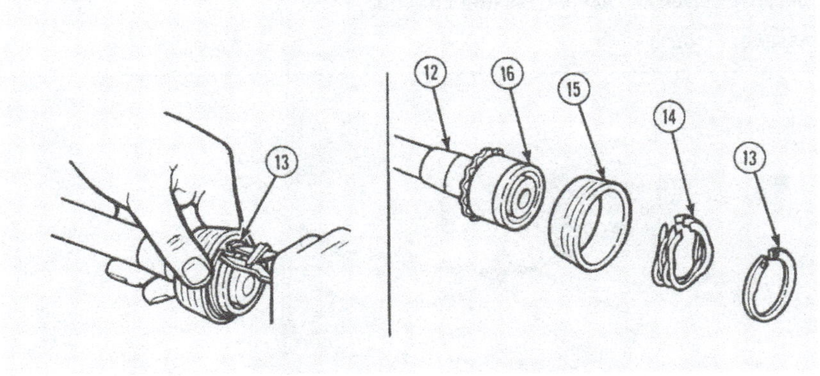

11. Remove retaining ring (13) using retaining ring pliers.

12. Remove slip ring spring (14) and handguard slip ring (15).

13. Do not remove barrel nut (16) from barrel assembly (12).

b. INSPECTION/CLEANING

WARNING

When using carbon removing compound, avoid skin contact. If carbon removing compound comes In contact with the skin, wash thoroughly with running water. Using a good lanolin base cream after exposure to compound is help ful. Using gloves and protective equipment is required.

1. Inspect gas tube for cracks. Replace if defective

2. Use carbon removing compound to remove carbon deposits from interior and exterior of gas tube. If a large amount of carbon is found and cannot be removed, replace the gas tube

NOTE

A small arms cleaning brush (bore) (item 4, app D) may be used to clean Interior of front sight assembly where gas tube is secured.

3-10. UPPER RECEIVER AND BARREL ASSEMBLY (CONT.).

b. INSPECTION/REPAIR (CONT.)

3. Inspect bore for burrs, cracks, rust, bulges, and pits using the following guidelines

(a) Pits no wider than a land or groove and no longer than 3'8 inch (0.95 cm) are al lowed in the bore.

(b) Lands that appear dark blue due to coating of gliding metal from projectiles are allowable.

(c) Definitely ringed bores or bores ringed sufficiently to bulge the outside surface of the barrel are cause for rejection. Replace barrel assembly if defective

4. If the upper receiver is separated from the barrel assembly, Inspect chamber for pits utilizing a flashlight. Pits 1'8 inch 10 32 cm) In length are cause for rejection Replace rifle barrel assembly If defective

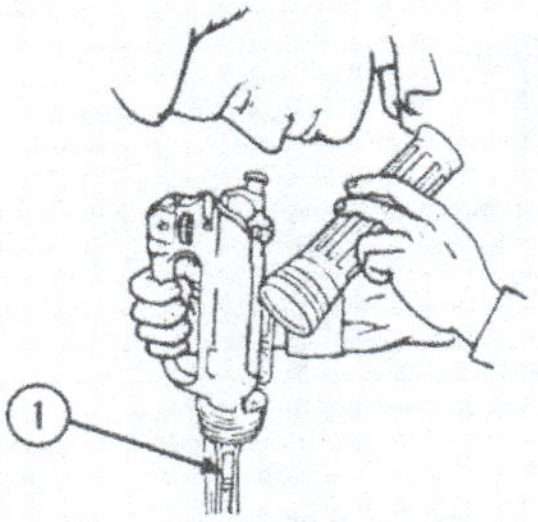

5. If upper receiver and barrel assembly is assembled, Inspect chamber using reflector tool (1) and flashlight. Pits 1 '8 inch (O 32 cm) In length are cause for rejection. Replace rifle barrel assembly If defective. If rifle barrel assembly Is replaced, Inspect headspace (p 3-45).

6. Inspect upper receiver assembly for cracks, corrosion, wear, or damage

(a) Small dents or gouges that do not affect functioning will not be cause for rejection.

(b) If upper receiver assembly contains cracks or holes, the upper receiver assembly will be replaced

7. Inspect all parts for damage and wear. Replace all defective parts.

NOTE

Damaged or missing teeth of the barrel nut is not cause for rejection provided the proper torque value can be obtained during installation using the tools depicted. If removal of the barrel is not possible with the combination tool, a pipe wrench or other such tool may be used during removal.

8. Inspect front sight guards for bends, if bent see page 3-37 for repair procedures.

c. REPAIR

1. Repair corroded upper receiver assembly surfaces as follows:

(a) Sand corroded area with abrasive cloth and make sure all corrosion has been removed.

WARNING

When using solid film lubricant or dichloromethane, be sure the area is well ventilated

(b) Wash area with technical dichloromethane (methylenechloride) to re move all dirt, grease, and foreign material

(c) Apply sealing compound, mixed In accordance with manufacturer's directions, to areas to be filled

(d) Spread sealing compound as smoothly as possible Into defective area using a putty knife or similar tool

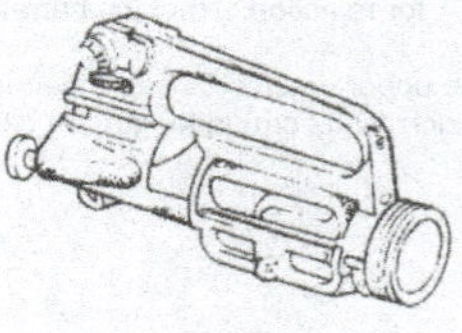

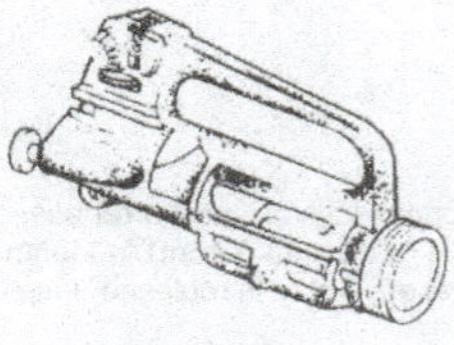

3-10. UPPER RECEIVER AND BARREL ASSEMBLY (CONT.).

c. REPAIR (CONT.)

NOTE
Do not feather edges

(e) Place a sheet of polyethylene (Item 32, app D), cut to size, over filled area Rub by hand to smooth.

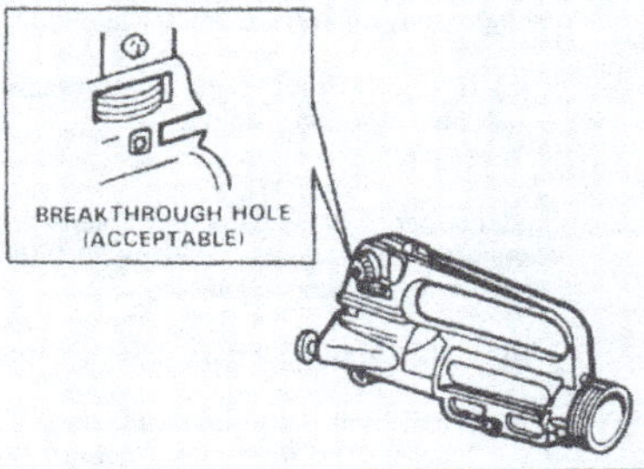

2. After curing, remove polyethylene sheet In accordance with Instructions by the manufacturer.

WARNING
When using solid film lubricant or dichloromethane, be sure the area is well ventilated.

CAUTION
Solid film lubricant is to be used only as an exterior surface protective finish and touchup If solid film lubricant comes in contact with recoiling parts or functional surfaces of the rifle, remove immediately by washing with technical dichloromethane.

3. Wash area with technical dichloromethane (methylenechloride) ; to remove dirt and grease, and foreign material.

4. Roughen area to be refinished with abrasive cloth and clean surface again Do not touch the area with fingers.

5. Repair shiny surfaces by spraying a coat of solid film lubricant in accordance with Instructions supplied by the manufacturer Dry 24 hours before handling.

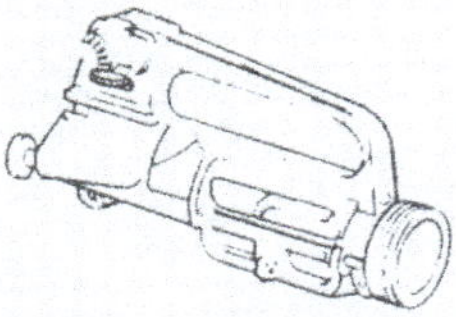

SHINY SURFACES (REPARABLE).

6. Straighten bent front sight guards (1) as follows:

NOTE

Remove spring before heating. (Heat will damage spring.) The sight post and plunger may be reused unless damaged.

(a) Remove front sight post, detent, and helical spring (see p 2-43).

NOTE

Use copper or brass caps (jaw Inserts on bench vise to prevent damage to front sight base (2) during clamping.

(b) Place front sight base (2) in a bench vise.

(c) Heat front sight guards (1) and bend with pliers. The front sight guards (1) should be put back as nearly as possible to the original position. Allow front sight housing to air cool.

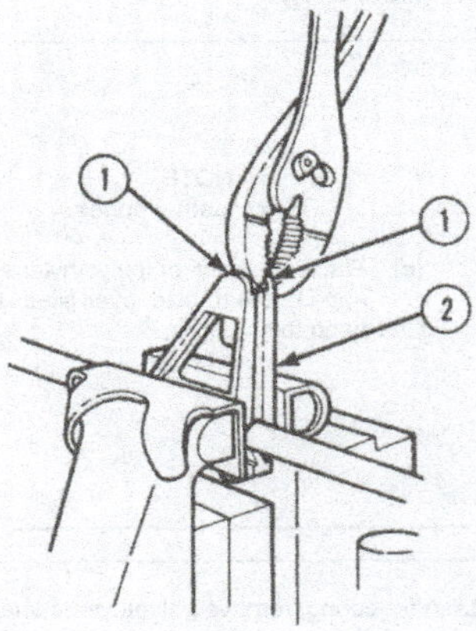

WARNING

Dry cleaning solvent is flammable and toxic and should be used in a well ventilated area. The use of rubber gloves is necessary to protect skin when washing rifle parts.

(d) Roughen any damaged surface of front sight guards with abrasive cloth and clean with dry cleaning solvent. Always wear rubber gloves when using dry cleaning solvent.

3-10. UPPER RECEIVER AND BARREL ASSEMBLY (CONT.).

c. REPAIR (CONT.)

CAUTION

Do not allow solid film lubricant to flow into front sight post threaded well.

(e) Apply solid film lubricant to cover the damaged finish.

(f) If front sight guards cannot be straightened utilizing the above procedures, reduce the rifle barrel assembly.

7. Slightly bent barrels may be straightened as follows'.

(a) Check straightness using straightness gage 8448202 (p 3 45) If the barrel falls the straightness test, and the gage remains in the barrel In the area of the front sight assembly, perform step (b) to determine If it may be straightened.

(b) With the gage remaining in the bore, hold the rifle in a vertical position with the end of the barrel into which the gage was inserted pointing up Ensure that if when the gage passes through the barrel it will not be damaged Using hand pressure' ONLY, flex the portion of the barrel between the front sight assembly and the compensator in all four directions (left, right, forward, and back) If the barrel is only slightly bent, the gage will drop through when the barrel is flexed in one of these directions Note the direction which allowed the gage to drop through the' barrel.

CAUTION

Remove the gage from the barrel before continuing

NOTE

If the gage does not pass through the barrel when it is flexed, replace the rifle barrel assembly.

(c) Place the barrel in a vise using appropriate protective laws . Clamp the barrel between the front sight assembly and the compensator approximately 1 inch (2.54 cm) from the front sight assembly The rifle barrel assembly should be in a horizontal position with the side noted in step (b) toward you.

CAUTION
Do not apply pressure to the receiver.

(d) Grasp the BARREL near the receiver so that when force is applied the barrel will flex In the same direction as noted In step (b).

(e) Give the barrel a sharp jerk of approximately 20 to 40 pounds of force (f) Remove the barrel from the vise and recheck straightness (step (al).

(g) If gage still will not pass through the barrel, perform step (b) to determine direction of bend If the barrel is still bent in the same direction as before, perform steps (c) through (f) using slightly more force. If the barrel Is now bent In the opposite direction, replace the rifle barrel assembly.

(h) If the gage passes freely through the barrel, the barrel shall be considered straight and continue In service.

(i) If the barrel has been straightened, the rifle must be targeted (p 3-45).

d. REASSEMBLY

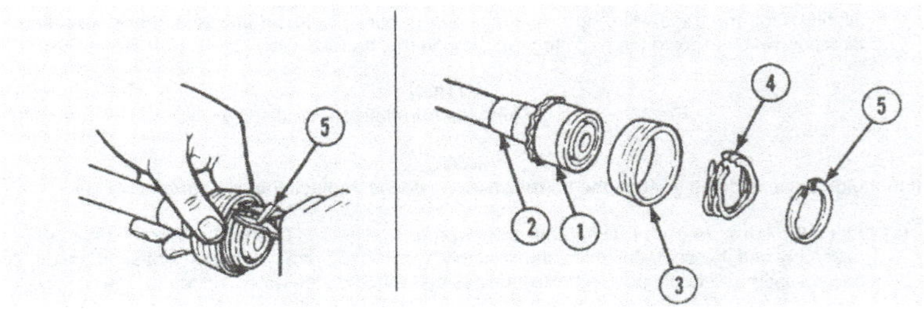

1. Position barrel nut (1) by sliding it to the rear of rifle barrel assembly (2) as far as possible
2. Slide handguard slip ring (3) over barrel nut (1).
3. Press slip ring spring (4) from both sides and insert It Into handguard slip ring (3)
4. Install retaining ring (5) against slip ring spring (4) using retaining ring pliers Snap retaining ring (5) to barrel nut (1).

NOTE
After cleaning, apply molybdenum disulfide grease to threads of barrel nut assembly before Installation.

3-10. UPPER RECEIVER AND BARREL ASSEMBLY (CONT).

d. REASSEMBLY (CONT)

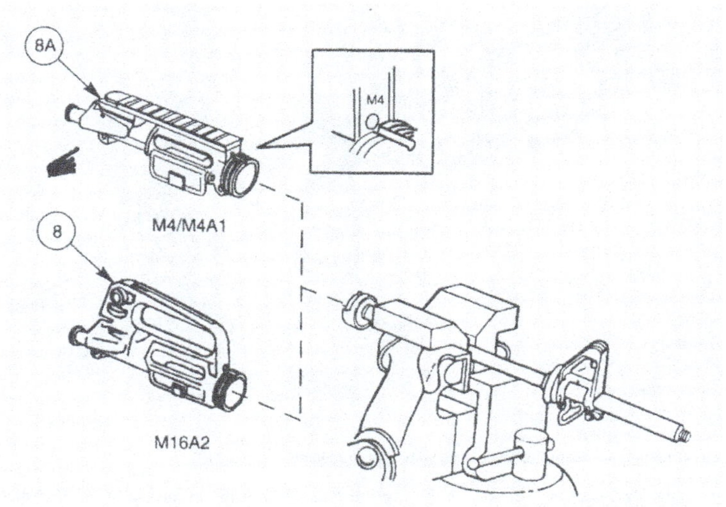

NOTE
The alignment pin (6) must not show any signs of looseness.

5. Position rifle barrel assembly (2) with alignment pin (6) up. Using barrel remover fixture (7), clamp rifle barrel assembly (2) in vise.

NOTE
The slot should fit the alignment pin perfectly with very little or no rotational play present. Note the play in a new rifle barrel assembly and new upper receiver assembly and use this as a guide.

6. Align upper receiver assembly (8) or (8A) using alignment pin (6) and the slot in upper receiver assembly (8) or (8A). Install over end of rifle barrel assembly (2).

7. Wipe upper receiver threads clean and ensure there are no burrs. Apply molybdenum disulfide grease to the threads before installation.

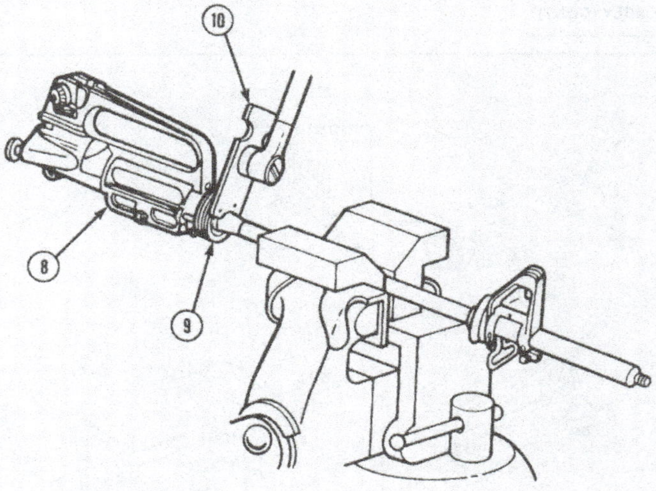

8. Engage threads of barrel nut assembly (9) with upper receiver assembly (8).

9. Using combination wrench (10) and torque wrench, torque barrel nut assembly (9) to 30 ft-lb (40.5 N-m). Torque is measured when both wrenches are used together.

NOTE

Three times torquing procedure provides for a better thread fit and prevents barrel nuts from becoming loose. Do not use the torque wrench for loosening.

10. Make certain all three drive pins on combination wrench are engaged with barrel nut assembly (91 Loosen and repeat torque operation. Then loosen the barrel nut again.

3-10. UPPER RECEIVER AND BARREL ASSEMBLY (CONT).

d. REASSEMBLY (CONT)

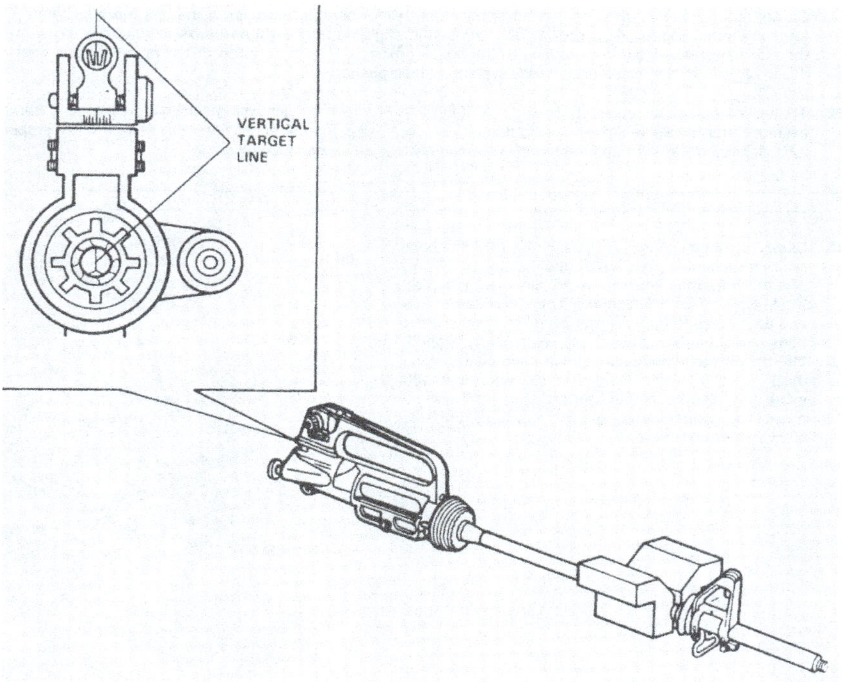

11. Front sight post must be Installed. Loosen the vise and align the bore on a distant vertical target Center the target In the bore from 12 o'clock through 6 o'clock The, front sight post should be on line and vertical with the target. Tighten vise Adjust the rear sight windage until a proper sight picture is obtained on the vertical target The rear sight aperture will be approximately In the center of the rear sight base If the rifle barrel assembly is properly aligned in the upper receiver assembly

NOTE

If rifle barrel assembly (usually new) is not properly aligned In the upper receiver assembly (usually an old part), excessive windage will be present and the upper receiver assembly will require replacement to obtain the proper fit between the alignment pin and slot.

CAUTION

Do not torque over 80 ft-lb (108 N-m) while tightening the barrel nut assembly to the next hole, to allow for proper alignment of gas tube.

NOTE

Do not attempt to hold the upper receiver assembly with a pry bar, however, If the rifle barrel assembly turns In the holding fixture, a pry bar may be used through the front sight assembly base to help prevent the rifle barrel assembly from turning In the holding fixture. Use care not to distort or bend front sight assembly or retaining pins. Use "buddy system" to hold pry bar.

12. Torque the barrel nut assembly again to 30 ft-lb (40.5 N-m) while maintaining sight alignment. The barrel nut assembly may be tightened beyond 30 ft-lb (40.5 N-m) to align the barrel nut assembly serrations for proper gas tube clearance Never loosen the barrel nut assembly to align for gas tube clearance.

13. Check alignment of barrel nut assembly (9) with upper receiver assembly (8). The front 8 inches (20.32 cm) of a gas tube may be used as an alignment tool (see Illustration). This Is Inserted Into the bolt carrier key and then Inserted into the rear of the receiver. If the parts of the barrel nut assembly are properly aligned, the tool will pass freely and lay top dead center along the top of the barrel. A number 15 twist drill (0.180 inch) may also be needed as an alignment tool. If necessary, tighten barrel nut assembly to next hole to allow proper alignment.

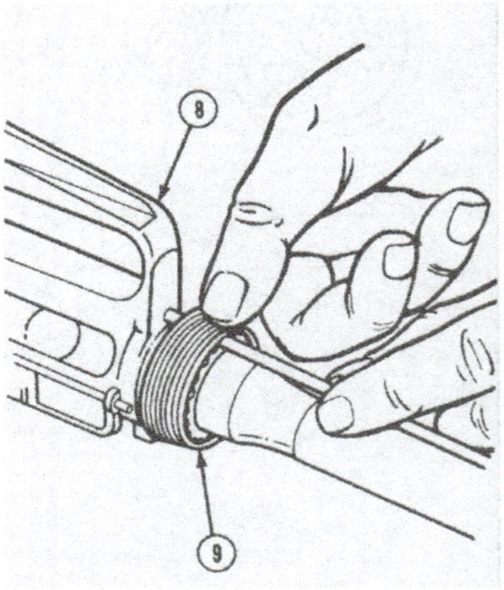

3-10. UPPER RECEIVER AND BARREL ASSEMBLY (CONT).

d. REASSEMBLY (CONT)

NOTE
The peel washer may be heated to remove thin sections. Always place thin section to rear.

14. Install peel washer (11) and compensator (12) on rifle barrel assembly (2).

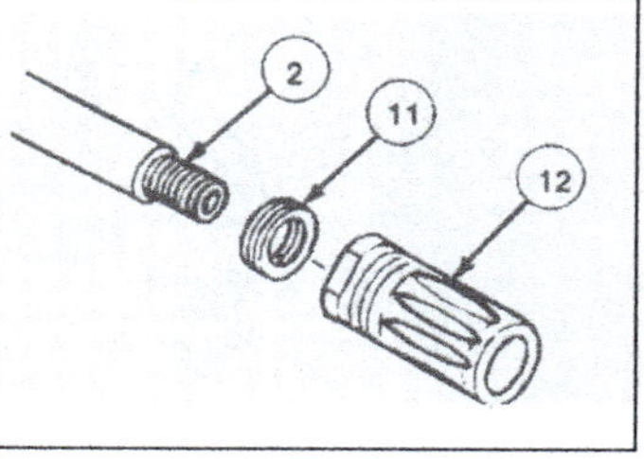

15. Torque compensator (12) to 15-20 ft-lb (20-27 N-m) using combination wrench (10) and torque wrench. Torque is measured when both wrenches are used together. Some portion of the openIng of the third or middle slot must be straight up (align with the front sight post) (12.1) at proper torque level. Thin sections of the peel washer may be removed or added as required. Save unused sections.

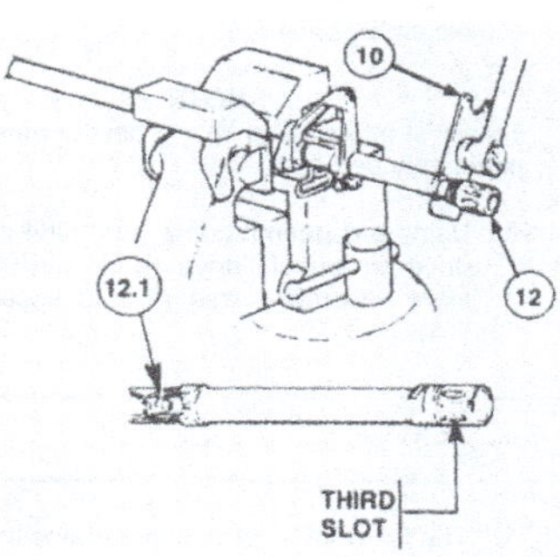

16. Slide gas tube (13) through the barrel nut assembly (9) and then slide forward, Inserting gas tube into hole In the front sight assembly base (14).

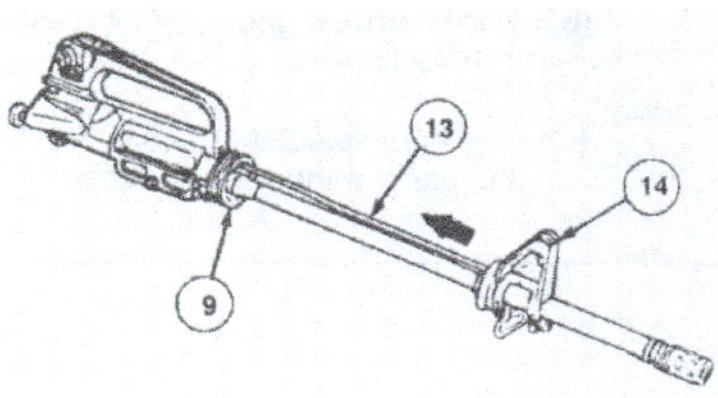

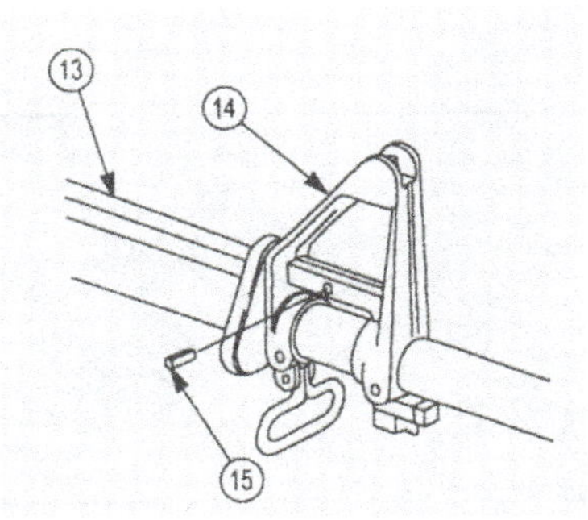

17. Align the holes in gas tube (13) and front sight assembly base (14).

NOTE

To assist in installing spring pin (1) modified nose pliers may be used.

18. Using ball-peen hammer and 5/64-inch diameter drive pin punch, drive spring pin (15) into front sight assembly base (14) to secure gas tube (13).

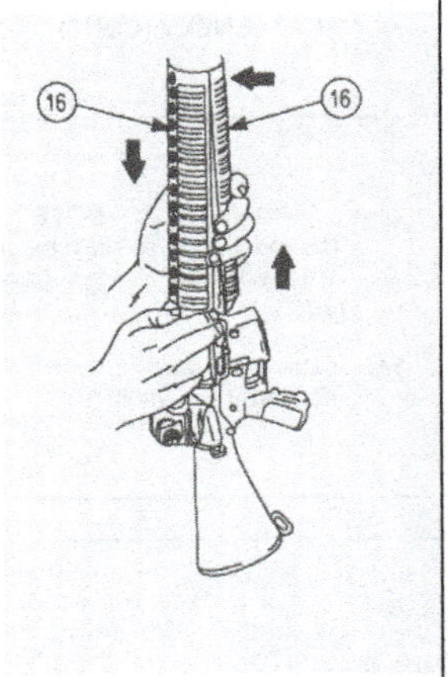

19. Install upper and lower handguard assemblies (16).

e. TEST

1. The following information pertains to the use of breech, bore, and other gages:

(a) All M1 6A2 barrels and chambers are chromed.

(b) Barrel erosion gage, PN 8448496 (normally used on M16A1 fully-chromed barrels), can be used to gage M16A2 barrels.

(c) The bore straightness gage, PN 8448202, is required for use on all barrels. The gage must pass through the barrel without being forced.

3-10. UPPER RECEIVER AND BARREL ASSEMBLY (CONT).

e. TEST (CONT)

2. Use barrel erosion gage (1) PN 8448496. Install key and bolt carrier assembly with bolt assembly and firing pin removed. Hold rifle vertical with receiver up. Insert gage into rear of key and bolt carrier assembly. The reject line must be read at the rear edge of the key and bolt carrier assembly.

3. If the reject mark passes beyond the rear surface of the key and bolt carrier assembly, the barrel is unserviceable and shall be replaced.

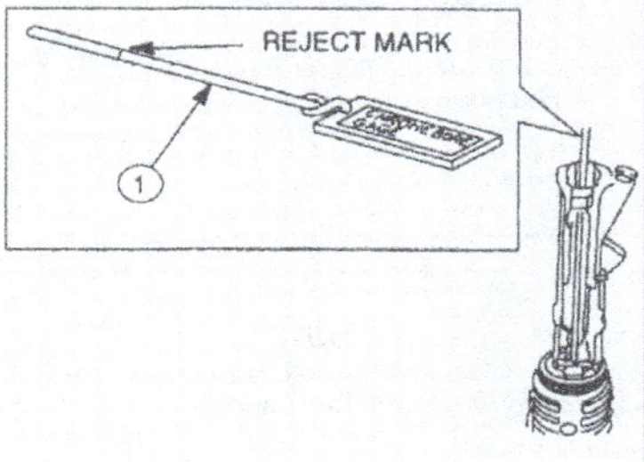

NOTE
Ensure barrel is clean prior to performing the following test.

4. Check straightness of bore using straightness gage (2) PN 8448202. Put gage in barrel. Tilt barrel and allow gage to fall through. Catch gage.

5. Gage must pass freely through barrel. If gage does not pass through barrel, recheck as follows. Hold upper receiver and barrel assembly (3) in vertical position with muzzle pointed down; insert gage into chamber end of barrel. Release gage and catch it as it exits muzzle end. If gage passes freely through the barrel, barrel is acceptable. If it does not, the barrel must be straightened or replaced. (See page 3-38 for straightening instructions/ procedures.)

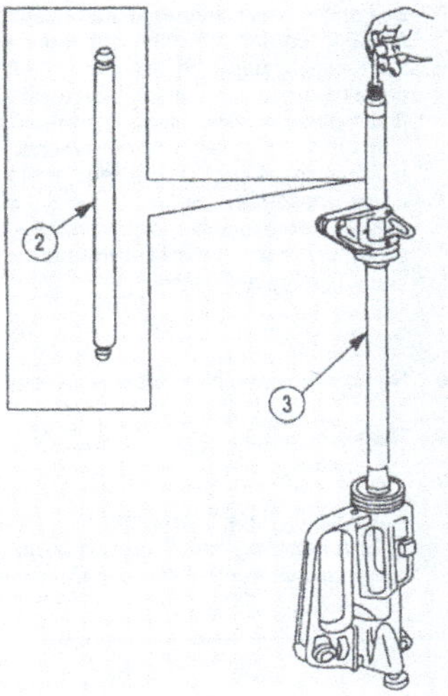

6. Assemble charging handle assembly (4), bolt assembly, and key and bolt carrier assembly (5) into upper receiver assembly (6).

7. Insert headspace gage (7) PN 7799734 in chamber.

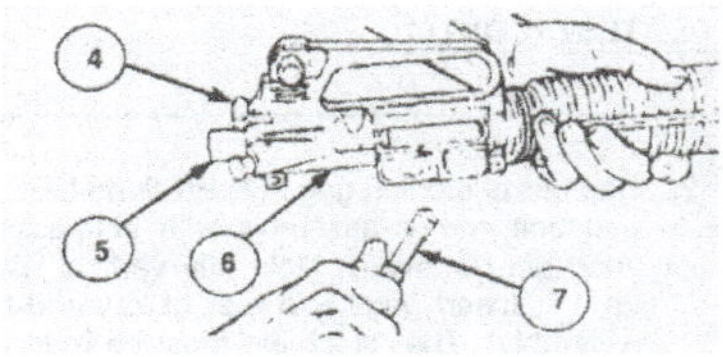

NOTE

For the purpose of this test "light finger pressure" is defined as 8 1/2 to 8 3/4 pounds.

8. Check headspace by pressing key and bolt carrier assembly (5) and charging handle assembly (4) forward using light finger pressure.

9. Bolt should not rotate to locked position. Key and bolt carrier assembly (5) must protrude from rear of upper receiver assembly (6) for proper headspace. If excessive headspace, first replace old bolt assembly with an unused bolt assembly and then recheck. If headspace is not corrected, replace rifle barrel assembly; then recheck with the original bolt assembly to determine if the bolt assembly is still good or if the bolt assembly should be replaced also.

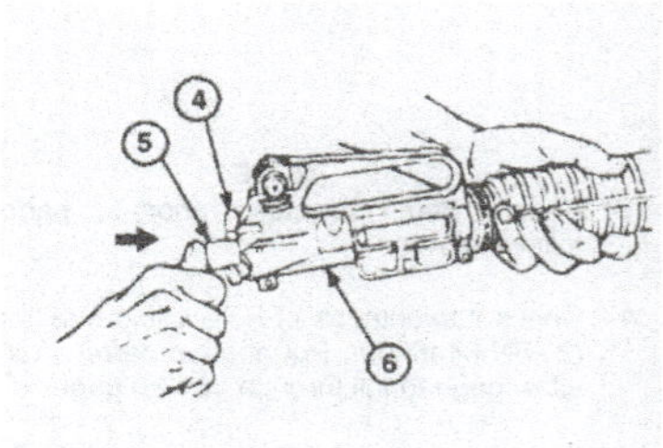

10. Remove key and bolt carrier assembly, bolt assembly, charging handle assembly, and headspace gage.

11. Reassemble rifle, refer to page 3-77.

NOTE

Rifles which have been rebarreled must be function-fired with seven rounds of 5.56mm ball ammunition. After rebarreling, the rifle must be targeted with three rounds of 5.56mm ball ammunition at 25 meter range using target. Refer to TM 9-1005-319-10 and FM 23-9

3-10. UPPER RECEIVER AND BARREL ASSEMBLY (CONT).

e. TEST (CONT)

12. (a) If the test fails using finger pressure, remove the gage and perform the test again as follows: With the muzzle down, stack 8 1/2 to 8 3/4 pounds of trigger weights (8) on a locally fabricated spacer/weight (9) on the bolt carrier assembly (10) Insert headspace gage (11) and test per above instructions.

(b) Remove trigger weights (8), spacer/weight (9), bolt carrier assembly (10), charging handle (12), and headspace gage (11).

(c) If excessive headspace, first replace bolt assembly and then recheck If headspace is not corrected, replace barrel assembly; then recheck with the original bolt to determine if the bolt is still good or If the bolt should be replaced also.

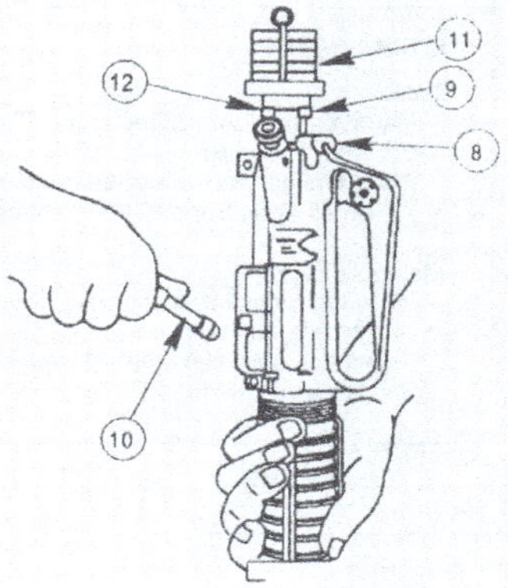

3-11. UPPER RECEIVER ASSEMBLY AND REAR SIGHT ASSEMBLY.

This task covers:

a. Disassembly
b. Inspection
c. Repair
d. Lubrication
e. Reassembly
f. Mechanical Zero Procedures (A.F. Only)

INITIAL SETUP:

Tools
- (ARMY) Small Arms Repairman Tool Kit (item 3, app B)
- Field Maintenance Basic Less Power Small Arms Shop Set (item 1, app B)

Materials/Parts
- Cleaner, lubricant, and preservative (CLP) (item 9, app D)
- Lubricant, solid film (item 21, app D)
- Index screw (9349065)

Equipment Conditions
- 3-29 Upper receiver assembly removed.

General Safety Instructions
- To avoid injury to your eyes, use care when removing and installing spring-loaded parts.
- When using solid film lubricant or dichloromethane, be sure the area is well ventilated.

a. DISASSEMBLY

WARNING

To avoid injury to your eyes, use care when removing and installing springloaded parts.

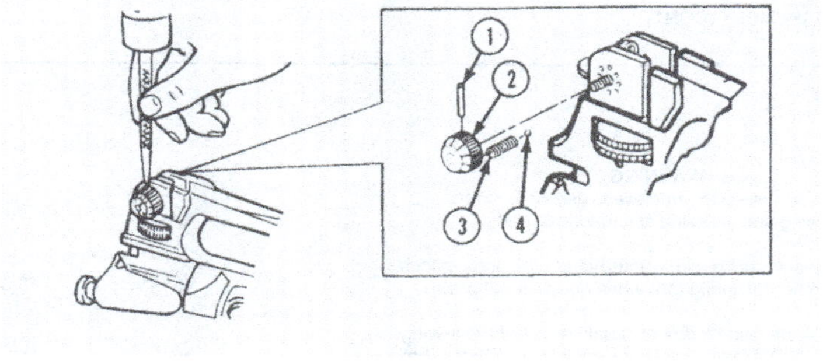

CAUTION

Be sure to catch small parts (1, 2, 3, and 4). AIR FORCE ONLY: Use magnet to keep from losing small parts (1, 3, and 4).

1. Drive out spring pin (1) using a hammer and 1/16 inch punch.
2. Catch rear sight assembly windage knob (2), helical spring (3), and ball bearing (4).

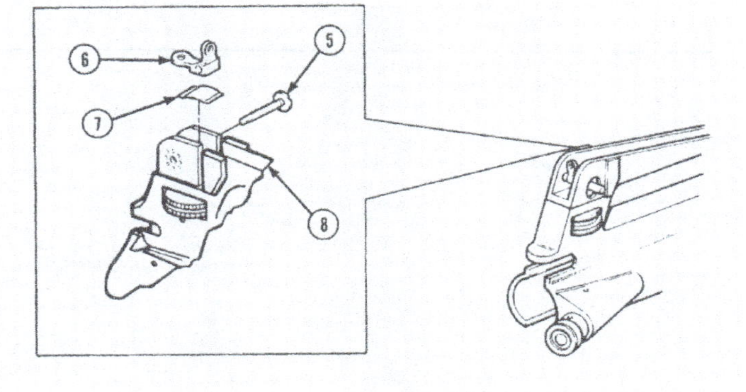

3. Using a flat-bladed screwdriver, remove rear sight assembly windage screw (5)
4. Remove sight aperture (6) and flat spring (7) from upper receiver (8).

3-11. UPPER RECEIVER ASSEMBLY AND REAR SIGHT ASSEMBLY (CONT).

a. DISASSEMBLY (CONT)

WARNING

To avoid injury to your eyes, use care when removing and installing spring-loaded parts.

5. Drive out spring pin (9) using a 3/32 Inch punch Catch helical spring (10) when punch is withdrawn.

6. Rotate elevation index (11) until rear sight assembly base (12) clears upper receiver (8). Catch ball bearing (13) and helical spring (14) as rear sight base clears.

7. Push elevation index (11) out with thumb using slight rotation motion. Catch ball bearing (15) and helical spring (16).

8. Use 1 16 Inch allen wrench to remove index screw (17). Discard Index screw (1 7) Separate elevation index (11) from elevation knob (18) by hand.

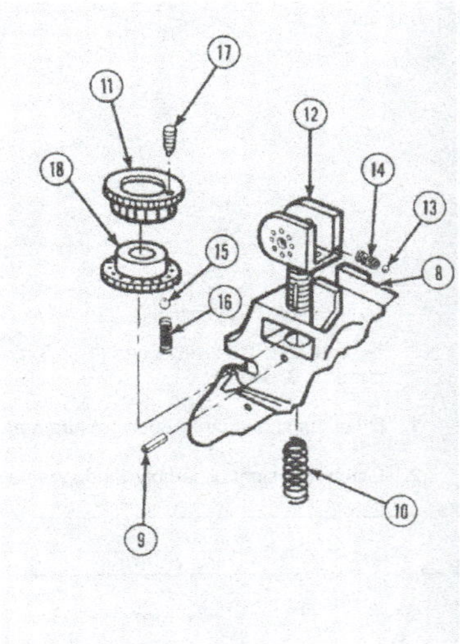

9. Remove pin (1 9) using 3,32 Inch drive pin punch and hand hammer .

WARNING

To avoid Injury to your eyes, use care when removing and installing spring-loaded parts.

10. Remove forward assist assembly (20) and helical spring (21) from upper receiver (8). For further disassembly of forward assist assembly see page 3-59.

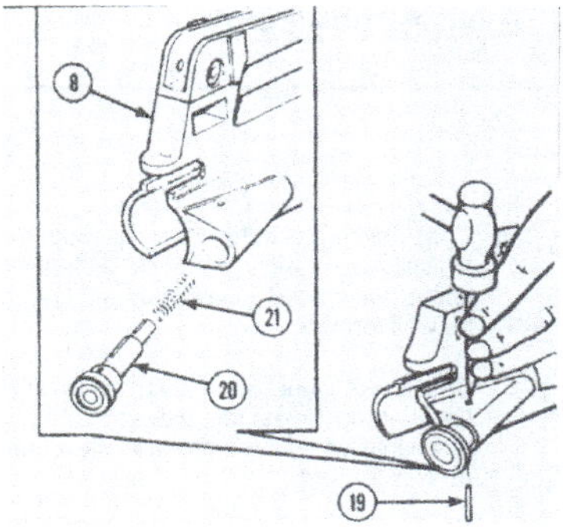

b. INSPECTION

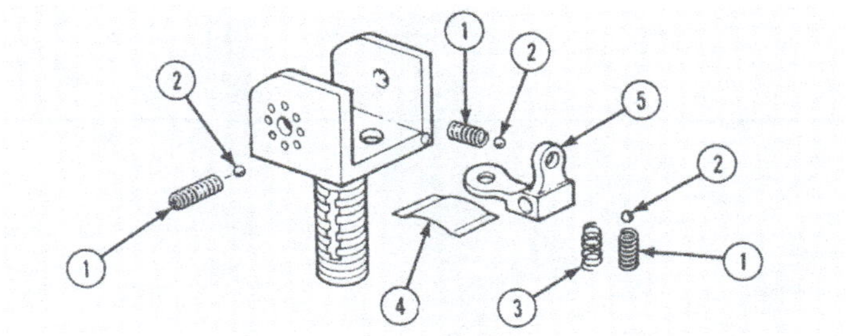

1. Check rear sight parts for serviceability. Inside of apertures should be round and distinct. Replace If defective
2. Visually inspect rear sight assembly helical springs (1) ball bearings (2), and helical spring (3) for breaks, bends, and missing parts Ball bearings should be smooth and round. Replace If defective
3. Check upper receiver for cracks, corrosion, and damage Clear drain hole with a piece of wire. Repair (p 3-29) or replace If defective.
4. Check that flat spring (4) retains sight aperture (5) firmly In either position. Replace flat spring (4) If sight aperture is not firm.

3-11. UPPER RECEIVER ASSEMBLY AND REAR SIGHT ASSEMBLY (CONT).

b. INSPECTION (CONT)

5. Check elevation Index 161 and windage knob (71 for legibility of markings Check underside of windage knobs for cracks. Detent indexing surfaces should be well formed.

6. Check rear sight assembly base (81 for serviceability Clear drain holes for springs Threaded portion of rear sight assembly base and elevation knob should be well formed.

7. Inspect rear sight guards for bends, if bent refer to the following page for repair procedures.

8. Inspect all parts for damage and wear Replace all defective parts.

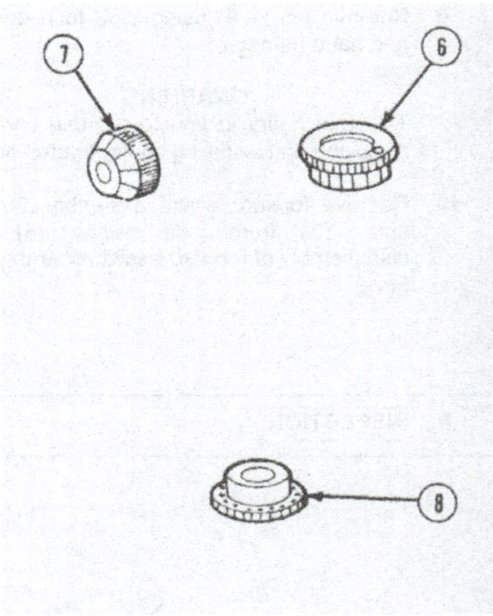

c. REPAIR

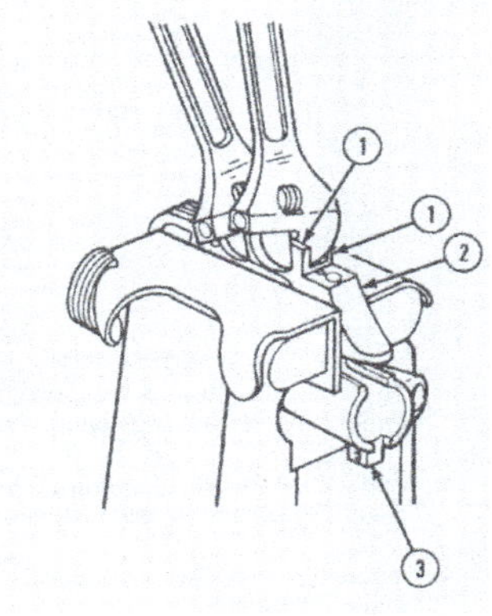

1. To straighten bent rear sight assembly guards (1), remove rear sight assembly components. Place carrying handle (2) In a vise using jaw clamps. Tighten vise to firmly hold upper receiver (3).

2. Using two adjustable wrenches, gradually bend guards (1) to straighten. When bending the guards (1), gradually bend beyond the straight point as the guard will partially return when bending pressure is stopped.

3. After straightening, use a flat file to remove any nicks, kinks, or burrs that remain on the inside of guards (1).

CAUTION

Do not use wire brush on aluminum surfaces

4. Apply solid film lubricant to brightened area for final protective coating.

5. Replace rear sight assembly components and check that rear sight assembly functions properly. If rear sight assembly functions check out, return upper receiver assembly to service.

6. If rear sight guards cannot be straightened utilizing the above procedures, replace the upper receiver

d. LUBRICATION

Lubricate upper receiver assembly and rear sight assembly. Apply CLP to helical springs and ball bearings (three each) and threaded portion of screws before installation. Lubricate helical springs and ball bearings through their respective drain holes.

3-11. UPPER RECEIVER ASSEMBLY AND REAR SIGHT ASSEMBLY (CONT).

e. REASSEMBLY

WARNING
To avoid injury to your eyes, use care when removing and Installing spring-loaded parts.

1. Apply CLP to helical spring (1) and forward assist assembly (2) and Install them into upper receiver (3).

2. Install spring pin (4) using 3, 32 Inch drive pin punch and hand hammer

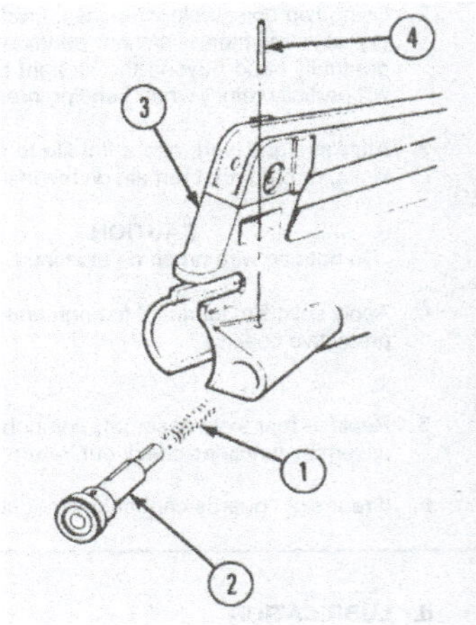

3. Assemble elevation knob (5), elevation index (6), and new index screw (7) using 1 16 Inch alien wrench. Do not overtighten Index screw as scale will require adjustment

4. Install ball bearing (8) and helical spring (9) using needle nose pliers or tweezers

5. Depress ball bearing (8) with a punch Inserted through access hole, and slide elevation knob assembly In upper receiver (3) from the side Center elevation knob assembly

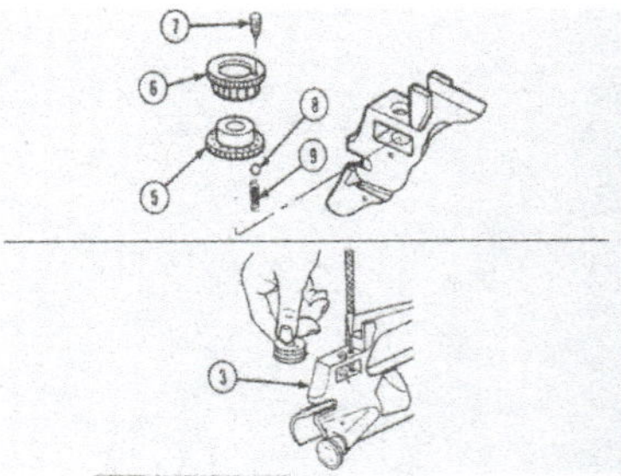

WARNING

To avoid injury to your eyes, use care when removing and installing spring-loaded parts.

NOTE

All springs are identical when new. Once disassembled from the rifle, their free length may vary due to different amounts of compression when installed. If the length of springs varies, use longest as #22, and the shortest as #9

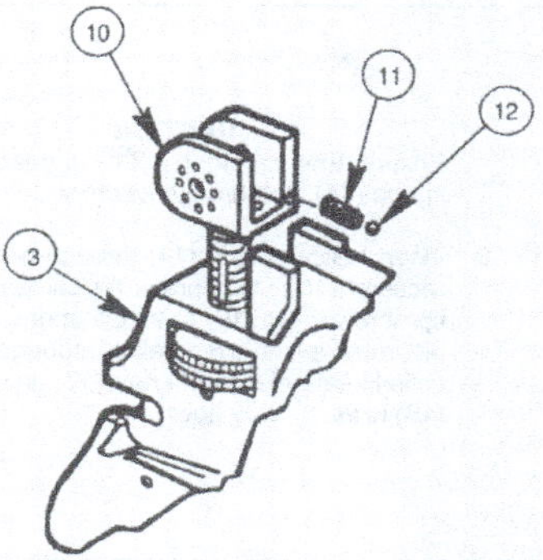

6. Insert threaded portion of rear sight assembly base (10) into upper receiver (3) and rotate elevation knob assembly until threads engage.

7. Insert helical spring (11) and ball bearing (12) in their hole as rear sight assembly base is lowered into upper receiver as elevation knob assembly is further rotated. Rotate elevation knob assembly until rear sight assembly base is all the way down. Then come up 22 clicks before installing spring pin. Check spring action of helical spring (11) on upper receiver.

3-11. UPPER RECEIVER ASSEMBLY AND REAR SIGHT ASSEMBLY (CONT).

e. REASSEMBLY (CONT)

CAUTION

Ensure that spring pin (1 3) passes over helical spring (14), not through its coils

8. Insert helical spring (14) through underside of upper receiver (15). Compress helical spring with a small tip screwdriver (16) to install spring pin (1 3) Spring pin must pass over helical spring, not through its coils Rotate elevation knob (17) until rear sight base (18) Is all the way down.

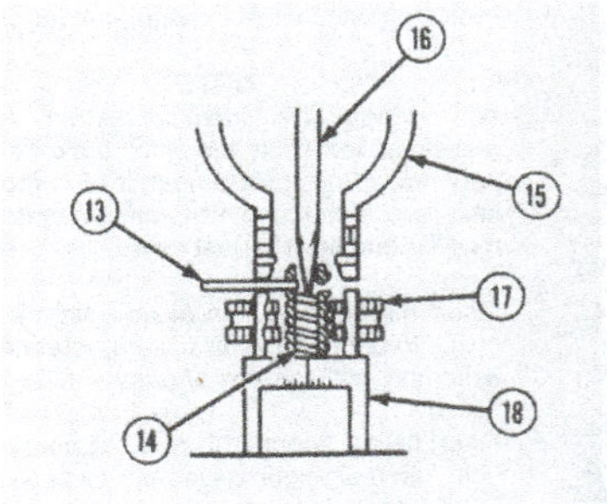

9. Install flat spring (19) and sight aperture (20) In rear sight assembly base (10). Install windage screw (21) with screwdriver.

10. Insert helical spring (22) and ball bearing 123) In windage knob (24)

NOTE

Tilt upper receiver toward wind age knob during positioning to prevent loss of ball bearing.

11. Position windage knob on shaft of windage screw (211 Align holes In windage knob with hole of shaft In windage screw and hold In alignment with a small screwdriver or punch Install spring pin (25).

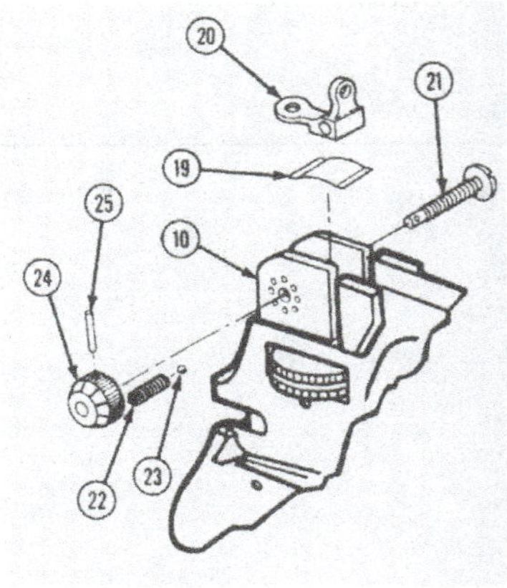

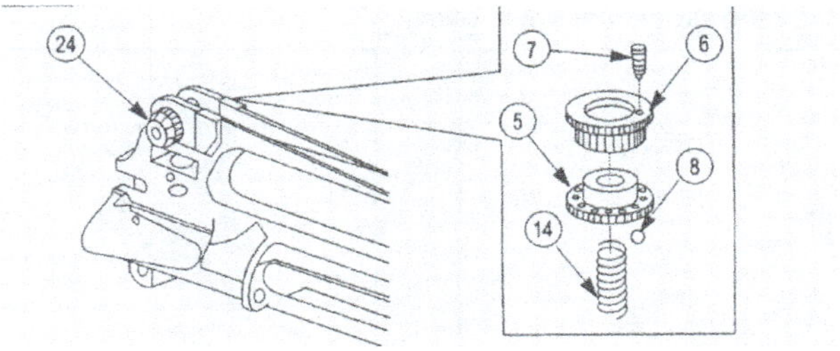

12. Rotate and test elevation index (6) and windage knob (24) for ease of functioning.

13. Inspect elevation knob zero as follows:

(a) Rotate elevation knob (5) counterclockwise until the rear sight assembly is all the way down. If a whole click is not felt as the rear sight assembly stops, the rear sight assembly has bottomed out and will not pivot freely.

(b) Position elevation knob back slightly to its last whole click so the rear sight assembly base is under tension of the ball bearing (8) and helical spring (14). The 300 meter mark should align with the mark on to receiver.

(c) If the 300 meter mark is not aligned with the mark on receiver, slip the range scale in the following manner:

(1) Position the 300 meter mark with the mark on the receiver.

(2) Insert a 1/16 inch allen wrench through the access hole of the rear sight assembly base and into the index screw (7).

(3) Loosen the index screw three turns and leave the wrench in place.

(4) Rotate lower portion of elevation knob

(5) counterclockwise until it stops (range scale should not have moved). Elevation knob should be positioned on its last whole click. (5) Tighten index screw (7) and remove wrench.

(6) Check for proper setting.

NOTE

After the rifle is assembled, the rear sight is centered, and placed at the 300 meter mark, perform the owing check: While looking at a light background, obtain good sight alignment. If the hole in the rear sight aperture appears oval Instead of round, the rear sight base or upper receiver should be replaced. To determine which part requires replacement, replace the rear sight base first. If this does not resolve the problem, replace the upper receiver.

14. Assemble rifle, refer to page 3-77

3-11. UPPER RECEIVER ASSEMBLY AND REAR SIGHT ASSEMBLY (CONT).

f. MEACHANICAL ZERO PROCEDURES (A.F. ONLY)

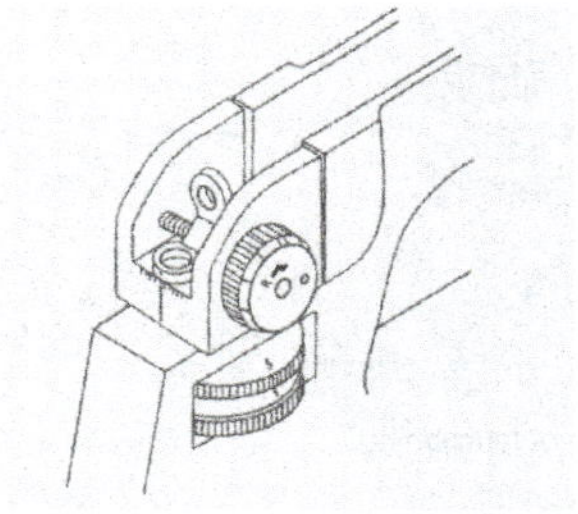

1. Center rear sight by moving windage knob in the appropriate direction.

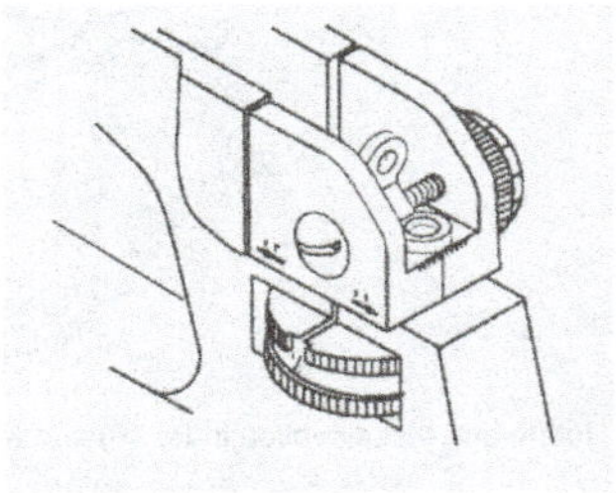

2. Always push in on windage screw head after making rear sight adjustments.

3. Visually check rear sight to ensure it is centered after making adjustments. Also, ensure the rear sight is set in the shortrange (unmarked aperture) position.

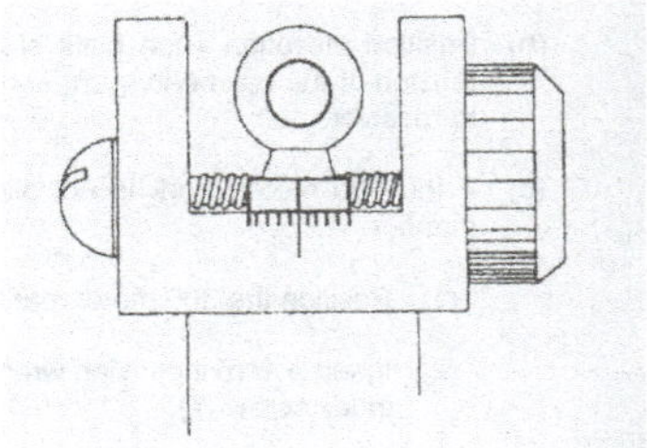

NOTE

This procedure, when used in conjunction with front sight mechanical zero adjustment (p 2-54), will give an approximate battle site zero to most M16A2 rifles. The above steps can also be used before firing a new or newly assigned rifle. Use the procedure to check rifles stored in preferred packaging during routine inspections. This will help ensure people armed with the rifles will stand a better chance of hitting an enemy if the rifles must be used before a live fire zero can be made. Whenever possible, zeroing of the rifle should be accomplished using ball ammunition on a 25 meter zeroing target using the "L" aperture.

After the rifle is assembled, the rear sight is centered, and placed at the 300 meter mark, perform the following check: While looking at a light background sight through the rear sight and obtain good sight alignment with the front sight placed in the center of the rear sight aperture. If while performing this check the hole in the rear sight aperture appears oval instead of round, the rear sight base or upper receiver should be replaced. To determine which part requires replacement, replace the rear sight base first. If this does not resolve the problem, replace the upper receiver.

3-12. FORWARD ASSIST ASSEMBLY.

This task covers:

a. Disassembly **c.** Repair **e.** Reassembly
b. Inspection **d.** Lubrication

INITIAL SETUP:

Tools
(ARMY) Small Arms Repairman Tool Kit (item 3, app B)
Field Maintenance Basic Less Power Small Arms Shop Set (item 1, app B)

Materials/Parts
Cleaner, lubricant, and preservative (CLP) (item 9, app D) | a

Equipment Conditions
3-48 Forward assist assembly removed

General Safety Instructions
To avoid injury to your eyes, use care when removing and installing spring-loaded parts

a. DISASSEMBLY

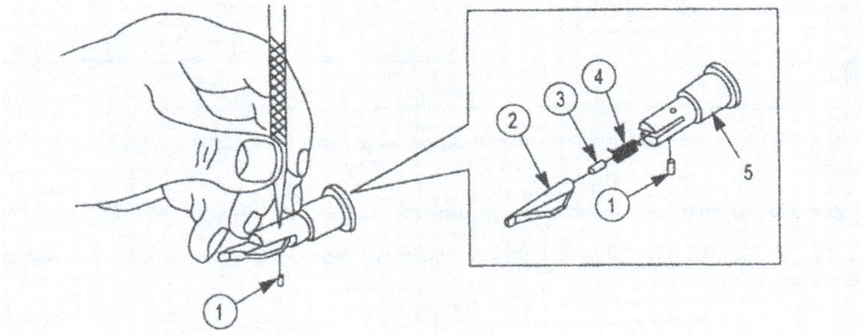

WARNING
To avoid injury to your eyes, use care when removing and installing spring-loaded parts.

1. Drive out spring pin (1) using 1/16 inch drive pin punch and hand hammer

2. Remove forward assist pawl (2), pawl detent (3), and helical spring (4) from plunger assembly (5)

3-12. FORWARD ASSIST ASSEMBLY (CONT)

b. INSPECTION

1. Inspect forward assist pawl (1) for burrs, chips, and cracks. Minor burrs may be removed using fine files or stones, as required. Replace forward assist pawl if defective.

2. Inspect pawl detent (2) for burrs and cracks. Minor burrs may be removed using fine files or stones, as required. Replace pawl detent if defective.

3. Inspect helical spring (3) for kinks, breaks, and wear. Replace helical spring if defective.

4. Inspect plunger assembly (4) for wear, burrs, chips, and breaks. Minor burrs may be removed using fine files or stones, as required. Replace forward assist assembly if defective.

5. Inspect spring pin (5) for wear. Replace if defective.

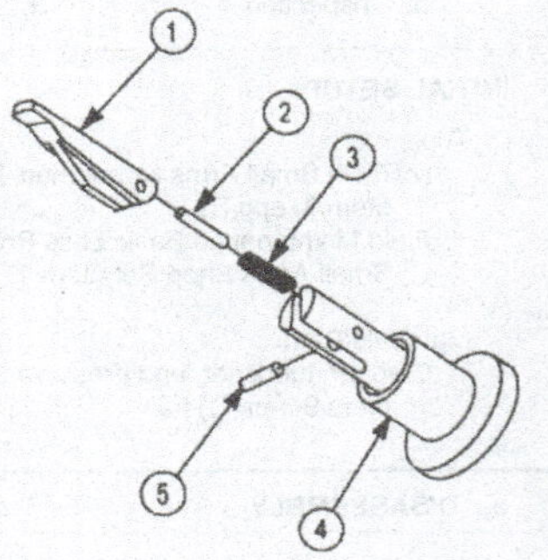

c. REPAIR

1. Repair forward assist pawl using fine files or stones, as required, to smooth burrs. Do not deform forward assist pawl.

2. Repair pawl detent using fine files or stones, as required, to smooth burrs. Do not deform pawl detent.

3. Repair plunger assembly using fine files or stones, as required, to smooth burrs. Do not deform plunger assembly.

d. LUBRICATION

Lubricate helical spring, pawl detent, and forward assist pawl with CLP (p 2-33) before installation.

e. REASSEMBLY

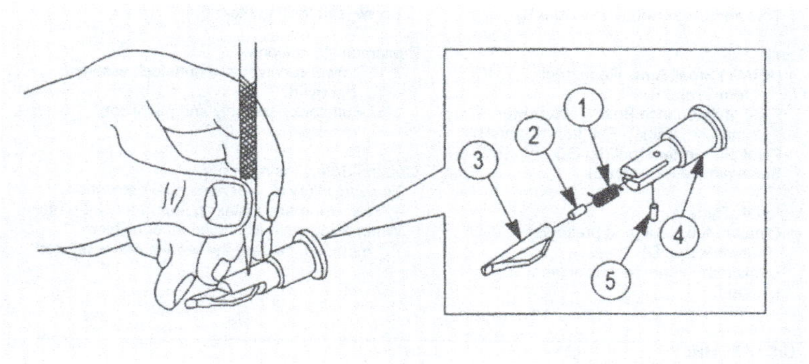

WARNING
To avoid injury to your eyes, use care when removing and installing spring-loaded parts.

1. Install helical spring (1), pawl detent (2), and forward assist pawl (3) into plunger assembly (4).
2. Align holes and install spring pin (5) using 1/16 inch drive pin punch and hand hammer. Spring pin must be flush or slightly below flush after reassembly.
3. Assemble rifle, refer to page 3-77.

3-13. LOWER RECEIVER AND BUTTSTOCK ASSEMBLY.

This task covers:

a. Disassembly
b. Inspection
c. Repair
d. Test
e. Reassembly

INITIAL SETUP

Test Equipment
Tool and Gage Set (item 2, app B)

Tools
(ARMY) Small Arms Repairman Tool Kit (item 3, app B)
Field Maintenance Basic Less Power Small Arms Shop Set (item 1, app B)
Pivot pin removal tool (fig E-3, app E)
Slave pin (fig E-5, app E)

Materials/Parts
Cleaner, lubricant, and preservative (CLP) (item 9, app D)
Dichloromehtane, technical (item 15, app D)
Lubricant, solid film (item 21, app D)
Screw, self-locking (item 6, p C-11)

Equipment Conditions
3-15 Lower receiver and buttstock assembly removed
2-57 Buttstock assembly and pistol grip removed

General Safety Instructions
To avoid injury to your eyes, use care when removing and installing spring-loaded parts.
When using solid film lubricant or dichloro methane, be sure the area is well ventilated.

a. DISASSEMBLY

WARNING

To avoid injury to your eyes, use care when removing and installing spring-loaded parts.

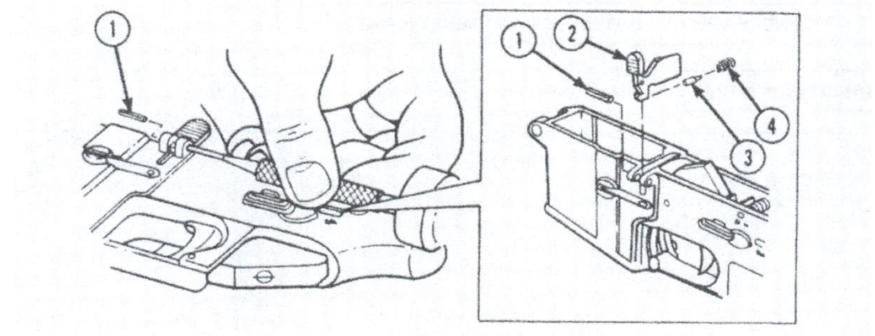

1. Remove spring pin (1) using 3/ 32 inch drive pin punch and hand hammer.
2. Remove bolt catch (2), bolt catch plunger (3), and bolt catch spring (4).

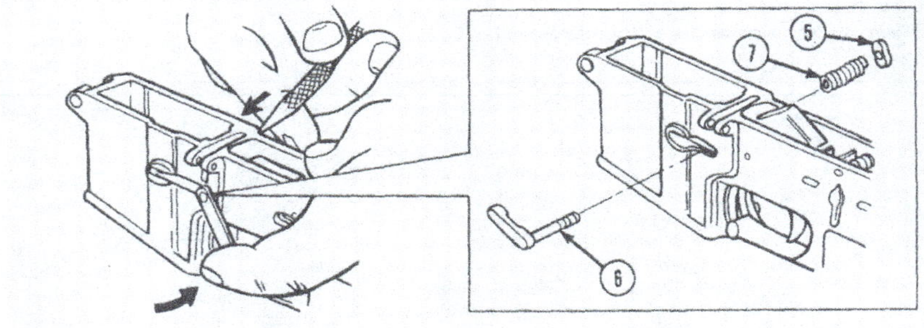

WARNING

To avoid injury to your eyes, use care when removing and installing spring-loaded parts.

3. Using drive pin punch, press in on magazine button (5) and turn magazine catch 16) counterclockwise to unscrew and remove.
4. Remove magazine button (5) and magazine catch spring (7).

3-13. LOWER RECEIVER AND BUTTSTOCK ASSEMBLY (CONT).

a. DISASSEMBLY (CONT)

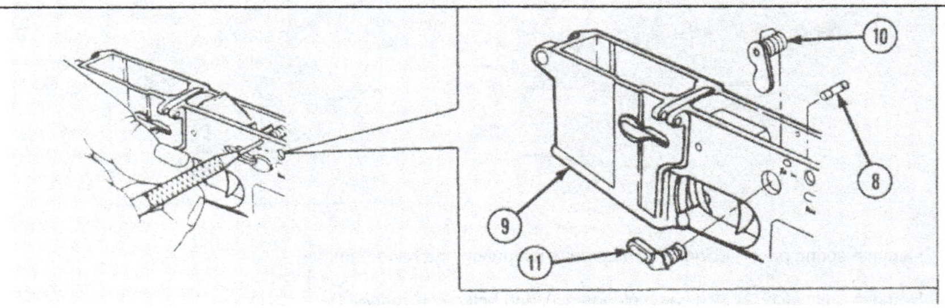

5. Use a rounded tip drive pin punch to push automatic sear pin (8) out of lower receiver (9).

NOTE
To remove automatic sear, selector lever must be positioned to BURST (if installed).

6. Remove automatic sear (10) and selector lever (11).

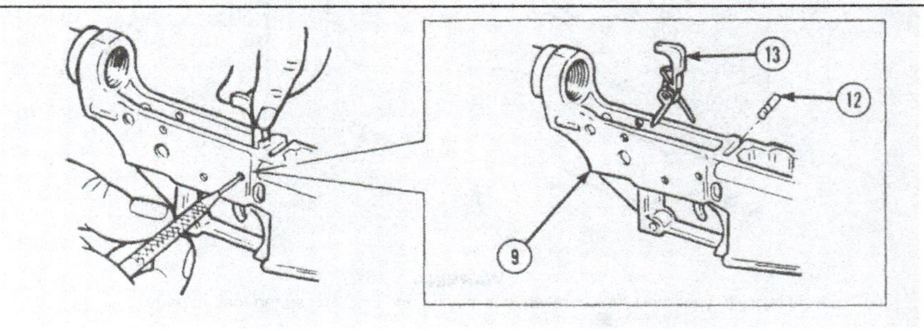

NOTE
To remove (hammer assembly should be forward), place selector lever (if installed) to SEMI position.

7. Use drive pin punch to push hammer pin (12) from lower receiver (9).

8. Remove hammer assembly (13). If further disassembly is required, see page 3-73.

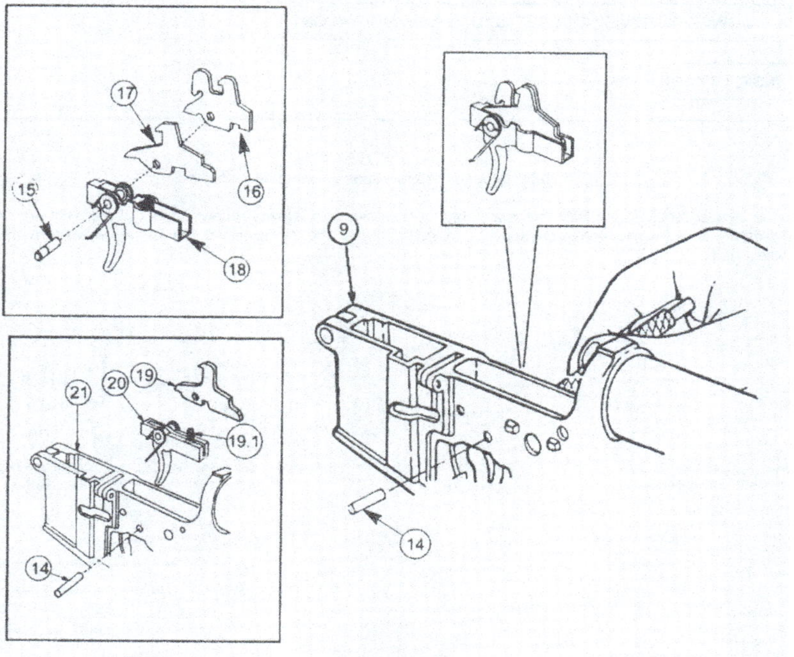

NOTE

Use of fabricated slave pin will allow removal of the following parts as a unit.

9. Remove trigger pin (14) by pushing from the left side of lower receiver (9) with fabricated slave pin (15) and a drive pin punch.

10. Remove semiautomatic disconnector (16), burst disconnector (17), and trigger assembly (18). If further disassembly of trigger assembly is required, see page 3-75.

M4A1 ONLY

11. Remove disconnector (19), disconnector spring (19.1), and trigger assembly (20) from lower receiver and extension subassembly (21).

3-13. LOWER RECEIVER AND BUTTSTOCK ASSEMBLY (CONT).

b. INSPECTION

NOTE

Refer to page 3-35 for repair of corroded surfaces.

Inspect lower receiver and buttstock assembly for corrosion in the lower receiver lobes of the pivot area or hinge pin area. If extensive corrosion appears in these areas, the receiver will not be repaired; turn rifle in for replacement.

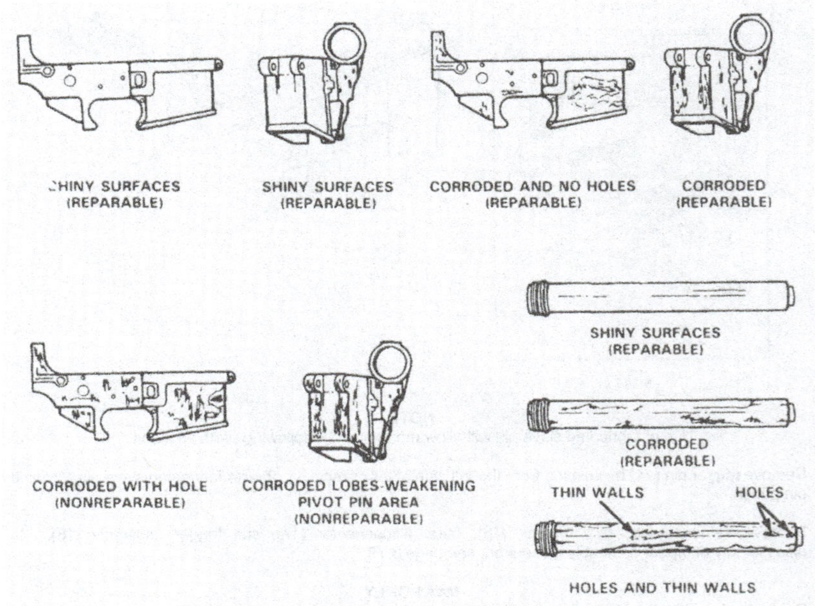

WARNING

When using solid film lubricant or dichloromethane, be sure the area is well ventilated.

CAUTION

Do not use a wire brush on aluminum surfaces.

NOTE

If a weapons lower receiver is missing one third or more of its exterior protective finish, resulting in an unprotected, light reflecting surface, it is candidate for overhaul. This missing finish will be considered a shortcoming. This shortcoming requires action to obtain a replacement weapon. Once a replacement has been received, evacuate the original weapon to depot for overhaul.

Solid Film Lubricant (SFL) is the only authorized touchup for the weapon and may be used on up to one third of the exterior finish of the weapon. **FOR ARMY CONUS USE ONLY AND AIR FORCE TRAINING WEAPONS ONLY:** SFL may be used as a touchup without limitation on the upper receiver and barrel assembly. This is to say that units which DO NOT fall under the category of Divisional Combat Units or rapid deployment type units may have up to 100 percent of the exterior surface of the upper receiver and barrel assembly protected with SFL if necessary.

c. REPAIR

Repair or replace all parts of lower receiver and buttstock assembly if defective.

3-13. LOWER RECEIVER AND BUTTSTOCK ASSEMBLY (CONT).

d. TEST

1. With the upper receiver attached to the lower receiver, and the pivot pin and takedown pins in place, perform the following test:

   (a) Apply hand pressure to push the upper receiver as far to one side as possible.

   (b) Attempt to insert a 0.020 inch thickness gage between the pivot pin lugs of the upper and lower receivers.

   (c) If the thickness gage penetrates to the pivot pin at all accessible locations, repair by replacement of the upper receiver (see (b) below) or replacement of rifle is required.

2. If the rifle fails the above test, remove the upper receiver and install a "NEW" upper receiver and perform the test again.

3. If the rifle now passes the above test, it shall be considered serviceable and continue in use.

4. If the rifle falls the test with a new upper receiver, this failure shall be considered a shortcoming. This shortcoming requires action to obtain a replacement rifle. Once a replacement has been received, evacuate the original rifle to depot for overhaul.

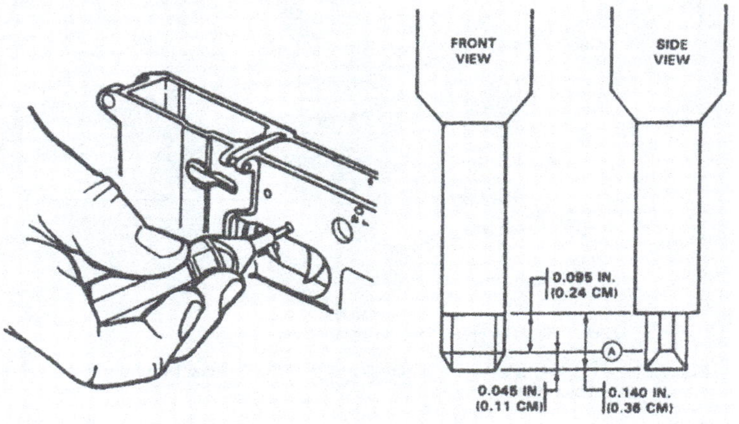

NOTE

If the lower receiver is not disassembled, visually inspect for broken or damaged parts, and to ensure that the hammer and trigger springs are correctly installed before beginning this test. It is not necessary to disassemble the lower receiver for the sole purpose of this visual Inspection. If broken or damaged parts are found, disassemble (p 3-62) and repair as authorized.

5. Test two hammer pin holes and two trigger pin holes using not-go plug gage PN 12006472. This test may be conducted by disassembly of the lower receiver (p 3-62) or by pushing the pin far enough to disengage the end of the pin from the side of the receiver which is being tested. If the lower receiver is not disassembled and the not-go plug gage enters any hole to first shoulder (A), the lower receiver must be disassembled and all four holes must be tested again.

6. Gently insert the not-go plug gage and rotate it 180 degrees. If the not-go plug gage passes through any one of the four pin holes, the rifle is unserviceable and will be turned in for replacement. The gage must extend through the wall thickness to be unserviceable.

7. After completion of gaging operation, visually inspect hammer and trigger springs to ensure proper location of spring legs.

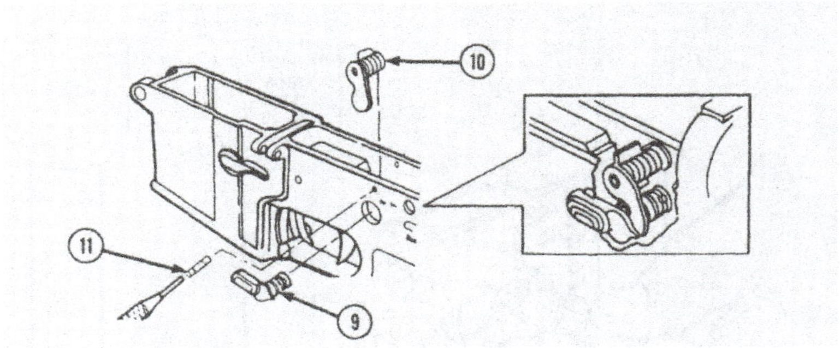

NOTE

Hammer assembly should be cocked prior to installing the selector lever.

Selector lever (9), if installed, must be positioned to BURST. Long leg of automatic sear (10) spring must rest on top of selector lever.

5. Install selector lever (9) and automatic sear (10).

6. Install automatic sear pin (11) (install on the right side) into receiver using drive pin punch. Push in until flush.

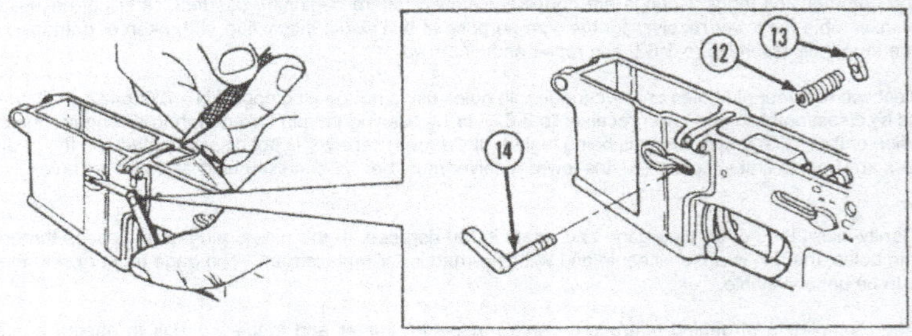

7. Install magazine catch spring (12) and magazine button (13).

NOTE

Drive pin punch should be larger than hole in magazine button.

8. Install magazine catch (14). Push in on magazine button (13) using a drive pin punch and turn magazine catch (14) clockwise until threaded end of magazine catch is flush with magazine button head.

3-13. LOWER RECEIVER AND BUTTSTOCK ASSEMBLY (CONT).

e. REASSEMBLY (CONT)

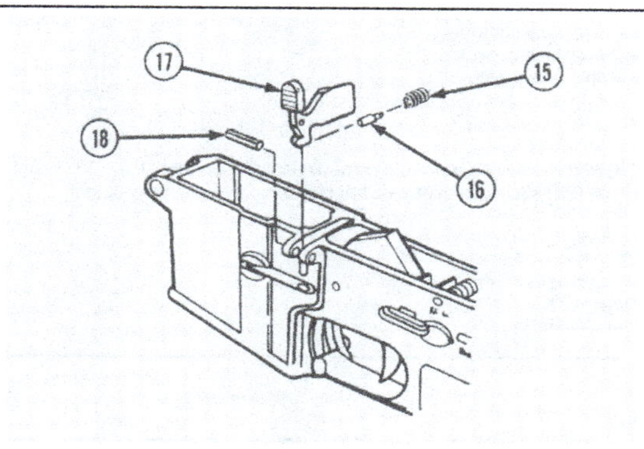

9. Install bolt catch spring (15), bolt catch plunger (16), and bolt catch (17).

10. Secure by installing spring pin (18) using 3/32 inch drive pin punch and hand hammer.

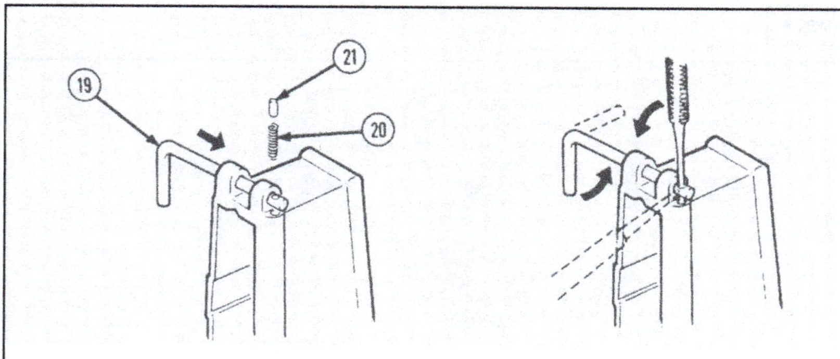

11. Install fabricated pivot pin installation tool (19). Insert helical spring (20) and pivot pin detent 21). Compress pivot pin detent in recess with punch and rotate tool. Remove punch.

WARNING
To avoid injury to your eyes, use care when removing or installing spring-loaded parts.

NOTE
Rounded end of pivot pin detent must be in the groove of the pivot pin (22) when assembly is complete.

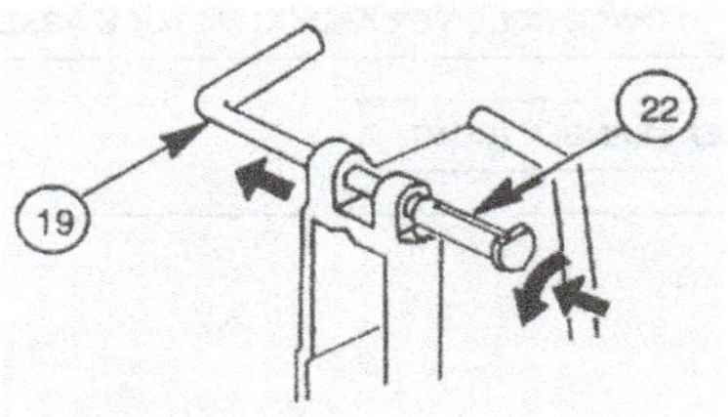

12. Position pivot pin (22) and removing fabricated pivot pin installation tool (19) while maintaining pressure, slide pivot pin (22) into hole. Rotate pivot pin to receive pivot pin detent.

13. Assemble rifle, refer to page 3-82.

3-14. HAMMER ASSEMBLY

This task covers:
a. Disassembly
b. Inspection/Repair
c. Reassembly

INITIAL SETUP

Equipment Conditions
3-62 Hammer assembly removed

a. DISASSEMBLY

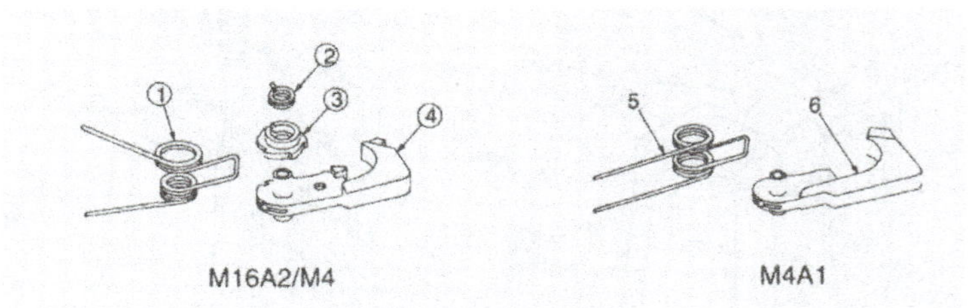

NOTE
M16A2 burst cam is black. M4 burst cam is nickel colored (shiny).

(M16A2/M4) remove hammer spring (1), cam clutch spring (2), and burst cam (3) from hammer and hammer pin retainer assembly (4). **(M4A1)** remove hammer spring (5) from hammer (6).

3-14. HAMMER ASSEMBLY (CONT).

b. INSPECTION/REPAIR

1. Inspect hammer spring for deformities, breaks, and bends. Pay special attention to the large coil. Replace hammer spring if defective.
2. Inspect cam clutch spring and burst cam for deformities, breaks, and bends; replace if defective.
3. Inspect hammer and hammer pin retainer assembly for chips and breaks. Hammer pin should click home under strong finger pressure. Install hammer pin into hole in hammer to check spring retention of the hammer pin. Replace hammer and hammer pin retainer assembly if defective.

c. REASSEMBLY

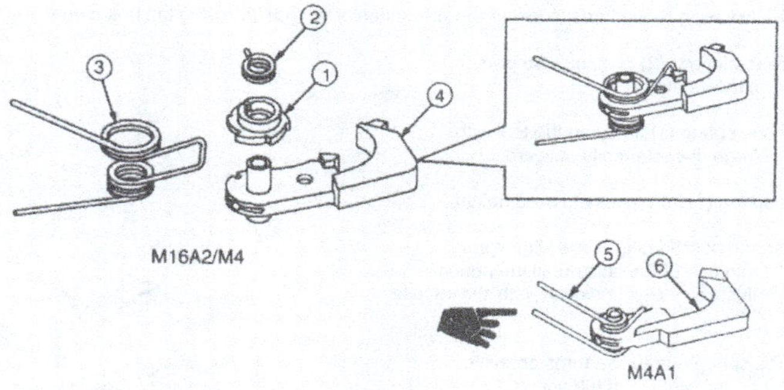

NOTE

M16A2 burst cam is black. M4 burst cam is nickel colored (shiny).

Cam clutch spring should be assembled with bend to the inside. The large loop of the hammer spring should be assembled over the burst cam.

(M16A2/M4) install burst cam (1), cam clutch spring (2), and hammer spring (3) on hammer and hammer pin retainer assembly (4). **(M4A1)** install hammer spring (5) on hammer (6).

3-15. TRIGGER ASSEMBLY AND TRIGGER SUBASSEMBLY (CONT).

b. INSPECTION/REPAIR

1. Inspect trigger spring for kinks, deformities, and weakness. Replace if defective.
2. Inspect disconnector springs for deformities, bends, breaks, and weakness. Replace if defective.
3. Inspect trigger for chips, wear, and cracks. Inspect for damaged searing surface on the trigger nose. Replace if defective.

c. REASSEMBLY

M16A2 AND M4 ONLY
NOTE
Use bolt carrier key tool to install disconnector springs.

1. Use the following procedures to install two disconnector springs (1) using the bolt carrier key tool:

(a) Secure the trigger (2) in a soft vise jaws or similar device.

(b) Place one spring (1) firmly on the tool with large diameter coils outward.

(c) Press spring (1) into recess to solid height.

(d) Hold spring at solid height and slide spring into one of the holes until the punch is flush and perpendicular with the recess wall.

(e) Turn the spring one to two turns opposite of coil winding of the spring.

(f) Discontinue winding when an audible click or snap is heard or felt. This indicates that the spring is seated.

(g) Hold the spring (1) in place when removing the tool to avoid unseating or damaging the spring.

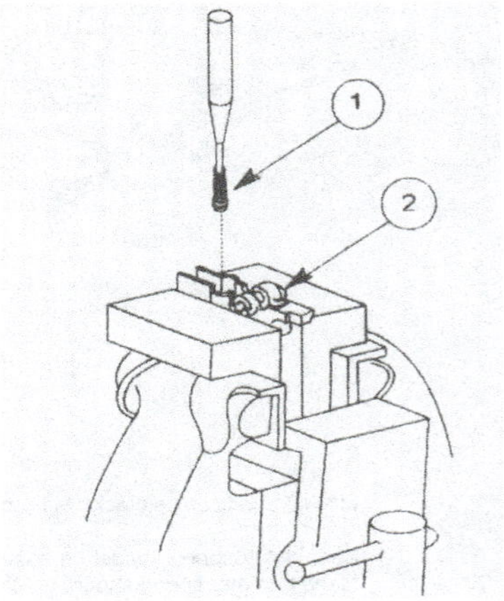

M4 CARBINE ONLY
NOTE
The semi and burst disconnector springs are not the same. The Semi disconnector spring (left side) (figure C-15, item 1B) is black while the burst disconnector spring (right side) (figure C-15, item 1A) is nickel (shiny). Ensure that the correct spring is installed on each side for proper functioning.

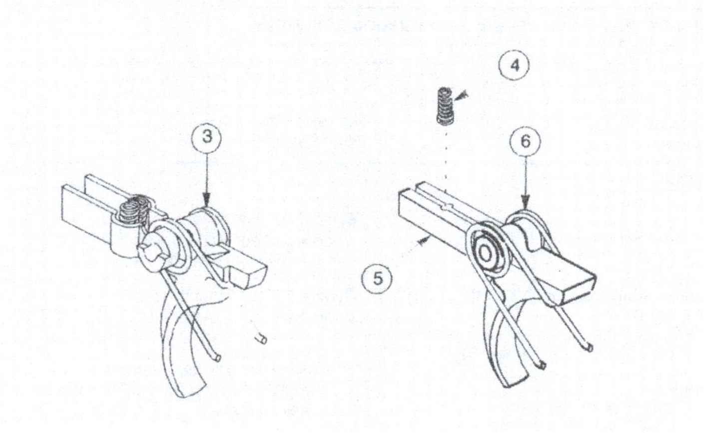

M16A2/M4 M4A1

M16A2 and M4

2. Install trigger spring (3).

M4A1 ONLY

3. Install disconnector spring (4) by inserting large end of spring into trigger (5).

4. Install trigger spring (6) on trigger (5).

3-16. LOWER RECEIVER AND RECEIVER EXTENSION ASSEMBLY.

This task covers:

a. Disassembly
b. Inspection
c. Repair/Modify
d. Reassembly

INITIAL SETUP

Tool Equipment
Tool and Gage Set (item 2, app B)

Tools
(ARMY) Small Arms Repairman Tool Kit (item 3, app B)
Field Maintenance Basic Less Power Small Arms Shop Set (item 1, app B)

Materials/Parts
Cloth, abrasive (item 13, app D)
Grease, molybdenum disulfide (item 19, app D)
Lubricant, solid film (item 21, app D)

Equipment Conditions
3-62 Lower receiver and receiver extension assembly removed

General Safety Instructions
To avoid injury to your eyes, use care when removing and installing spring-loaded parts.
When using solid film lubricant or dichloromethane, be sure the area is well ventilated.

a. DISASSEMBLY

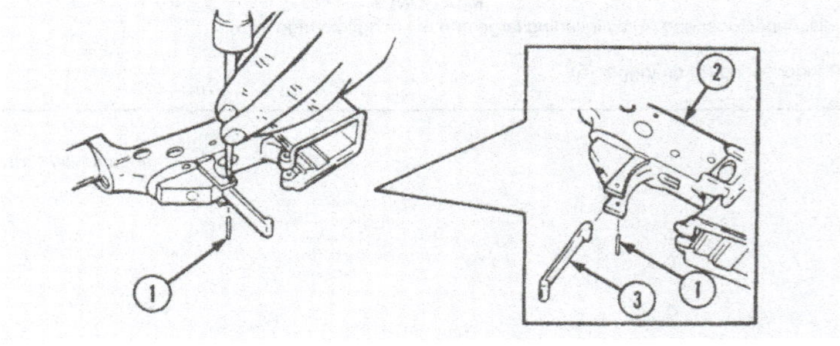

1. Remove spring pin (1) from lower receiver (2) using 1/8 inch drive pin punch and hand hammer.

2. Remove trigger guard (3).

3-16. LOWER RECEIVER AND RECEIVER EXTENSION ASSEMBLY (CONT).

a. DISASSEMBLY (CONT)

3. Use padding between lower receiver and brass vise laws. Grip the solid portion of the lower receiver with brass vise jaws which conform to the shape of the lower receiver in this area.

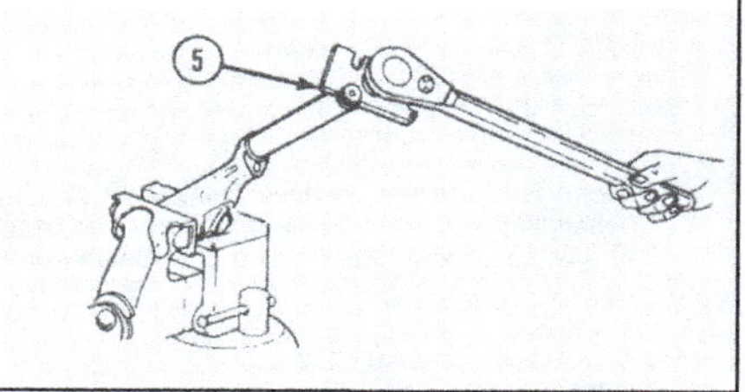

RIFLE ONLY

4. Clamp lower receiver (2) in a machinist's vise using vise jaw caps and tighten on solid portion just tight enough to hold.

WARNING

To avoid injury to your eyes, use care when removing and installing spring-loaded parts.

NOTE

As lower receiver extension is removed, catch buffer retainer and helical spring. Lower receiver is a serial number controlled item.

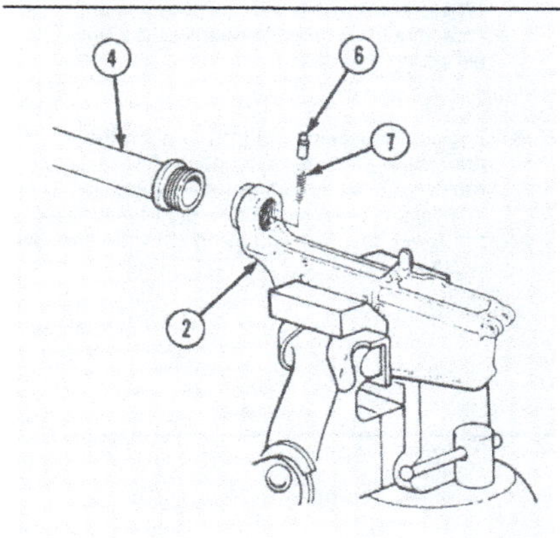

5. Remove lower receiver extension (4) from lower receiver (2) using combination wrench (5) and socket wrench handle. Catch buffer retainer (6) and helical spring (7).

CARBINE ONLY

6. Clamp lower receiver (8) in vise and tighten on solid portion just tight enough to hold.

NOTE

Use wooden vise jaws in place of brass vise jaw caps.

7. Remove the lower receiver extension (9), by loosening the extension locking nut (10) using the special tool (item 12, app C). Catch buffer retainer and spring (11).

CAUTION

While performing the following step, care should be taken to restrain the pivot pin spring and detent.

8. Loosen the locking nut (10) to allow the receiver end plate (12) to disengage from the lower receiver. Hold the buffer retainer and spring (11) in position with your index finger and unscrew the lower receiver extension (9) from the lower receiver (8).

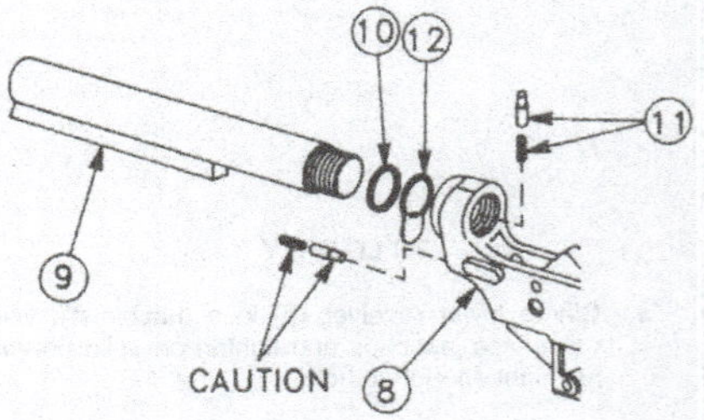

b. INSPECTION

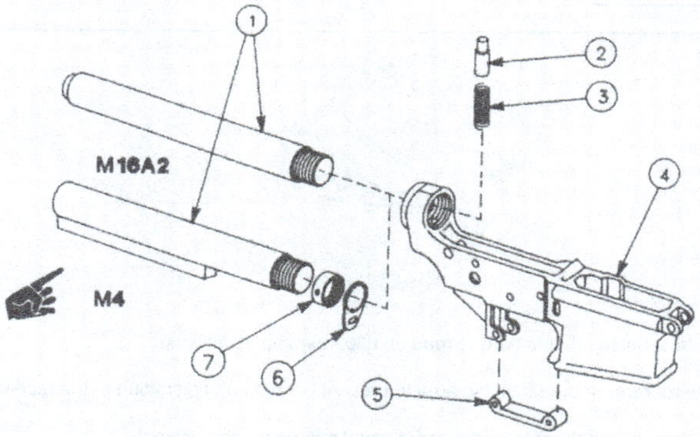

1. Inspect lower receiver extension (1) for corrosion, dents, and wear. Repair or replace if defective.
2. Inspect buffer retainer (2) for wear. Replace if defective.
3. Inspect helical springs (3) for deformities and breaks. Replace if defective.
4. Inspect lower receiver (4). See page 3-66, INSPECTION.
5. Inspect trigger guard (5) for deformities and check operation of plunger and spring. Replace trigger guard if defective.
6. **CARBINE ONLY:** Inspect receiver end plate (6) and locking nut (7) for damage. Replace if damaged.

c. REPAIR/MODIFY

1. Repair lower receiver extension by using abrasive cloth to remove light corrosion. Retouch using solid film lubricant.

NOTE

AIR FORCE ONLY: Only depot maintenance is authorized to restamp the serial number on rifle.

ARMY ONLY: Only direct support level is authorized to restamp serial number.

3-16. LOWER RECEIVER AND RECEIVER EXTENSION ASSEMBLY (CONT).

c. REPAIR/MODIFY (CONT)

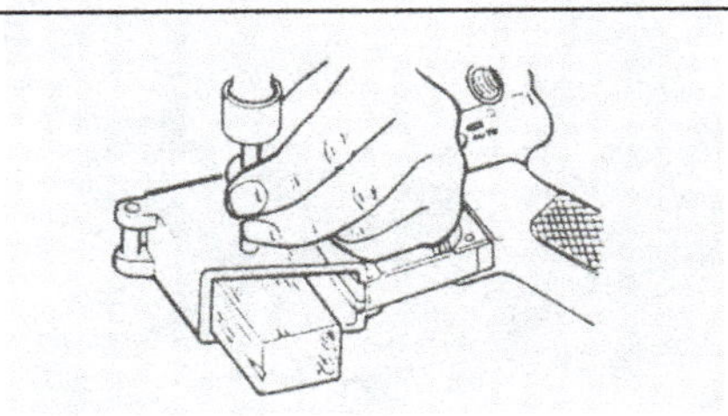

2. **ARMY ONLY:** If serial number is hard to read on rifle, restamp as follows:

   (a) Support the receiver In the stamping area to prevent bending and distortion of the receiver.

   (b) Exercise extreme care to restamp the same serial number as the original.

   (c) Restamp the serial number the same size as the original serial number.

NOTE

Most rifle serial numbers are 1/8 inch (0.31 cm) in height, or close enough that this size is acceptable for such restamping. In the event that a rifle has a serial number that cannot be reproduced by the use of the die sets contained In the Set D Field Maintenance Post, Camp, and Station Small Arms Shop Set, local purchase of an appropriate size die set is authorized.

d. REASSEMBLY

RIFLE ONLY

WARNING

To avoid injury to your eyes, use care when removing and installing spring-loaded parts.

1. Install helical spring (1) and buffer retainer (2) into lower receiver (3).

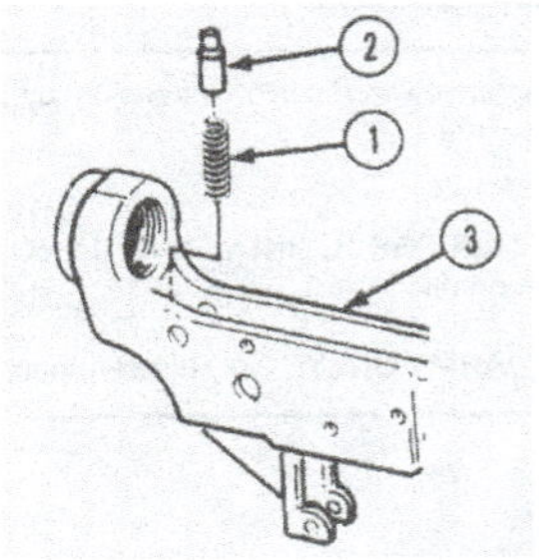

RIFLE ONLY

2. Lubricate threads of lower receiver (3) and lower receiver extension (4) with molybdenum disulfide grease (item 19, app D) before reassembly.
3. Install lower receiver extension (4) into lower receiver (3) while depressing buffer retainer.

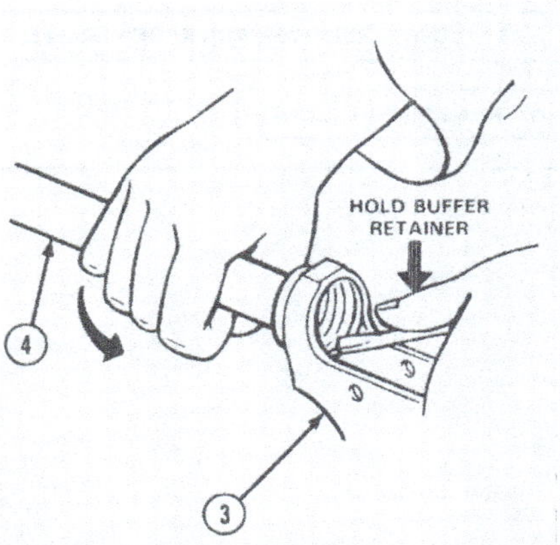

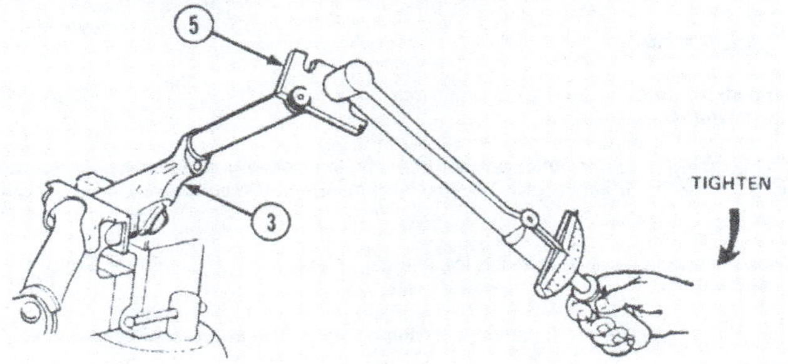

RIFLE ONLY

4. Use padding between lower receiver and brass vise jaws. Use vise jaws in vise and brass vise jaw caps, if available.
5. Clamp solid portion of lower receiver (3) in a machinist's vise using vise jaws. Grip the solid portion of the lower receiver with vise jaws which conform to the shape of the lower receiver in this area.
6. Using combination wrench (5) and torque wrench, torque lower receiver extension to 35-39 ft-lb (47 25 52 65 N•m). Torque is read when both wrenches are used to gether.

3-16. LOWER RECEIVER AND RECEIVER EXTENSION ASSEMBLY (CONT).

d. REASSEMBLY (CONT)

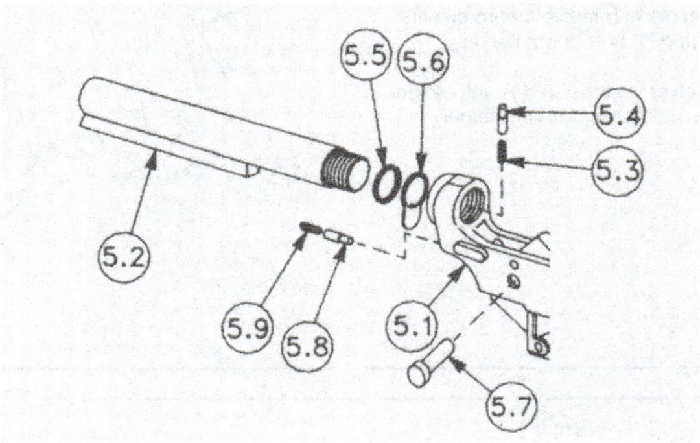

CARBINE ONLY

6A. Lubricate threads of lower receiver (5.1) and lower receiver extension (5.2) with molybdenum disulfide grease (item 19, app D) before reassembly.

6B. Pre-position the spring (5.3) and buffer retainer (5.4) into the retaining hole of the lower receiver (5.1). Screw the locking nut (5.5) onto the lower receiver extension (5.1) with the three notches on the locking ring (5.5) facing forward.

6C. Align the receiver end plate (5.6) onto the lower receiver extension (5.2) with the lug of the receiver end plate (5.6) facing forward.

6D. Pre-position the takedown pin (5.7), detent (5 8), and spring (5.9) in lower receiver assembly (5.1).

6E. Push down on the buffer retainer (5.4) and spring (5.3) and at the same time, screw the lower receiver extension (5.2) in until it retains the buffer retainer (5 4) in position.

6F. Align the lug of the receiver end plate (5.6) into the rear of the lower receiver (5.1). Screw the locking nut (5.5) forward until it contacts the receiver end plate (5.6).

6G. Using the special tool (Item 12, app C) tighten the locking nut (5.5) until snug.

6H. Using the special tool (item 12, app C), and torque wrench, torque locking nut (5.5) to 40 ±2 inch pounds.

6J. Stake the receiver end plate (5.6) in 2 places across from the notches in the locking nut (5.5).

3-16. LOWER RECEIVER AND RECEIVER EXTENSION ASSEMBLY (CONT).

d. REASSEMBLY (CONT)

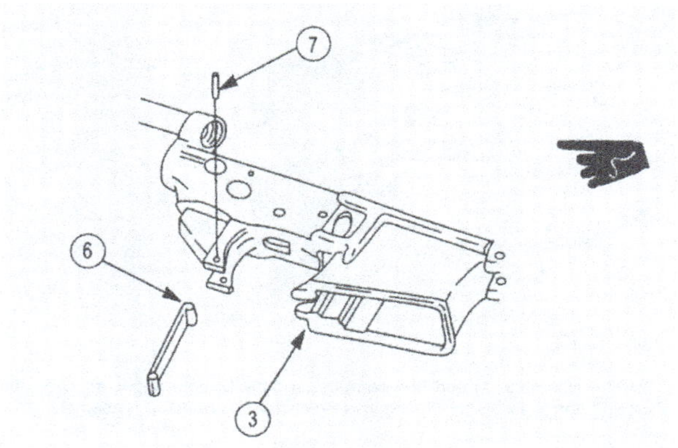

7. Install trigger guard (6) into lower receiver (3).

8. Install spring pin (7) using 1/8 inch drive pin punch and hand hammer.

9. Deleted.

3-17. MAJOR COMPONENTS OF M16A2 RIFLE

This task covers:

a. Reassembly   b. Test   c. Inspection

INITIAL SETUP

Test Equipment
Tool and Gage Set (item 2, app B)

Tools
(ARMY) Small Arms Repairman Tool Kit (item 3, app B)

Reference
TM 9-1005-319-10

General Safety Instructions
To avoid injury to your eyes, use care when removing and installing spring-loaded parts.
Live ammunition must not be near the work area.

a. REASSEMBLY

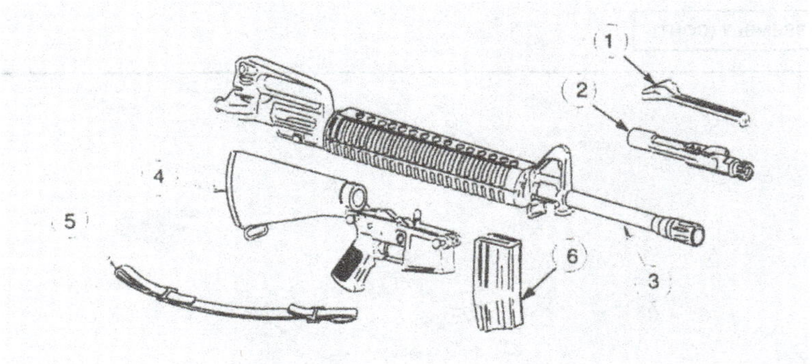

1. Refer to TM 9-1005-319-10.

2. Install charging handle assembly (1) and bolt carrier assembly (2) into upper receiver and barrel assembly (3). Join upper receiver and barrel assembly (3) and lower receiver and buttstock assembly (4).

3. Snap on sling (5) and install magazine (6).

b. TEST

Test trigger pull as follows:

M16A2 and M4 ONLY

(a) Clear the weapon. Place selector to BURST. Pull the trigger and hold it to the rear. Pull the charging handle to the rear and return to the bolt closed position three times. (This will place the BURST disconnector in the deep notch of the BURST cam.)

(b) Release the trigger. Place the selector to SEMI. Hold the rifle in the vertical position. Using trigger pull measuring fixture, 7274758, add weights until hammer trips. Determine weight applied.

(c) Hammer must not trip when 5.5 lbs (2.49 kg) have been applied, hammer must trip on applying 9.5 lbs (4.31 kg).

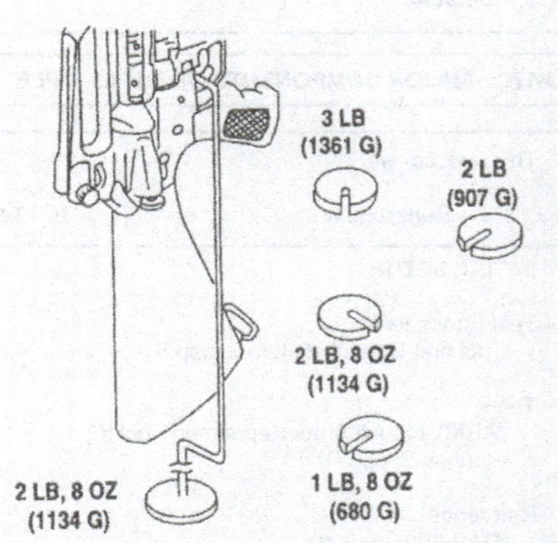

3-17. MAJOR COMPONENTS OF M16A2 RIFLE (CONT)

b. TEST (CONT)

M4A1 ONLY

(d) Clear the weapon. Cock weapon. Place weapon in SEMI position and hold weapon in vertical position.

(e) Using trigger pull measuring fixture 7274758 as shown, add weights until hammer trips. Determine weight applied.

(f) Hammer must not trip when 5.5 lbs (2.49 kg) have been applied. Hammer must trip on applying 8.5 lbs (3.86 kg).

ALL WEAPONS

(g) If weapon fails trigger pull test or excessive creep is present, replace trigger and/or hammer.

NOTE

Always gage trigger and hammer pin holes with not go plug gage 12006472 before replacing parts.

c. INSPECTION

Perform final inspection procedures below.

3-18. M16A2 RIFLE FINAL INSPECTION.

This task covers:

a. Final inspection
b. Test
c. Functional theory of three-round burst control

INITIAL SETUP

Test Equipment
Tool and Gage Set (item 2, app B)

References
TM 9-1005-319-10

General Safety Instructions
Live ammunition must not be near the work area.

Air Force rifles will be inspected in accordance with AFM 36-2227, Vol. 1.

a. FINAL INSPECTION I

1 Visually inspect general appearance of rifle should look almost new. All metal surfaces are to have a dull, rustor corrosion-resistant finish with no burrs or deep scratches.

2 Visually inspect barrel for serviceability. Barrels must be straight, clean, free of rust, powder fouling, and free of bulges and rings. Fine pitting is allowable.

Using moderate hand pressure, check for rotational movement of the front sight in relation to the barrel. Also, using moderate hand pressure, check for rotational movement of the barrel in relation to the upper receiver. If movement between the front sight and the barrel exist, the barrel assembly must be replaced. If movement between the barrel and the upper receiver exist, the barrel assembly must be aligned and tightened (page 3-29).

3 Visually inspect rifle for missing parts. All parts must be attached and all modifications must be applies. Steel parts must be rust free. Spring pins must be secure and screws must be tight.

4 Functionally inspect key and bolt carrier assembly and gas tube alignment. Refer to TM 9-1005-319-10 and use the following procedures:

(a) Disengage the takedown pin and open the receiver.

(b) Remove bolt carrier assembly.

(c) Remove bolt assembly from bolt carrier assembly.

(d) Insert key and bolt carrier assembly into upper receiver and barrel assembly. The bolt assembly must not be installed while performing test.

(e) Slide key and bolt carrier assembly forward to detect binding between key and bolt carrier assembly and gas tube by feel. Badly bent gas tube could cause damage to both the key and bolt carrier assembly and the gas tube. A slightly bent gas tube will cause unnecessary wear of the key and bolt carrier assembly and gas tube.

(f) Correct slight binding by removing handguard assemblies and slightly bending gas tube in the handguard area while repeating step (e) above until no binding is detected. Badly bent gas tubes will be replaced and realigned.

(g) Remove key and bolt carrier assembly from upper receiver and barrel assembly.

(h) Reassemble bolt assembly into key and bolt carrier assembly.

(i) Reinstall key and bolt carrier assembly into upper receiver and barrel assembly.

5 Make a functional check on an assembled weapon with selector lever in SAFE, SEMI, and BURST/AUTO positions Any portion of this check may be used alone to determine the operational condition of any specific firing position selected

3-18 M16A2 RIFLE FINAL INSPECTION (CONT)

a. FINAL INSPECTION (CONT)I

6 Check rear sight assembly as follows:

(a) Rotate elevation knob counterclockwise until the rear sight assembly is all the way down. If a whole click Is not felt as the rear sight assembly stops, the rear sight assembly has bottomed out and will not pivot freely.

(b) Position elevation knob back slightly to its last whole click as the rear sight assembly base is under tension of the ball bearing and helical spring. The 300 meter mark should align with the mark on the receiver (c) If the 300 meter mark is not aligned with the mark on receiver, slip the range scale in the following manner:

(1) Position the 300 meter mark with the mark on the receiver.

(2) Insert a 1/16 inch alien wrench through the access hole of the rear sight assembly base and into the index screw.

(3) Loosen the index screw three turns and leave the wrench in place.

(4) Rotate lower portion of elevation knob counterclockwise until it stops (range scale should not have moved). Elevation knob should be positioned on its last whole click.

(5) Tighten index screw and remove wrench.

(6) Check for proper setting.

7 Pull charging handle assembly to rear. Check that chamber is clear. Leave hammer in cocked position.

8. Place selector lever is SAFE position and pull trigger. Hammer should not fall.

WARNING

If rifle falls any of the following inspections, continued use of the rifle could result In injury to, or death of, personnel.

NOTE

For the purpose of the following tests, "SLOW" is defined as 1/4 to 1/2 the normal rate of trigger release

9. Place selector lever in SEMI position.

10 Pull trigger. Hammer should fall.

11 Hold trigger to the rear, charge rifle and release the trigger with a slow, smooth motion without hesitations or stops, until the trigger is fully forward; an audible click should be heard. Hammer should not fall.

12 Repeat the above selector lever SEMI position test (steps 10 and 11) five times. The rifle must not malfunction during any of these five repetitions. If the rifle malfunctions during any of these five tests, refer to page 3-7.

M16A2 and M4 ONLY

13 Place selector lever in BURST position. Charge rifle and pull trigger. Hammer should fall.

14 While holding the trigger to the rear, pull charging handle assembly to the rear and release it three times. Hammer should not fall the third time the charging handle assembly is released. It may or may not fall the first and second time the charging handle assembly is released. When the burst disconnector reached the deep notch of the burst cam, the burst disconnector should have held the hammer to the rear while the trigger was in the pulled position. Release the trigger with a slow, smooth motion without hesitations or stops, until the trigger is fully forward. The hammer should not fall.

15 Pull trigger. Hammer should fall. This would be the first round of a three-round burst.

NOTE

A detailed explanation of the three-round burst control can be found in the following pages.

M4A1 ONLY

15.1 Place selector lever in AUTO position. Charge weapon and pull trigger. Hammer should fall.

15.2 Hold trigger to the rear, charge weapon, and release trigger. Pull trigger. Hammer should not fall. Automatic assembly sear should have released the hammer as the bolt closed.

ALL WEAPONS

16 With the hammer in the forward position, using moderate finger/thumb pressure, attempt to place the selector lever in the SAFE position. Selector lever should not go in the SAFE position.

17 Perform the following additional functional checks and adjustments on assembled weapon:

- **(a)** Press magazine catch button. Make sure it functions properly.
- **(b)** Press bolt catch. Make certain it operates smoothly and holds bolt in open position.
- **(c)** Inspect front sight and rear sight assembly. Make certain they can be adjusted properly.
- **(d)** Actuate forward assist assembly It must work freely.

3-18 M16A2 RIFLE FINAL INSPECTION (CONT)

a. FINAL INSPECTION (CONT)

(e) Inspect upper receiver and barrel assembly. Barrel assembly should not rotate within upper receiver assembly.

(f) Check that third or middle slot of compensator is straight up (TDC).

b. TEST

1 Check headspace using headspace gage PN 7799734. See page 3-45, TEST.

2 Check firing pin protrusion using firing pin protrusion gage PN 7799735. See page 3-45, TEST.

3 Check extent of barrel erosion using barrel erosion gage PN 8448496. See page 3-45, TEST.

4 Check barrel straightness using barrel straightness gage PN 8448202. See page 3-45.

c. FUNCTIONAL THEORY OF THREE-ROUND BURST CONTROL (M16A2 AND M4 ONLY).

NOTE

First become familiar with the functioning of the firing mechanism especially when in the SAFE and SEMI positions. You should also understand the role that the automatic sear plays when firing in the BURST position. Functioning of the mechanism is explained below In a step by step manner. This actually will seem to complicate something that is really very simple and happens in less than 1 second. The diagrams below and on the following pages do not show the associated springs for the sake of simplicity The positioning of the burst cam is shown in detail

Functional check of three-round burst is as follows:

NOTE

Assume the rifle is fully loaded with a live round in the chamber and the selector lever on BURST.

(a) Hammer is cocked

(b) Front hook of burst lever is in stop notch

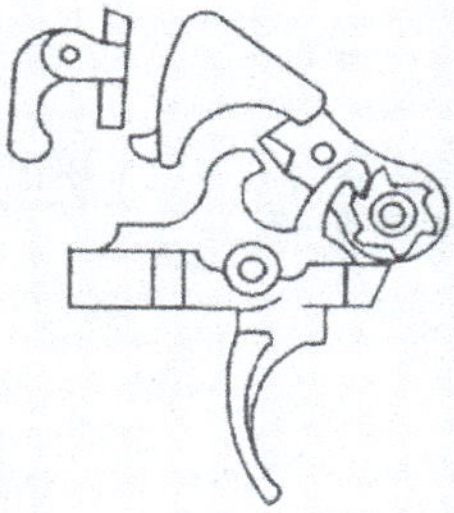

(c) Trigger Is pulled.

(d) Trigger nose drops and hammer falls firing the FIRST ROUND.

(e) Front hook of burst disconnector holds burst cam In place as hammer falls.

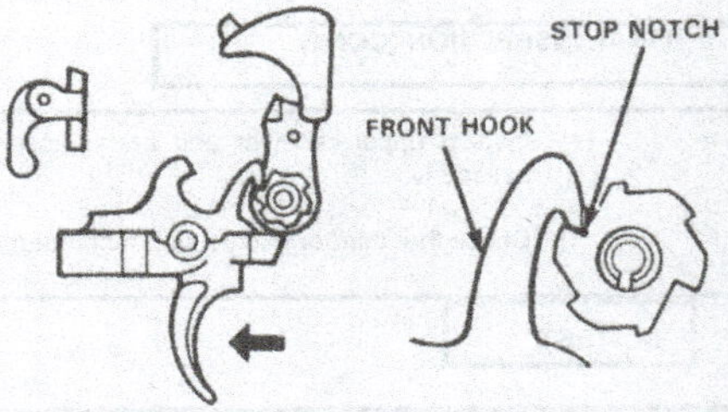

NOTE

Anytime the hammer falls forward, the clutch spring releases the burst cam and allows the front hook of the burst disconnector to keep It In place.

(f) As the key and bolt carrier assembly moves to the rear, the hammer is forced to the rear.

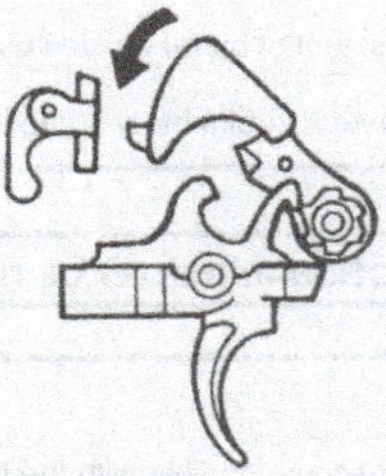

(g) The clutch spring of the burst cam clutches the burst cam and causes It to rotate one notch as the hammer is forced back.

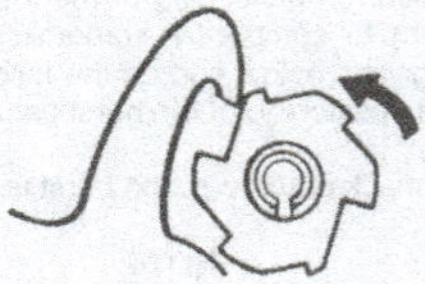

(h) When hammer is fully to the rear, the automatic sear catches It.

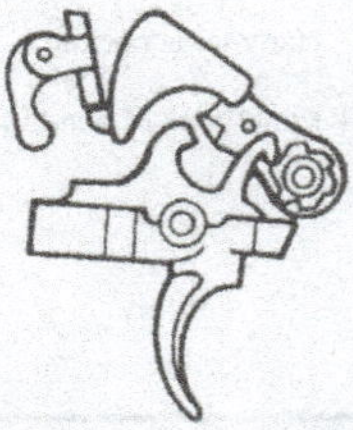

3-18 M16A2 RIFLE FINAL INSPECTION (CONT).

c. FUNCTIONAL THEORY OF THREE-ROUND BURST CONTROL (CONT)

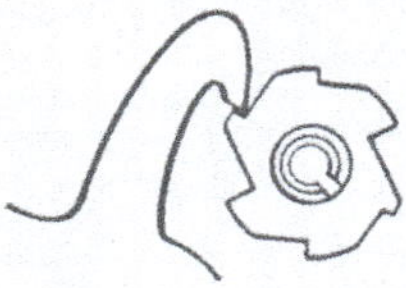

(i) The front hook of the burst disconnector is now fully In the second notch.

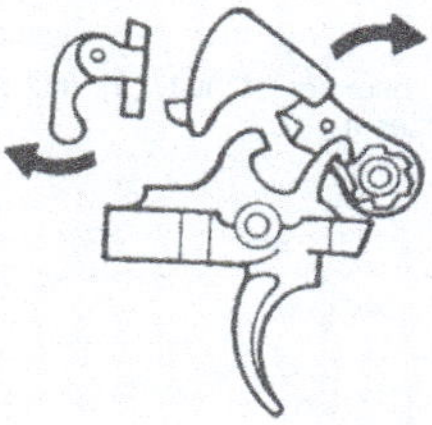

(j) As the key and bolt carrier assembly travels forward, the automatic sear releases the hammer and the hammer falls.

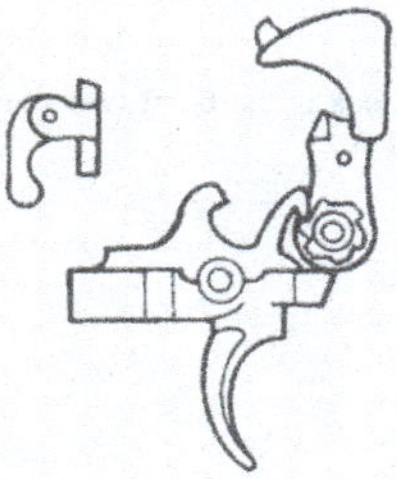

(k) When the hammer falls, the SECOND ROUND is fired.

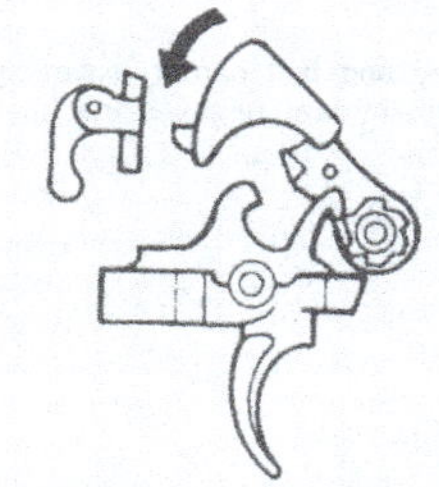

(l) As the key and bolt carrier assembly moves to the rear, the hammer is forced back to the rear.

(m) The clutch spring of the burst cam clutches against the burst cam and causes It to rotate one notch as the haiimmer is forced back

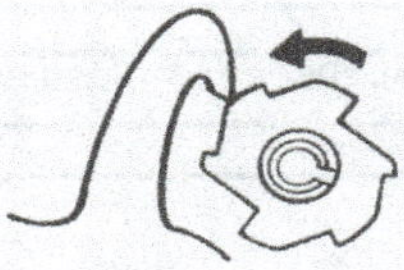

(n) When the hammer is fully to the rear, the automatic sear catches It

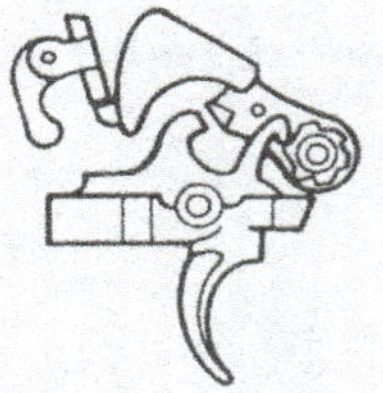

(o) The front hook of the burst disconnector IS now fully In the third notch

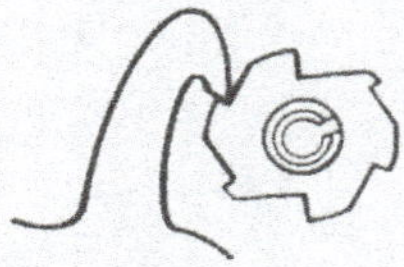

(p) As the key and bolt carrier assembly) travels forward, the automatic sear releases the hammer and the hammer falls

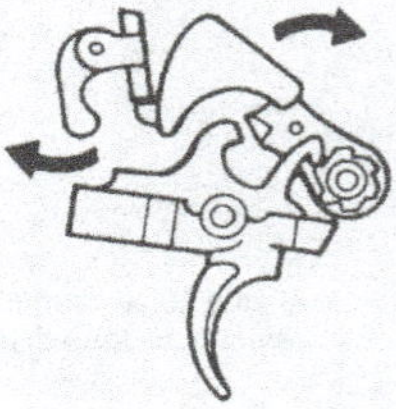

3-18. M16A2 RIFLE FINAL INSPECTION (CONT).

c. FUNCTIONAL THEORY OF THREE-ROUND BURST CONTROL (CONT)

(q) When the hammer falls, the THIRD ROUND Is fired

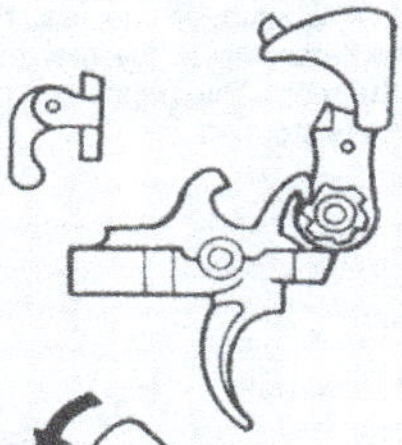

(r) As the key and bolt carrier assembly moves to the rear, the hammer is forced back to the rear

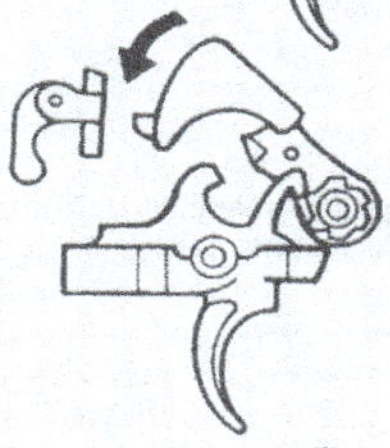

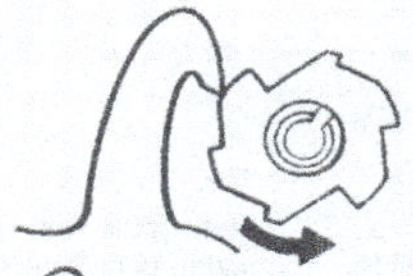

(s) The clutch spring of the burst cam clutches against the burst cam and causes it to rotate one notch as the hammer is forced back.

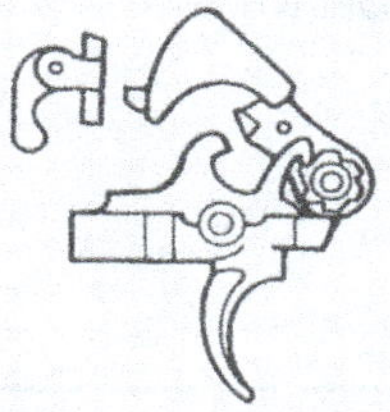

(t) When the hammer Is fully to the rear, It Is initially caught the, ;automatic sear However, the front hook of the first disconnector Is now fully in the next stop notch which is deeper than the others

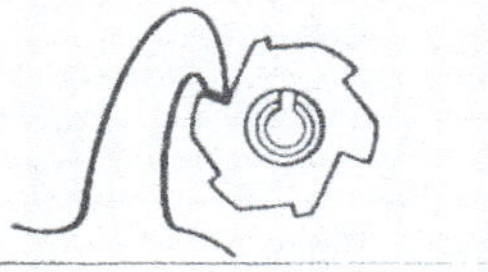

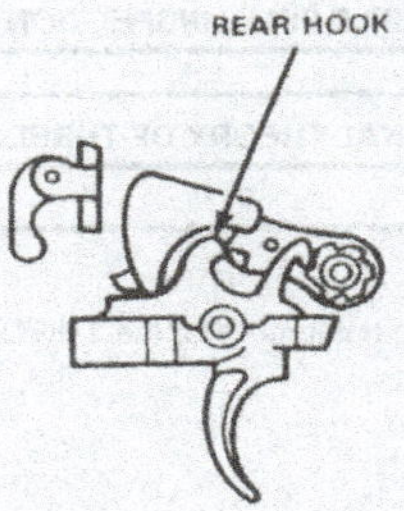

(u) Because a stop notch is deeper than the others, It allows the front hook of the burst disconnector further forward than before. This allows the rear hook of the burst disconnector to latch on the rear hammer notch. This holds the hammer fully to the rear even though the trigger is still to the rear. This happens when the burst Is over and the firing Is stopped.

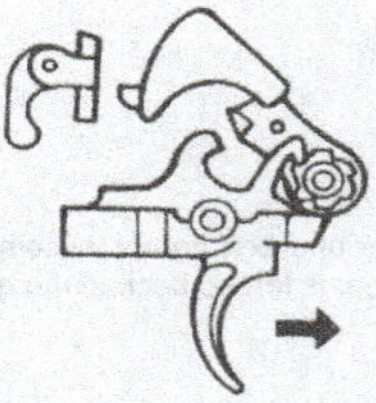

(v) Once the trigger is released, the trigger nose comes up and holds the hammer back.

NOTE

Pulling the trigger to the rear and holding it back again will fire another three round burst This will continue until the magazine is empty However the trigger must be released between each burst

3-19 MI 6A2 RIFLE AND M4/M4A1 CARBINE ANNUAL GAGING REQUIREMENTS.

This task covers.

a Inspection b Gaging

INITIAL SETUP

Test Equipment
Tool and Gage Set (item 2, app B)

Tools
(ARMY) Small Arms Repairman Tool Kit (item 3, app B)
Field Maintenance Basic Less Power Small Arms Shop Equipment (item 1, app B)

References
DA PAM 738-750
TM 9-1005-319-10
AFR 50-36
AFTO Form 105

General Safety Instructions
- To avoid injury to your eyes, use care when removing and installing spring-loaded parts
- Initial gaging is required 1 year from receipt of the weapons.
- All rifles and carbines must be gaged at least once annually for safety.
- All Army Reserve and Army National Guard M16A2 rifles must be inspected and gaged at least once every 2 years, after the initial inspection/gaging procedures have been accomplished. This 2 year interval may be maintained unless preventive maintenance checks and services (PMCS), or other physical evidence, indicates that an individual unit's rifles require inspection/gaging at a more frequent interval. If it is determined that a yearly inspection is necessary for an individual unit, only that unit will be affected. This will not affect other units in regard to the interval of inspection.
- It is recommended that training units inspect/gage all rifles at the end of each training cycle. Training units will inspect/gage all rifles at least once annually.
- Air Force M16 rifles will be inspected in accordance with AFR 50-36, Vol 1, Chapter 5.

a. INSPECTION

1 Visually inspect general appearance of weapon Overall appearance will be approximately that of a new weapon. For inspection criteria refer to final inspection, page 3-84 All visual and functional Inspection requirements must be met.

NOTE

To perform the following tests, disassemble weapon only as far as allowed in chapter 2 (Unit Maintenance Instructions). unless a deficiency is uncovered

2 Perform a general inspection of weapon per section III of this chapter Repair as required and authorized

b. GAGING

1	Gage bolt carrier assembly for firing pin protrusion using firing pin protrusion gage PN 7799735. See page 3-20, TEST.
2.	Gage bolt carrier assembly for firing pin hole wear using not-go plug gage PN 12620101. See page 3-24, TEST.
3	Inspect chamber in upper receiver and barrel assembly using chamber reflector tool PN 8448201 See page 3-33, INSPECTION/CLEANING.
4	Gage barrel in upper receiver and barrel assembly using barrel erosion gage PN 8448496 and bore straightness gage PN 8448202. See page 3-45, TEST.
5	Check headspace in upper receiver and barrel assembly by inserting headspace gage PN 7799734 in chamber. See page 3-45, TEST.
6	Gage pivot pin lug area clearance in lower receiver assembly using 0.020 thickness gage. See page 3-68.
7	Gage hammer and trigger pin holes in lower receiver assembly using taper plug gage PN 12006472. See page 3-68, TEST.
8	Gage trigger pull using trigger pull measuring fixture PN 7274758. See page 3-83, TEST.
9	Document inspection with DA PAM 738-750, AFTO Form 105, or NAVMC 11003 when completed.

Section IV. PREEMBARKATION INSPECTION OF MATERIEL IN UNITS ALERTED FOR OVERSEAS MOVEMENT

NOTE

This section refers to all the weapons (MI 16A2 rifle, M4, and M4A1) unless otherwise stated.

3-20 PURPOSE: This section establishes standards for overseas shipment (pre-embarkation in section criteria) for all weapons. These standards are provided to ensure that the user is furnished equipment which will perform its mission without early failure or major maintenance problems.

3-21 SCOPE.

a. The standards prescribed provide for a high percentage of remaining life in affected rifles; therefore, rifles designated for overseas shipment must qualify under the standards contained in the following paragraph, table, and in referenced DA publications, before they can be approved for shipping action.

b. Provisions of this standard apply to all US Army agencies/activities selecting or preparing rifles for shipment to US troops overseas. It also applies to CONUS troops preparing rifles for shipment overseas. Provisions do not apply either to rifles being prepared for shipment to MAP/MAS recipients unless specifically prescribed by MAP/MAS transaction for the materiel or to rifle being returned to CONUS from overseas. The maintenance instructions and standards contained herein do not apply to rifles once the material has arrived at the overseas destination. At that time, maintenance instructions contained in the applicable TM's will be used.

c. This applies to rifles which are the logistic responsibility of the US Army Armament, Munitions and Chemical Command.

3-22. GENERAL.

a. Only rifles which have been classified as serviceable condition code A, B, or C under AR 725-50 will be considered for overseas shipment. All items of equipment for which equipment serviceability criteria have been published must, as a minimum, be rated green under the ESC as a prerequisite to overseas shipment. In addition to the condition code standard, as enumerated above, and the required ESC rating prescribed herein, the rifle being considered for overseas shipment must meet the requirements of this section. The ESC will be discontinued as new operator manuals are revised which will be used to determine serviceability condition of rifle.

b. Waivers to provisions can only be granted by the gaining command of any particular end Item being considered for issue, deployment, or shipment. The issuing services may recommend issue or shipment of rifles not meeting the provisions when all the following conditions exist:

(1) Repair parts in required quantities cannot be obtained from the supply system prior to delivery of the end item.

(2) The gaining command concurs in the receipt of the end item for storage until required repair parts become available. The gaining command must also state that capability, facilities, and funds are available to perform the necessary work when parts become available.

(3) Department of the Army approval is obtained on a case-by-case basis.

(4) Required repair parts are requisitioned by the issuing command for delivery to the Framing command.

c. All Department of the Army MWO's applicable to the specific rifle being considered for shipment overseas must have been applied.

d. Refer to SB 746-1 for pertinent publications relating to equipment processing and marking information

e. Refer to AMC-P 310-9 for publications containing applicable overhaul standards.

3-23. SHIPMENT OR ISSUE.

a. **Organizational Repair Parts, Tools, and Equipment**. Rifles must be complete with all Items required by applicable Department of the Army publications, including those In the basic Issue items list of the appropriate operator's manual.

b. **Publications**. Operator publications applicable to the equipment log book must accompany the equipment. All log book entries must be complete and up-to-date including those covering any repairs, replacements, or adjustments made to the rifle In complying with this section.

c. **Documentation**. Prepare DA Form 2408-9 (Equipment Control Record) at time of overseas shipment or Issue to another stock record or property book account, in accordance with the provisions of DA PAM 738-750.

d. **Preparation**. Process rifles for shipment as required by shipping documents and pertinent regulations.

3-24. DISPOSITION. Disqualified rifles which do not qualify for shipment will either be redistributed within the camp, post, or station, be repaired, or become candidates for overhaul, cannibalization, or other disposition as required by existing regulations.

3-25. GENERAL INSPECTION CRITERIA.

WARNING

Before starting an Inspection, be sure to clear the rifle Do not actuate the trigger until the rifle has been cleared. Inspect the bore and chamber to ensure that It is empty and free from obstructions, and check to see that no ammunition Is In position to be Introduced.

a. Before Inspection, the materiel must be thoroughly cleaned of all grease, dirt, or other foreign matter that might Interfere with Its proper function or the use of gages and tools during inspection.

b. Materiel must be free of burrs, rust, or corrosion on functional surfaces.

3-25. GENERAL INSPECTION CRITERIA (CONT).

c. Parts must not be cracked, bent, distorted, or damaged and must be free of detrimental wear or looseness

d. Minor defects In metal components do not normally affect their acceptability For example, scratches and tool marks are ordinarily of no Importance

e. Inspect finish of metal surface

(1) General. Satisfactory metal surfaces for rifles range from black to light gray. A worn shiny metal surface is objectionable only when It is capable of reflecting light No rifle will be reelected unless exterior parts have a shine All rear sights must have a dull gray or black finish of all surfaces that would cause a glare

(2) M16A2 Rifle. Minor loss of finish (shiny spots, nicks, scratches) on exterior surfaces of the barrel and flash suppressor shall not be cause for rejection of M16A2 rifles located In hands of troops at training centers Large shiny surfaces, nicks, scratches, etc., can be restored by the use of solid film lubricant (Item 21, app D) Rifles (small arms) missing in excess of one-third or more of the exterior finish resulting In an unprotected, light reflecting surface, are considered candidates for overhaul The only authorized level of maintenance to phosphate finish small arms Is depot.

f. Plastic components must not be cracked or damaged In such a way as to Interfere with their structural strength Surface cracks, bruises, or dents that do not affect their strength will not be cause for rejection. Cracks will be cause for rejection. Criteria for determining which cracks are repairable are on page 2-1 4.

g. Barrels must be clean and free of corrosion such as that caused by moisture and powder fouling Standards of serviceability are indicated In (1) through (9).

(1) Pits In the chamber are allowable If they do not cause extraction difficulties

(2) Pits as wide as a land and 3'8 Inch (O 95 cm) or less in length are allowable for 5 56-mm barrels Pits not greater than the width of a land and less than 3 8-inch (O 95 cm) long are permissible

(3) Scattered or uniformly fine pits, or fine pits In a densely pitted area are allowable

(4) Tool marks are acceptable regardless of length They will appear as lines running laterally In the grooves, or may run spirally across the top of lands

(5) Ringed bores or bores ringed sufficiently to bulge the outside surface of the barrel are cause for rejection. However, faint rings or shadowy depressions do not indicate an unserviceable barrel and will not be cause for rejection. Gap in lined barrels will not be classified as a ringed bore.

(6) Lands that appear dark due to coating of gilding metal from projectiles will not be cause for rejection.

(7) Breech bore diameter will be checked on unlined barrels using the appropriate breech bore gage.

(8) Barrel erosion gages are provided for lined barrels. Bore wear will be checked using barrel erosion gage for the M16A2. For detailed instructions in the use of the above gage and for serviceability limits, refer to page 3-46.

(9) Flaking or checking (fine cracks) of chromium plate in barrels or chambers will not be cause for rejection, unless accompanied by pitting to the degree that extraction difficulties are encountered or accuracy is unacceptable.

h. Springs must be free of distortion and broken coils. Springs must have sufficient tension to perform their intended function.

i. Screw heads must be in serviceable condition and threads must not be stripped. Internal threads must not be stripped.

j. The sear, hammer, and/or cocking notches must be in good condition. Chipped engaging corners will be cause for rejection. Slight wear on functional surfaces, including engaging corners, shall be acceptable, providing the minimum trigger pull requirements and selector lever checks are met in accordance with instructions on page 3-83.

k. Chips, flat spots, or bent striker points on firing pins will be cause for rejection.

l. The cartridge engaging surfaces on extractors must not be chipped or deformed.

m. Evidence of any damage to sights will be cause for a sight alignment check. Rear sight bases should have no movement.

n. Rear sight elevating and windage mechanisms must operate with distinct clicks, without binding. Sights must have sufficient tension to retain their setting during firing. Graduations and numerals must be legible. Graduation filler Is not required.

o. Safeties must positively position in both the ON and OFF position. When in the ON or safe position, the rifle must not fire when the trigger is squeezed; when in the OFF or fire position, the rifle must fire when the trigger is squeezed.

3-25. GENERAL INSPECTION CRITERIA (CONT).

p. All locking devices such as latches, magazine latches, or detents must be positive in action and must not become disengaged due to normal handling and firing. Retaining pins and similar devices must not be subject to accidental loss during use or transportation.

q. Each rifle must be hand functioned to check for unusual binding, positive cocking action, and general operation. Dummy ammunition must be used to assure positive feeding, chambering, extraction, and ejection action.

Table 3-1. 5.56MM Rifle M16A2

Item	Standard
RIFLE:	
General	Clear rifle of any ammunition and Inspect In accordance with paragraph 3-25.
Barrel and barrel extension	Check barrel erosion. Use barrel erosion gage 8448496 for chrome lined barrels. Stripping of lands and grooves shall not be cause for rejection unless so determined by barrel erosion gage. Visually inspect, using chamber reflector tool 8448201. Pits 1/8 inch (0.31 cm) In length and those pits large enough to extend from the body of the chamber into the shoulder stop area and forcing cone area are cause for rejection. Large pits are defined as those 1/8 inch (0.31 cm) or more in diameter as determined by visual Inspection. Only closed flash suppressors are acceptable. Check barrel for straightness using bore straightness gage 8448202. Gage must pass freely through the bore to be acceptable, either dropped from the muzzle or chamber end.
Front sight and gas tube	Inspect gas tube for proper alignment with carrier key. Gas tube must not bind when mating with the key. Evidence of gas leaks around the front sight connection of the gas tube shall be cause for rejection until rifle has been function fired to determine if the loss of gas is sufficient to cause malfunction. If function firing malfunctions occur, repairs are necessary. Inspect front sight for damage.

Table 3-1. 5.56MM Rifle M16A2, M4/M4A1 Carbine - Cont

Item	Standard
Bolt carrier group	Inspect bolt for elongated or oversized firing pin hole using plain cylinder gage 12620101. Firing pinholes which permit the plain cylinder plug gage to fully penetrate at any position on the circumference will be rejected. Bolt face with a cluster of pits which are touching or tightly grouped, covering an area measuring approximately 1/8 inch (0.31 cm) across will be rejected. Bolts which contain pits extending into the firing pinhole will not be rejected unless firing pin hole gaging check determines rejection. Bolts which contain individual pits or scattered pits will not be cause for rejection. Only phosphated bolt carriers are acceptable. Both phosphated and chrome plated bolts are acceptable.
Bolt locking lugs and bolt cam pin hole	Inspect for cracks in the locking lugs and cam pin hole area. Use a black light, if available; otherwise, use a glass of no more than 3X magnification or use inspection penetrant (item 25, app D). Use instructions contained in kit for application. If cracks are detected, the bolts will be replaced. **NOTE** Particular attention must be given to the area where the lugs meet the bolt body and around the side walls of the cam pin hole. Bolt rings must not be broken. Ring gaps must be properly spaced approximately 1/3 turn apart and not in line. Firing pin protrusion must be not less than 0.028 inch (0.071 cm) or more than 0.036 inch (0.091 cm). (Use firing pin protrusion gage 7799735.) Socket head capscrews must be staked. Carrier key must not be dented where end mates with gas tube. Repair or replace damaged carrier keys.
Headspace	Inspect headspace using headspace gage 7799734. Excessive headspace will be cause for rejection.
Trigger pull	Inspect trigger pull using trigger measuring fixture 7274758. Trigger pull must be minimum 5.5 pounds (2.49 kg), maximum for M16A2 and M4 is 9.5 pounds (4.31 kg) and maximum for M4A1 is 8.5 pounds (3.86 kg). Test trigger pull, refer to page 3-83, steps a, b, c, and d.

3-25. GENERAL INSPECTION CRITERIA (CONT).

Table 3-1. 5.56MM Rifle M16A2, M4/M4A1 Carbine - Cont

Item	Standard
Lower receiver group	Inspect hammer and trigger pin holes using plain cylinder plug gage 12006472 Penetration of the gage in any one or more of the four holes will be cause for rejection. Inspect for cracks, corrosion, or mutilation which would affect functioning. Small dents or gouges will not be cause for rejection. Inspect receiver for corrosion in the lobes of the pivot or hinge pin area. Width between lobes shall not exceed 0 515 inch (1.30 cm). Inspect receiver for break through of metal. Inspect receiver and receiver extension for initial loss of protective coating.
Action spring	Free length of spring shall be between 11-3/4 and 13-1/2 inches (29.84 and 34.29 cm) **M16A2 ONLY** and 10-1/16 and 11-1/4 inches (25.56 and 28.58 cm) **M4/M4A1 ONLY**.
Handguard	Inspect handguard assembly for breaks, separations, and cracks. Breaks and separations of material which prevent proper retention or interfere with functioning of the weapon will be cause for handguard rejection and replacement. Cracks up to 1 inch (2.54 cm) in length are acceptable provided they do not extend into the handguard retaining flange (1) (critical area). **M16A2 ONLY** Each handguard assembly may have up to two of the three front retaining tabs (2) missing. If all three front tabs are missing, the handguard assembly must be replaced. **M4/M4A1 ONLY** No tabs can be missing. **ALL WEAPONS** Replace severely cracked handguards. Handguards that have a heat shield loose enough to rattle when installed on the weapon must be replaced.
Stock assembly	Inspect buttstock assembly for dents, cracks, and chips. Check for breaks and separation of material which could prevent proper functioning of weapon. M4/M4A1 inspect for proper functioning. Repair as required

Table 3-1. 5.56MM Rifle M16A2, M4/M4A1 Carbine - Cont

Item	Standard
	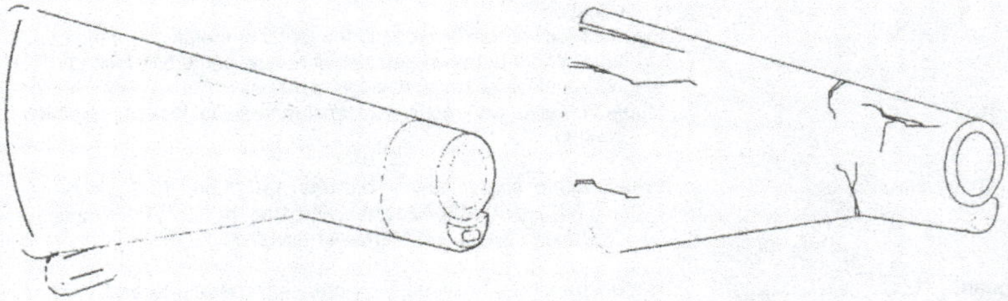
	M16A2 ONLY
	Under the following conditions, hairline cracks originating from buttplate end of buttstock are acceptable. No chipped away material is allowed. a. One hairline crack, not to exceed 1 inch (2.54 cm) in length, per side of buttstock. b. Two additional hairline cracks up to 0.22 inch (0.55 cm) in length, per side of buttstock. Buttstocks with unauthorized markings stamped into their surfaces will be replaced. Unauthorized markings, scratched, etched, carved, etc., are acceptable if they do not extend into the fiber of the buttstock which may weaken it. These marks may lie at any location on the buttstock. Cracks in the critical area at the front end of the buttstock are not acceptable and these buttstocks must be replaced.

CHAPTER 4

MAINTENANCE OF AUXILIARY EQUIPMENT

CHAPTER OVERVIEW

This chapter contains information and instructions to keep auxiliary equipment used with your weapon in good repair.

Section I. AUXILIARY EQUIPMENT REPAIR

4-1 GENERAL.

a The following items of auxiliary equipment are used in conjunction with the Weapons:

(1) 40mm Grenade Launcher M203, NSN 1010-00-179-6447, (M16A2 ONLY)

(2) Lock Plate, NSN 1005-00-233-9031.

(3) Top Sling Adapter, NSN 1005-00-406-1570.

(4) Blank Firing Attachment M15A2, NSN 1005-00-118-6192, (M16A2 ONLY), Blank Firing Attachment M23, NSN 1005-01-361-8208 (M4 ONLY)

Change 4**(5)** ARMY ONLY: Conversion Kit, M261 (caliber .22 rimfire adapter), NSN 1005-01-010-1561.

(6) Bayonet-Knife M7, NSN 1005-00-073-9238.

(7) Bayonet-Knife Scabbard M10, NSN 1095-00-223-7164.

(8) M9 Multi-Purpose Bayonet System, NSN 1005-01-227-1739.

(9) Night Vision Sight, Individual Served Weapon, AN/PVS-4, NSN 5855-00-629-5334 (M16A2 ONLY)

(10) M2 Practice Bolt, NSN 1005-01-184-4041 (M16A2 ONLY)

b Refer to TM 9-1010-221-23&P for unit maintenance for the Grenade Launcher M203.

c ARMY ONLY: Refer to TM 9-6920-363-12&P for unit maintenance for the M261 Conversion Kit (caliber rimfire adapter) M16, M16A1, and M16A2 Rifles.

d Refer to TM 9-1005-237-23&P for repair instructions and repair parts for Bayonet-Knife M7 and Bayonet Knife Scabbard M10 and M9 Multi-Purpose Bayonet System.

e Refer to TM 11-5855-213-23&P for maintenance for the Night Vision Sight, Individual Served Weapon, AN/PVS-4.

f Refer to TM 9-6920-746-12&P for unit maintenance for the M2 Practice Bolt.

4-2. LOCK PLATE

This task covers:

a Installation
b Removal
c Inspection

INITIAL SETUP

Tools
(ARMY) Small Arms Repairman Tool Kit
(item 3, app B)

WARNING

The lock plate prevents the selector lever from being placed in BURST and will be installed at the discretion of the unit commander. It is mandatory for use in civil disturbance (riot control). Do not keep live ammunition near work area.

a. INSTALLATION

1 Using a screwdriver, reach inside pistol grip (1) and remove screw (2) and lockwasher (3).

WARNING

To avoid injury to your eyes, use care when removing and installing spring-loaded parts.

2 Carefully remove pistol grip (1). Hold detent helical spring (4) in place.

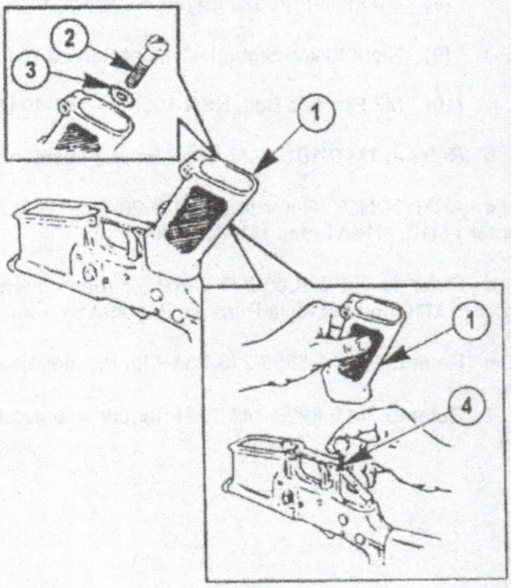

3 Install lock plate (5) with the detent helical spring (4) passing through the hole In the right side of the lock plate and the arm on the outside of the receiver The selector lever just point to the SAFE position

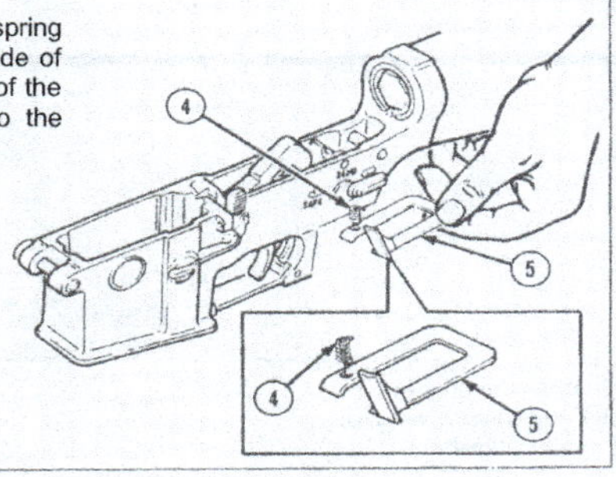

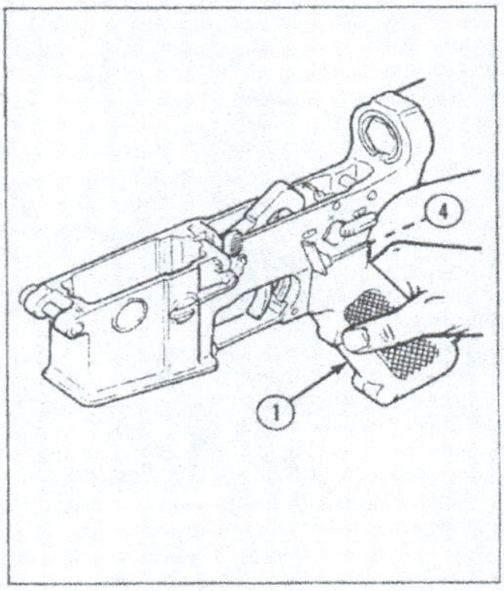

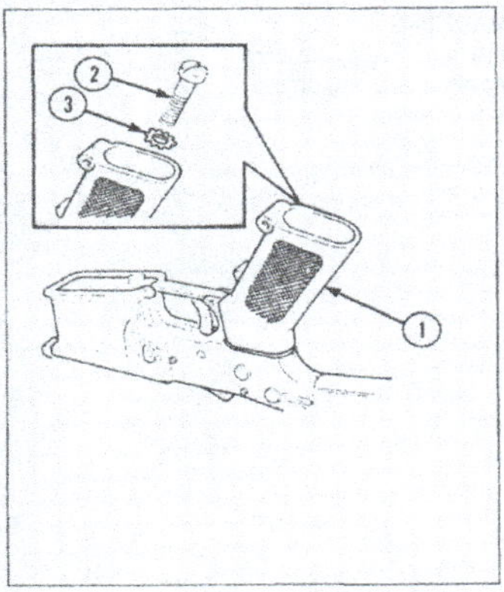

4 Carefully compress detent helical grip (1)

5. Using a screwdriver secure pistol by Installing lockwasher 131 spring (4) and position pistol grip (1) and screw 12)

4-2. LOCK PLATE (CONT).

b REMOVAL

1. Using a screwdriver, reach inside pistol grip (l) and remove screw (2) and lockwasher (3)

WARNING

To avoid Injury to your eyes, use care when removing and installing spring loaded parts

2 Carefully remove pistol grip (1) Hold detent helical spring (4) In place

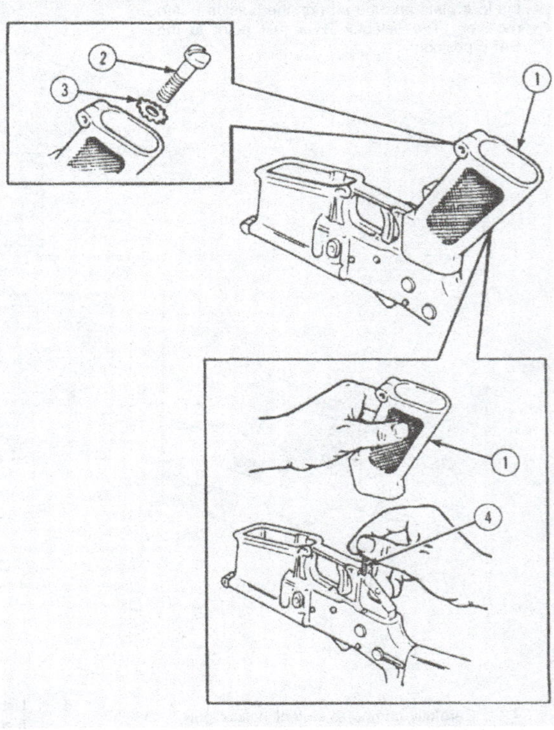

3 Remove lock plate (51 while continuing to hold detent helical spring (4) In place The selector lever must point to the SAFE position

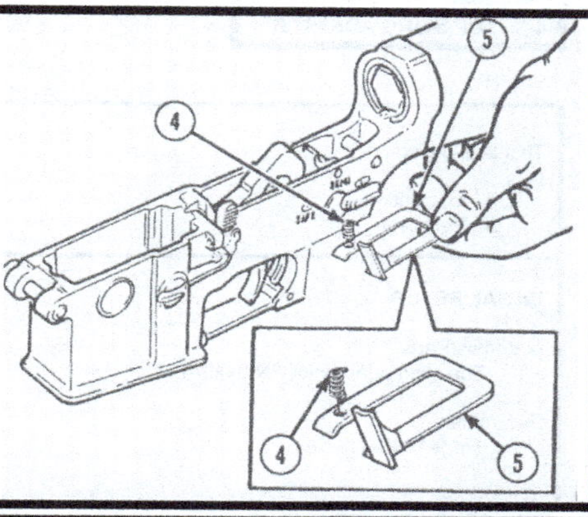

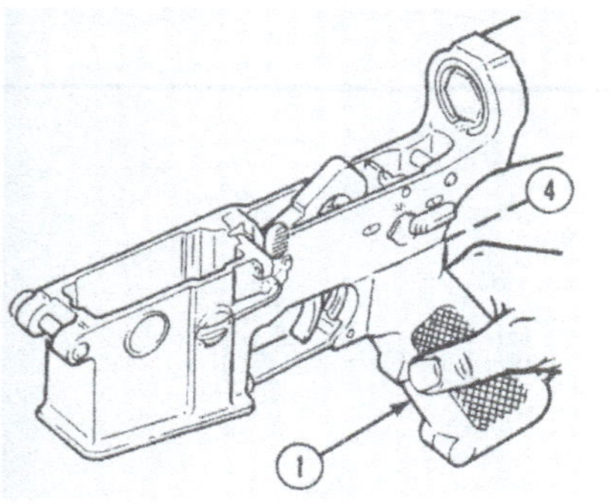

4 Carefully compress detent helical grip (l)

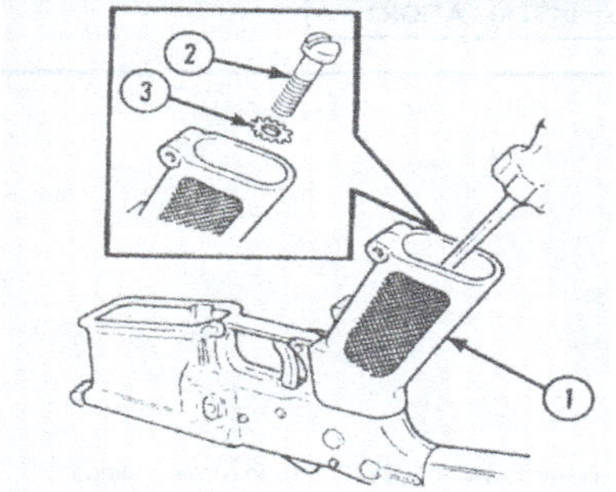

5. Using a screwdriver, secure pistol by installing lockwasher (3) spring (4) and position pistol grip (1) and screw (2)

c. INSPECTION

Inspect lock plate for serviceability and broken arm Replace If unserviceable or if arm is broken off

4-3. TOP SLING ADAPTER

This task covers:

a Installation
b Removal
c Inspection

INITIAL SETUP

Materials/Parts
Top sling adapter kit PN 8448471

References
TM 9-1005-319-10

a. INSTALLATION I

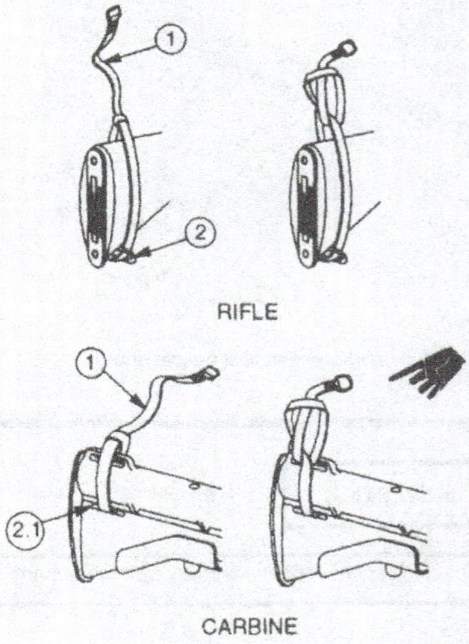

1 Refer to TM 9-1005-319-10 to remove sling.

RIFLE ONLY

2 Install top sling adapter strap (1) through sling swivel (2) and tie.

CARBINE ONLY

2.1. Install top sling adapter strap (1) through sling opening (2.1) and tie.

3 Remove upper handguard assembly.

4 Use pliers to install clamp (3) on front sight base (4).

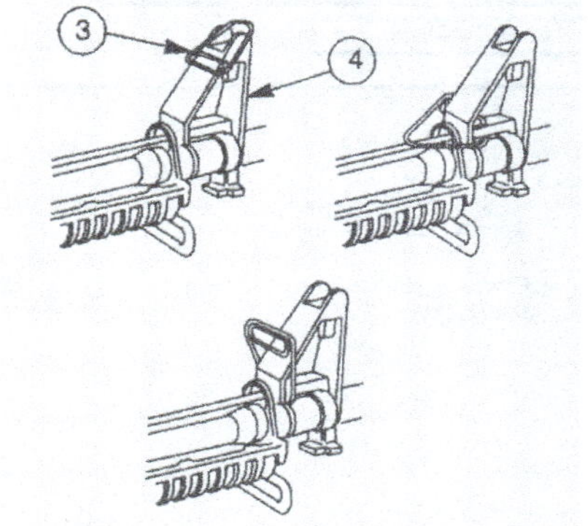

5 Attach sling (5) to top sling adapter strap (1) and to clamp (3).

6 Reinstall upper handguard assembly (6).

7 Adjust sling (5).

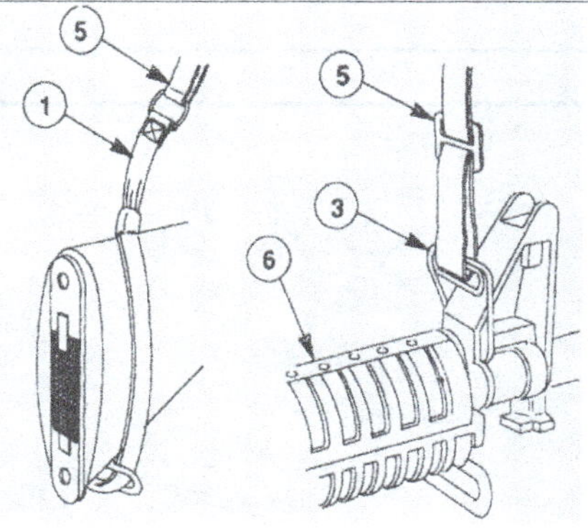

4-3. TOP SLING ADAPTER (CONT).

b. REMOVAL

1 Remove upper handguard assembly.

2 Remove sling (1) from clamp (2) and top sling adapter strap (3).

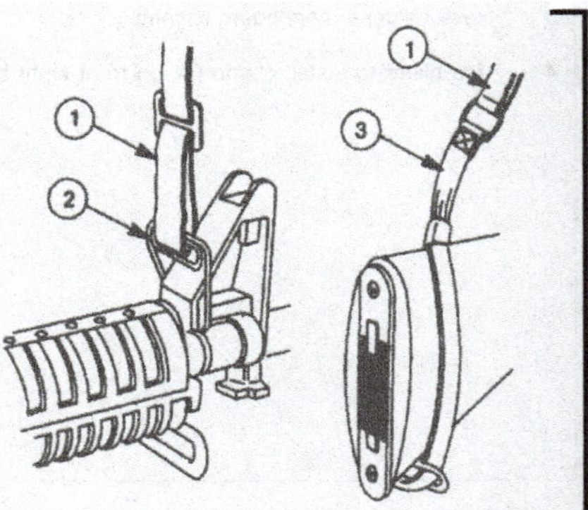

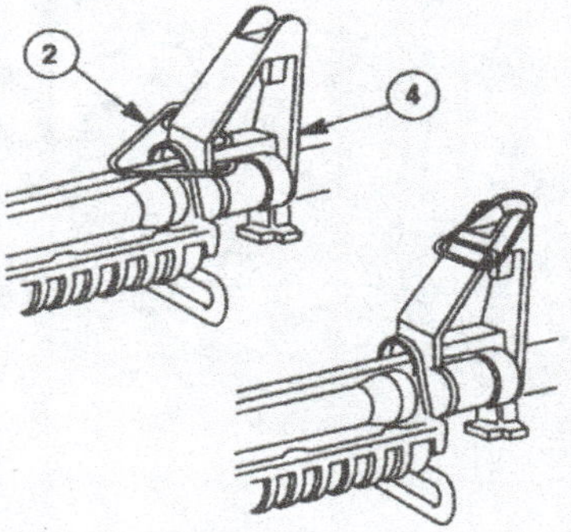

3 Use pliers to remove clamp (2) from front sight base (4). (5).

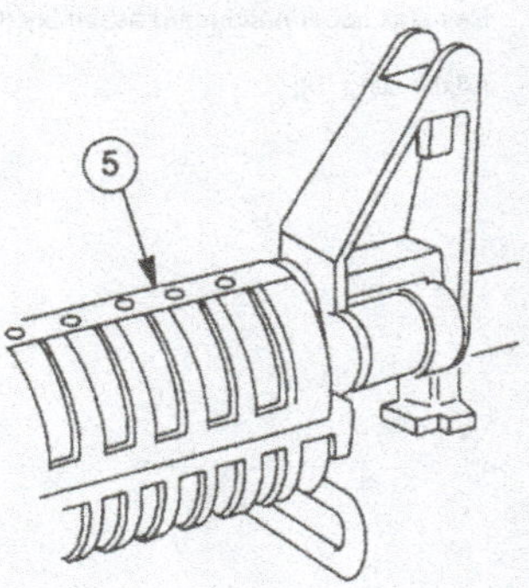

4 Reinstall upper handguard assembly

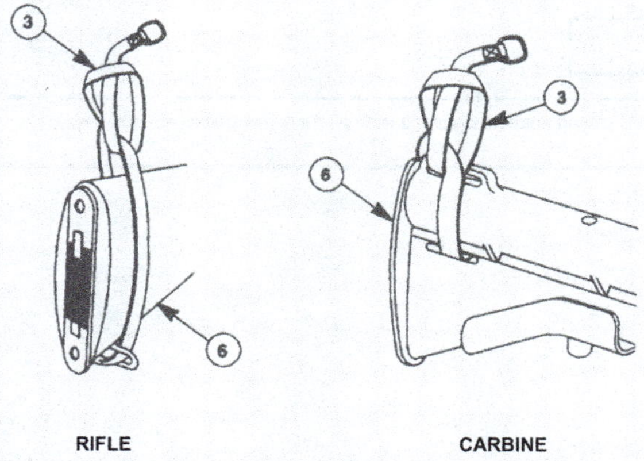

5 Untie top sling adapter strap (3) and remove from buttstock (6).

6 Refer to TM 9-1005-319-10 and install sling.

4-3. TOP SLING ADAPTER (CONT)

c. INSPECTION

Visually Inspect top sling adapter strap and replace If it is badly worn or damaged

4-4. BLANK FIRING ATTACHMENTS M15A2 (M16A2) AND M23 (M4/M4A1).

This task covers:

a Installation
b Removal
c Cleaning
d Inspection
e Repainting
f Replacement

INITIAL SETUP

Materials/Parts
Cleaner, lubricant, and preservative (CLP) (item 9, app D)
Coating compound, enamel (Red - M16A2) (item 23.2, app D)
Coating compound, enamel (Yellow - M4/M4A1) (item 23.3, app D)

General Safety Instructions
Do not keep live ammunition near the work area.

Only blank cartridge M200 is to be used when the blank firing attachment is attached to the weapon.
Do not fire blank ammunition at a representative enemy at distances of less than 20 feet (6.10m).
The unburned propellant grains can cause injury within this distance.

a. INSTALLATION

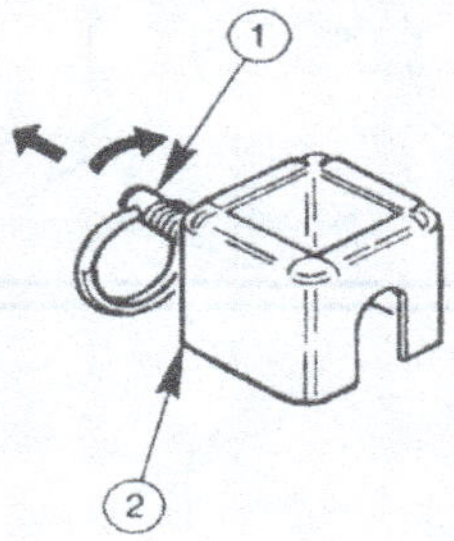

NOTE
M23 blank firing attachment is stamped "M4 Carbine Only" painted yellow and may be used on the M4 and M4A1 carbines. M15A2 BFA is painted red and is used on the M16A2 rifle.

1 Unscrew and pull slide (1) all the way out on blank firing attachment (2)

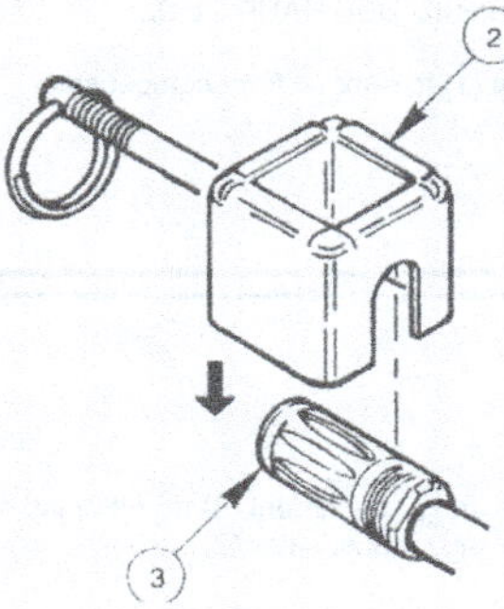

2 Hook blank firing attachment (2) behind the first groove of the compensator (3).

4-4 BLANK FIRING ATTACHMENTS M15A2 (M16A2) AND M23 (M4/M4A1)(CONT).

a. INSTALLATION (CONT)

CAUTION

Do not use tools to tighten the blank firing attachment. **USE HANDS ONLY.**

3 Push slide (2) into compensator (3) and hand tighten.

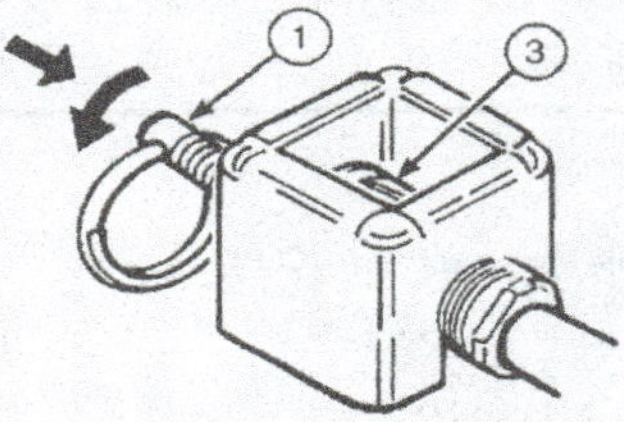

NOTE

Check for tightness after firing approximately 50 blank rounds.

b. REMOVAL

CAUTION

Do not use tools to tighten the blank firing attachment. USE HANDS ONLY.

1 Unscrew slide (1) to remove from compensator (2).

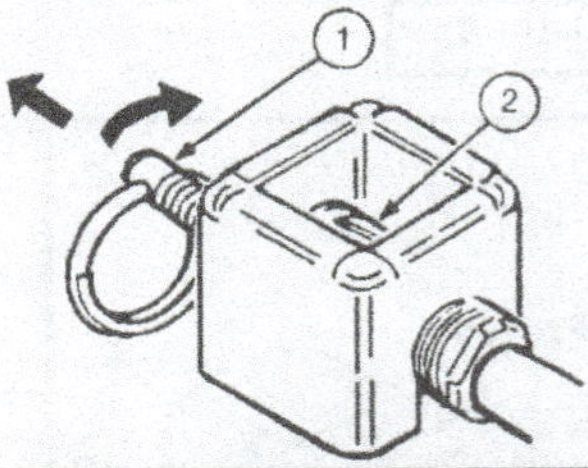

2. Unhook blank firing attachment (3) from behind the first groove of the compensator (2)

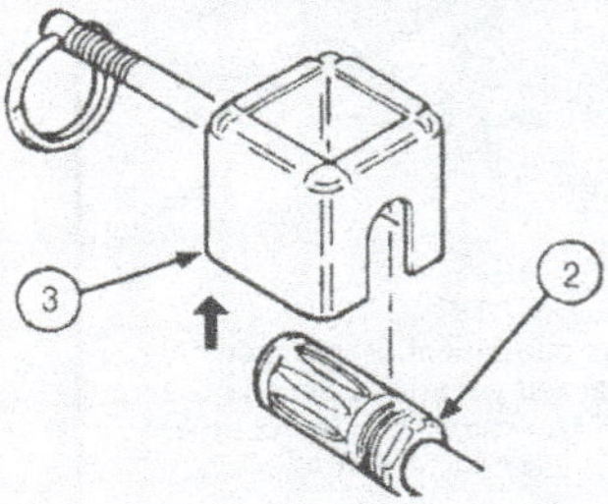

3 Screw slide (1) all the way in on blank firing attachment (3).

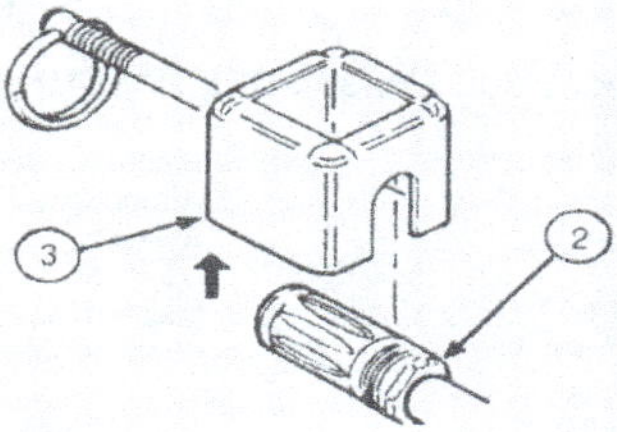

c. CLEANING

Clean blank firing attachment with CLP, wipe dry, and coat with CLP.

d. INSPECTION

Inspect blank firing attachment for cracks or distortion. Be sure the parts in the slide are clear and clean. If blank firing attachment is cracked or distorted, it is unserviceable.

e. REPAINTING

Repaint blank firing attachment using enamel coating compound (Red for M16A2 rifle or Yellow for M4/M4A1 carbine). Painting is the only repair authorized.

f. REPLACEMENT

Replace blank firing attachment if unserviceable.

APPENDIX A
REFERENCES

A-1. TECHNICAL BULLETINSINSTRUCTIONS/MANUALS/ORDERS.

TB 9-1000-247-34 ....Standards for Overseas Shipment or Domestic Issue of Small Arms, Aircraft Armament, Toward Howitzers, Mortars, Recoilless Rifles, Rocker Launchers and Associated Fire Control Equipment

TB 43-0002-73 ....Maintenance Expenditures Limits for FSC Group 10; FSC Classes 1005, 1010, 1015, 1030, 1055, 1090, and 1095

TI 05538A-15/2A ....Early Warning Indications of Malfunctions

TI 05538A-35/12A ....Checking Muzzle and Breech Erosion in Barrel

TI 8005-24/20 ....Prefire Inspection, Small Arms Weapons, Ordnance Material

TM 9-1005-237-23&P.............................Organizational and Direct Support Maintenance Manual (Including Repair Parts and Special Tools List) for Bayonet-Knife M6 and M7, with Bayonet-Knife Scabbard M10 and M9 Multipurpose Bayonet System

TM 9-1005-319-10..................................Operator's Manual for 5.56mm, M16A2 Rifle and M4/M4A1 Carbine

TM 9-1010-221-23&P.............................Organizational, Direct Support and General Support Maintenance Manual (Including Repair Parts and Special Tools List) for Launcher, Grenade: 40mm M203 W/E (NSN 1010-00-179-6447)

TM 9-6920-363-12&P.............................M261 Conversion Kit

TM 9-6920-746-12&P.............................M2 Practice Bolt

TM 11-5855-213-23&P...........................Night Vision Sight, Individual Served Weapon, AN/PVS-4

TM 750-244-7...Procedures for Destruction of Equipment in Federal Supply Classifications 1000, 1005, 1015, 1020, 1025, 1030, 1055, 1090, and 1095 to Prevent Enemy Use

TM 4700-15/1 ....Equipment Record Procedures

TO 00-35D-54 ....Materiel Deficiency Reporting and Investigating System

TO 11 A-13-10-7 ....Storage and Maintenance Procedures for Small Arms Ammunition

TO 11W-1-10 ....Historical Data Recording of Inspection, Maintenance, and Firing Data for Ground Weapons

TO 33K-1-100...(TMDE) Interval Calibration and Repair Reference Guide and Work Unit Code Manual

A-2. FIELD MANUALS.

FM 21-11 ....First Aid for Soldiers

FM 23-9 ....M16A1 and M16A2 Rifle and Rifle Marksmanship

A-3 PAMPHLETS.

DA PAM 25-30..Consolidated Index of Army Publications and Blank Forms

DA PAM 738-750...................................The Army Maintenance Management System (TAMMS)

A-4 RELATED PUBLICATIONS.

AFM 36-2227, Vol 1................................Combat Arms Training Management and Range Operation

AMC-P 310-9..Equipment Publications Listing

AR 725-50 ....Requisitioning, Receipt, and Issue System

AR 750-1 ....Army Materiel Policies

CTA 8-100 ....Army Medical Department Expendable/Durable Items

CTA 50-970 ....Expendable/Durable Items (Except: Medical, Class V, Repair Parts and Heraldic Items)

DOD 4160.21-M-1 ....Defense Demilitarization Manual

FMFM 1-3A...Field Firing Techniques

SB 746-1...Publications for Packing Army General Supplies

SPI 00-856-6885 ....Special Packaging Instructions for M16 Rifle

A-5 FORMS.

AFTO Form 22 ....Technical Order System Publications Improvement Report

AFTO Form 105 ....Inspection, Maintenance, and Firing Data for Ground Weapons

DA Form 2028 ....Recommended Changes to Publications and Blank Forms

DA Form 2028-2.....................................Recommended Changes to Equipment Technical Manuals

DA Form 2404 ....Equipment Inspection and Maintenance Worksheet

DA Form 2408-9.....................................Equipment Control Record

DD Form 314 ....Preventive Maintenance Schedule and Record

NAVMC 10772 .... Recommended Changes to Technical Publications

NAVMC 11003 .... Ordnance Serialized Items Subsidiary Records

SF 364 .... Report of Discrepancy (ROD)

SF 368 .... Product Quality Deficiency Report

A-6 SUPPLY CATALOGS/SUPPLY LISTS.

SC 4933-95-CL-A11 .... Shop Set, Small Arms Field Maintenance

SC 5180-95-CL-A07 .... Tool Kit, Small Arms Repair

APPENDIX B

MAINTENANCE ALLOCATION CHART

Section I. INTRODUCTION

B-1. GENERAL.

a. This section provides a general explanation of all maintenance and repair functions authorized at various maintenance levels

b. The Maintenance Allocation Chart (MAC) In section II designates overall authority and responsibility for the performance of maintenance functions on the identified end Item or component. The application of the maintenance functions to the end Item or component will be consistent with the capacities and capabilities of the designated maintenance levels

c. Section III lists the tools and test equipment (both special tools and common tool sets) required for each maintenance function as referenced from section II.

d. Section IV contains supplemental instructions and explanatory notes for a particular maintenance function.

B-2. MAINTENANCE FUNCTIONS.

Maintenance functions will be limited to and defined as follows: (except for ammunition MAC')

a. Inspect. To determine the serviceability of an Item by comparing Its physical, mechanical, and or electrical characteristics with established standards through examination leg , by sight, sound, or feel).

b. Test. To verify serviceability by measuring the mechanical, pneumatic, hydraulic, or electrical characteristics of an Item and comparing those characteristics with prescribed standards

c. Service. Operations required periodically to keep an Item In proper operating condition, I e to clean (includes decontaminate, when required), to preserve, to drain, to paint, or to replenish fuel, lubricants, chemical fluids. or gases.

d. Adjust. To maintain or regulate, within prescribed limits, by bringing Into proper or exact position, or by setting the operating characteristics to specified parameters

'Exception is authorized for ammunition MAC to permit the redesignation/redefinition of maintenance function headings to more adequately identify ammunition maintenance functions The heading designations and definitions will be Included In the appropriate technical manual for each category of ammunition

B-2. MAINTENANCE FUNCTIONS (CONT).

e. Align. To adjust specified variable elements of an item to bring about optimum or desired performance

NOTE

Air Force Precision Measurement Equipment Laboratory (PMEL) person will calibrate small arms Inspection gages In accordance with TO 33K-1-100

f. Calibrate. To determine and cause corrections to be made or to be adjusted on Instruments or test, measuring, and diagnostic equipments used In precision measurement Consists of comparisons of two Instruments, one of which Is a certified standard of known accuracy, to detect and adjust any discrepancy in the accuracy of the Instrument being compared

g. Remove/Install. To remove and Install the same item when required to perform service or other maintenance functions. Install may be the act of emplacing, seating, or fixing Into position a spare, repair part, or module (component or assembly) In a manner to allow the proper functioning of an equipment or system.

h. Replace. To remove an unserviceable Item and Install a serviceable counterpart In Its place "Replace"' is authorized by the MAC and is shown as the 3d position code of the SMR code.

i. Repair. The application of maintenance services2, including fault location troubleshooting3, removal 'installation, and disassembly'assembly4 procedures, and maintenance actions" to Identify troubles and restore serviceability to an Item by correcting specific damage, fault, malfunction, or failure In a part, subassembly, module (component or assembly), end Item, or system.

j. Overhaul. That maintenance effort (service/action) prescribed to restore an item to a completely serviceable operational condition as required by maintenance standards In appropriate technical publications (i.e , DMWR). Overhaul is normally the highest degree of maintenance performed by the Army Overhaul does not normally return an Item to like new condition

k. Rebuild. Consists of those services/actions necessary for the restoration of unserviceable equipment to a like new condition In accordance with original manufacturing standards Rebuild is the highest degree of materiel maintenance applied to Army equipment The rebuild operation Includes the act of returning to zero those age measurements (hours miles, etc) considered In classifying Army equipment/components.

---

[2]Servlces--Inspect, test, service, adjust, align, calibrate, and or replace

[3]Fault locate troubleshoot The process of investigating and detecting the cause of equipment malfunctioning; the act of isolating a fault within a system or unit under test (UUT).

[4]Disassemble assemble encompasses the step-by-step taking apart (or breakdown) of a spare/functional group coded Item to the level of Its least competency identified as maintenance significant (.e , assigned an SMR code) for the level of maintenance under consideration '

[5]Actionswelding, grinding, riveting, straightening, facing, remachinery, and or resurfacing

B-3. EXPLANATION OF COLUMNS IN THE MAC, SECTION II.

a. **Column 1, Group Number**. Column 1 lists functional group code numbers, the purpose of which is to identify maintenance significant components, assemblies, subassemblies, and modules with the next higher a assembly. End item group number shall be "OO."

b. **Column 2, Component Assembly**. Column 2 contains the names of components. Assemblies, subassemblies, and modules forwhich maintenance is authorized.

c. **Column 3, Maintenance Function**. Column 3 lists the functions to be performed on the it em listed in Column 21. (for detailes explanation of these functions, see paragraph B-2.)

d. **Column 4, Maintenance Level**. Column 4 specifies, by the list ing of a work time figure in the appropriate subcolumn(s), the level of maintenance. If the number of complexity of the t asks within the listed maintenance function vary at different maintenance levels, appropriate work time figures will be shown for each level. The work time figure represents the average time required to restore an item (assembly, subassembly, component, module, end it4m, or system) or a serviceable condition under typicalfield operating conditions. This time includes preparation time (including any necessary disassembly assembly time), troubleshooting fault location time, and quality assurance quality control time in addition to the time required to perform the specific tasks identified for the maintenance functions authorized i the maintenance allocation chart. The symbol designations for the varies maintenance levels are as follows:

C..Operator or Crew
O..Unit Maintenance
F..Direct Support Maintenance
H..General Support Maintenance
L..Specialized Repair Activity (SRA)[6]
D..Depot Maintenance

e. **Column 5, Tools and Equipment**. Column 5 specifies, by code, those common tool sets (not individual tools) and special tools., TMDE, and support equipment required to perform the designated function.

f. **Column 6, Remarks**. This column shall, when applicable, contain a lett4er code, in alphabetic order, which shall be keyed to the remarks contained in Section IV.

[6]This maintenance level is not included in Section II, column (4) of the Maintenance Allocation Chart. To identify functions to this level of maintenance, enter a work time figure in the "H" column of section II, column (4), and use an associated reference code in the Remarks column (6). Key the c ode to Section IV . Remarks, and explain the SRA complete repair application there. The explanatory remarks)s) shall reference the specific Repair Parts and Special Tools List (RPSTL) TM which contains additional SRA criteria and the authorized spare/repair parts.

B-4. EXPLANATION OF COLUMNS IN TOOL AND TEST EQUIPMENT REQUIREMENTS, SECTION III.

a. Column 1, Reference Code. The tool and test equipment reference code correlates with a code used In the MAC, Section II, Column 5.

b. Column 2, Maintenance Level. The lowest level of maintenance authorized to use the tool or test equipment.

c. Column 3, Nomenclature. Name or identification of the tool or test equipment.

d. Column 4, National Stock Number. The National stock number of the tool or test equipment.

e. Column 5, Tool Number. The manufacturer's part number.

B-5. EXPLANATION OF COLUMNS IN REMARKS, SECTION IV.

a. Column 1, Reference Code. The code recorded In column 6, Section II.

b. Column 2, Remarks. This column lists information pertinent to the maintenance function being performed as indicated In the MAC, Section II.

Section II. MAINTENANCE ALLOCATION CHART FOR M16A2 RIFLE AND M4/M4A1 CARBINES

(1) GROUP NUMBER	(2) COMPONENT ASSEMBLY	(3) MAINTENANCE FUNCTION	(4) MAINTENANCE LEVEL UNIT C	UNIT O	DS F	DS H	DEPOT D	(5) TOOLS AND EQPT	(6) REMARKS CODE
00	M16A2 5.56MM RIFLE, AND M4/ M4A1 5.56MM CARBINES	Inspect	0.1	0.3	0.7			2	
		Test			0.2			2,3	
		Service	0.2	0.3	0.3				
		Replace		0.1					
		Overhaul							
00.1	CARRYING HANDLE ASSEMBLY M4/M4A1	Inspect	0.1	0.1	0.1				
		Service	0.1						
		Remove/ Install	0.1	0.1	0.1				
		Replace			0.3			1,3	
		Repair		0.1	0.2			1,3	
00.101	Rear Sight Assembly M4/ M4A1	Inspect			0.1				
		Replace			0.3			1,3	
		Repair			0.2			1,3	
01	BOLT CARRIER ASSEMBLY	Inspect	0.1	0.1	0.1				
		Service	0.1	0.1					
		Remove/ Install	0.1						
		Replace			0.1			1,2	
		Repair	01	0.1				3	
0101	Bolt Assembly	Inspect	0.1	0.1	0.1				
		Test			0.1			2	
		Service	0.1						
		Remove/ Install	0.1						
		Replace			0.1			2	
		Repair		0.1	0.1			3	

*'Worktimes are included in DMWR 9-1005-249

MAINTENANCE ALLOCATION CHART (CONT)

(1) GROUP NUMBER	(2) COMPONENT ASSEMBLY	(3) MAINTENANCE FUNCTION	(4) MAINTENANCE LEVEL UNIT C	UNIT O	DS F	DS H	DEPOT D	(5) TOOLS AND EQPT	(6) REMARKS CODE
0102	Key and Bolt Carrier Assembly	Inspect	0.1		0.1				
		Service	0.1						
		Remove/ Install	0.1						
		Replace			0.1			1	
		Repair		0.1	0.1			1,3	
02	HANDLE ASSEMBLY	Inspect	0.1						
		Service	0.1						
		Remove/ Install	0.1						
		Replace			0.1		**		
		Repair			0.1			3	
03	UPPER RECEIVER AND BARREL ASSEMBLY	Inspect	0.1	0.2	0.1				
		Test			0.2			2	
		Service	0.2		0.1			1	
		Remove/ Install	0.1						
		Replace			0.5			1,2	
		Repair		0.2	0.5			1,2,3	A,B
0301	M16A2 Barrel Assembly and M4/ M4A1 Replacement Barrel	Inspect		0.1	0.1				
		Replace			0.3			1,2	
		Repair	0.1					3	
0302	Upper Receiver Assembly	Inspect			0.1				
		Replace			0 5			1.2	
		Repair			0.3			1.3	
030201	Forward Assist Assembly	Inspect			0.1				
		Replace			0.1				
		Repair			0 1			1.3	
030202	Rear Sight Assembly M16A2	Inspect			0 1				
		Replace			0.3			1,3	
		Repair			0 2			1,3	

MAINTENANCE ALLOCATION CHART (CONT)

(1) GROUP NUMBER	(2) COMPONENT ASSEMBLY	(3) MAINTENANCE FUNCTION	(4) MAINTENANCE LEVEL UNIT C	O	DS F	DS H	DEPOT D	(5) TOOLS AND EQPT	(6) REMARKS CODE
04	LOWER RECEIVER AND BUTTSTOCK ASSEMBLY	Inspect	0.1	0.1	0.1				
		Test			0.1			2	
		Service	0.2	0.2					
		Repair		0.2	0.3			1,2,3	C,D,E
0401and 0401A	Buttstock Assembly	Inspect		0.1					
		Remove/ Install		0.1				3	
		Replace		0.1				3	
		Repair		0.1				3	
0402	Hammer Assembly	Inspect			0.1				
		Remove/ Install			0.1				
		Replace			0.1				
		Repair			0.1				
0403	Trigger Assembly	Inspect			0.1				
		Remove/ Install			0.1			1,3	
		Replace			0.1			1,3	
		Repair			0.1			1,3	
040301	Trigger Subassembly M16A2 and M4	Inspect			0.1				
		Replace			0.1			1,3	
		Repair			0.1			1,3	
0404 and 0404A	Lower Receiver and Receiver Extension Assembly	Inspect			0.1				
		Test			0.1			2	
		Repair			0.2			1,3	

Section III. TOOL AND TEST EQUIPMENT REQUIREMENTS FOR M16A2 RIFLE AND M4/M4A1 CARBINES

TOOL OR TEST EQUIPMENT REF CODE	(2) MAINTENANCE CATEGORY	(3) NOMENCLATURE	(4) NATIONAL/NATO STOCK NUMBER	(5) TOOL NUMBER
1 (ARMY ONLY)	F	Shop Set, Small	4933-00-754-0664	SC 4933-95-CL-A11
2 (ARMY ONLY)	F	Tool and gage Set, DS/GS Maintenance for 5.56mm Rifle, M16 Series	4933-00-056-7106	8426685
3 (ARMY ONLY)	O	Tool Kit, Small Arms Repairman	5180-00-357-7770	SC 5180-95-CL-A07
4 (A.F ONLY)	F	Torque Wrench, ft-lb	5120-00-640-6365	A-A-411
5 (A.F ONLY)	F in.-lb	Torque Wrench,	5120-00-230-6380	T-E-12A
6 A.F ONLY)	F	Trigger Weights	4933-00-647-3696	7274758

Section IV. REMARKS

Reference Code	Remarks
A	Tool, Front Sight Post Removal and Installation
B	Depressor, Front Sight Detent
C	Tool, Pivot Pin Removal
D	Tool, Pivot Pin Installation
E	Pin, Slave

APPENDIX C

REPAIR PARTS AND SPECIAL TOOLS LIST

Section I. INTRODUCTION

C-1. SCOPE.

This RPSTL lists and authorizes spares and repair parts; special tools; special test, measurement, and diagnostic equipment (TMDE); and other special support equipment required for performance of unit and direct support maintenance of the M16A2 rifle and M4/M4A1 carbines. It authorizes the requisitioning, issue, and disposition of spares, repair parts, and special tools as indicated by the source, maintenance, and recoverability (SMR) codes.

C-2. GENERAL.

In addition to Section I, Introduction, this Repair Parts and Special Tools List is divided into the following sections:

a. Section II Repair Parts List. A list of spares and repair parts authorized by this RPSTL for use in the performance of maintenance. The list also includes parts which must be removed for replacement of the authorized parts. Parts lists are composed of functional groups in ascending alphanumeric sequence, with the parts in each group listed in ascending figure and item number sequence. Bulk materials are listed by item name in FIG BULK at the end of the section. Repair parts kits or sets are listed separately in their own functional group within Section II. Repair parts for repairable special tools are also listed in the section.

b. Section III Special Tools List. A list of special tools, special TMDE, and other special support equipment authorized by this RPSTL (as indicated by Basis of Issue (BOI) information in DESCRIPTION AND USABLE ON CODE (UOC) column) for the performance of maintenance.

c. Section IV Cross-Reference Indexes. A list, in National item identification number (NIIN) sequence, of all National stock numbered items appearing in the listings, followed by a list in alphanumeric sequence of all part numbers appearing in the listing. National stock numbers and part numbers are cross-referenced to each illustration figure and item number in alphanumeric sequence and cross-references NSN, CAGEC, and part number.

C-3. EXPLANATION OF COLUMNS (SECTIONS II AND III).

a. ITEM NO. (Column (1)). Indicates the number used to identify items called out in the illustration.

b. SMR CODE (Column (2)). The Source, Maintenance, and Recoverability (SMR) code is a 5-position code containing supply/requisitioning information, maintenance level authorization criteria, and disposition instruction, as shown In the following breakout:

C-3. EXPLANATION OF COLUMNS (SECTION II AND III) (CONT).

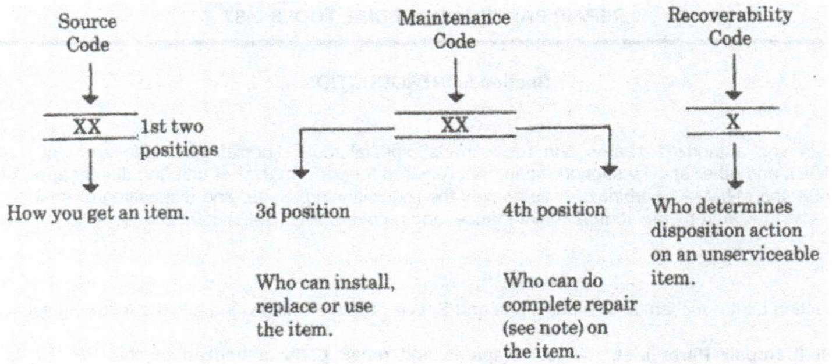

*Complete Repair: Maintenance capacity, capability, and authority to perform all corrective maintenance tasks of the "Repair" function in a use/user environment In order to restore serviceability to a failed item.

(1) Source Code. The source code tells you how to get an item needed for maintenance, repair, or overhaul of an end item/equipment. Explanations of source codes follow:

Code	Explanation
PA PB PC* * PD PE PF PG	Stocked items; use the applicable NSN to request/requisition Items with these source codes They are authorized to the level indicated by the code entered In the 3d position of the SMR code. * *NOTE: Items coded PC are subject to deterioration.
KD KF KB	Items with these codes are not to be requested/requisitoned Individually They are part of a kit which is author->zed to the maintenance level indicated In the 3d positon of the SMR code The complete kit must be requisitioned and applied.
MO-(Made at Unit/ AVUM Level) MF- (Made at DS/ AVIM Level) MH-(Made at GS Level) ML- (Made at Specialized Repair Act (SRA)) MD-(Made at Depot)	Items with these codes are not to be requested/requisitioned Individually They must be made from bulk material which Is identified by the part number In the DESCRIPTION AND USABLE ON CODE (UOC) column and listed In the Bulk Material group of the repair parts list in this RPSTL If the Item Is authorized to you by the 3d position code of the SMR code, but the source code Indicates It is made at a higher level, order the Item from the higher level of maintenance.

Code	Explanation
AO- (Assembled by unit/AVUM Level) AF- (Assembled by DS/AVIM Level) AH- (Assembled by GS Level) AL- (Assembled by SRA) AD- (Assembled by Depot)	Items with these codes are not to be requested/requisitioned Individually. The parts that make up the assembled Item must be requisitioned or fabricated and assembled at the level of maintenance indicated by the source code. If the 3d position code of the SMR code authorizes you to replace the Item, but the source code Indicates the Item Is assembled at a higher level, order the Item from the higher level of maintenance.

XA-Do not requisition an "XA"-coded item. Order Its next higher assembly. (Also, refer to the NOTE below.)

XB--If an "XB" item is not available from salvage, order It using the CAGEC and part number given.

XC-installation drawing, diagram, instruction sheet, field service drawing, that is identified by manufacturer's part number.

XD-Item is not stocked. Order an "XD"-coded Item through normal supply channels using the CAGEC and part number given, If no NSN Is available.

NOTE

Cannibalization or controlled exchange, when authorized, may be used as a source of supply for items with the above source codes, except for those source coded "XA" or those aircraft support Items restricted by requirements of AR 750-1.

(2) Maintenance Code. Maintenance codes tell you the level(s) of maintenance authorized to USE and REPAIR support items. The maintenance codes are entered In the third and fourth positions of the SMR code as follows:

(a) The maintenance code entered in the third position tells you the lowest maintenance level authorized to remove, replace, and use an Item. The maintenance code entered In the third position will indicate authorization to one of the following levels of maintenance

Code	Application/Explanation
C	-Crew or operator maintenance done within unit or aviation unit maintenance.
O	-Unit or aviation unit level can remove, replace, and use the Item.
F	-Direct support or aviation intermediate level can remove, replace, and use the Item.
H	-General support level can remove, replace, and use the Item.
L	-Specialized repair activity can remove, replace, and use the Item.
D	-Depot level can remove, replace, and use the Item

C-3. EXPLANATION OF COLUMNS (SECTIONS II AND III) (CONT).

(b) The maintenance code entered In the fourth position tells whether or not the Item is to be repaired and Identifies the lowest maintenance level with the capability to do complete repair (I e, perform all authorized repair functions). (NOTE: Some limited repair may be done on the Item at a lower level of maintenance, If authorized by the Maintenance Allocation Chart (MAC) and SMR codes I This position will contain one of the following maintenance codes

Code	Application/Explanation
O	-Unit or aviation unit is the lowest level that can do complete repair of the item
F	-Direct support or aviation Intermediate is the lowest level that can do complete repair of the Item
H	-General support is the lowest level that can do complete repair of the item .
L	-Specialized repair activity is the lowest level that can do complete repair of the item
D	-Depot is the lowest level that can do complete repair of the Item
Z	-Nonreparable. No repair Is authorized
B	-No repair is authorized. (No parts or special tools are authorized for the maintenance of a "B" coded item) However, the Item may be reconditioned by adjusting, lubricating, etc , at the user level

(3) Recoverability Code. Recoverability codes are assigned to Items to Indicate the disposition action on unserviceable items The recoverability code is entered In the fifth position of the SMR code as follows.

Recoverability Codes	Application/Explanation
Z	Nonreparable item unserviceable, condemn and dispose of the item at the level of maintenance shown In 3d position) of SMR code
O	Reparable Item. When uneconomically reparable condemn and dispose of the item at unit or aviation unit level
F	Reparable Item. When uneconomically reparable, condemn and dispose of the Item at the direct support or aviation level
H	Reparable Item. When uneconomically reparable, condemn and dispose of the Item at the .general support level

Recoverability Codes	Application/Explanation
D	-Reparable item. When beyond lower level repair capability, return to depot. Condemnation and disposal of item not authorized below depot level.
L	-Reparable item. Condemnation and disposal not authorized below Specialized Repair Activity (SRA).
A	-Item requires special handling or condemnation procedures because of specific reasons (e.g., precious metal content, high dollar value, critical material, or hazardous material). Refer to appropriate manuals/directives for specific instructions.

c. NSN (Column (3)). Indicates the National stock number (NSN) assigned to the item and which will be used for requisitioning.

d. CAGEC (Column (4)). The Contractor and Government Entity Code (CAGEC) is a 5-digit numeric code which is used to identify the manufacturer, distributor, or Government agency, etc., that supplies the Item.

e. PART NUMBER (Column (5)). Indicates the primary number used by the manufacturer (individual, company, firm, corporation, or Government activity), which controls the design and characteristics of the item by means of its engineering drawings, specifications standards, and inspection requirements to identify an item or range of items.

NOTE

When you use an NSN to requisition an item, the item you receive may have a different part number from the part ordered.

f. DESCRIPTION AND USABLE ON CODE (UOC) (Column (6)). This column includes the following Information:

(1) The Federal item name and, when required, a minimum description to identify the Item.

(2) The physical security classification of the item is indicated by the parenthetical entry which is a physical security classification abbreviation, e.g., Phy Sec C1 (C) Confidential, Phy Sec C1 (S) Secret, Phy Sec C1 (T) Top Secret.

(3) Items that are included in kits and sets are listed below the name of the kit or set.

(4) Spare/repair parts that make up an assembled item are listed immediately following the assembled item line entry.

(5) Part numbers for bulk materials are referenced in this column In the line item entry for the Item to be manufacture red/fabricated.

(6) When the item is not used with all serial numbers of the same model, the effective serial numbers are shown on the last line(s) of the description (before UOC).

(7) The usable on code, when applicable (see paragraph 5, Special Information).

C-3. EXPLANATION OF COLUMNS (SECTION II AND III) (CONT).

(8) In the Special Tools List section, the basis of issue (BOI) appears as the last line(s) in the entry for each special tool, special TMDE, and other special support equipment. When density of equipment supported exceeds density spread indicated in the basis of Issue, the total authorization is increased proportionately.

(9) The statement "END OF FIGURE" appears just below the last item description in Column 5 for a given figure In both Section II and Section III.

g. Oty (Column (7)) . The Qty (quantity per figure column) indicates the quantity of the item used in the breakout shown on the illustration figure, which is prepared for a functional group, subfunctional group, or an assembly. A "V" appearing In this column in lieu of a quantity indicates that the quantity is variable and the quantity may vary from application to application.

C-4. EXPLANATION OF COLUMNS (SECTION IV).

a. NATIONAL STOCK NUMBER (NSN) INDEX.

(1) STOCK NUMBER column. This column lists the NSN by National Item Identification Number (NIIN) sequence. The NIIN consists of the last nine digits of the NSN.

NSN
(i.e., 5305-01-674-1467). When using this column to locate an item, ignore the first 4 digits NIIN of the NSN. However, the complete NSN should be used when ordering items by stock number.

(2) FIG. column. This column lists the number of the figure where the item Is Identified/located. The figures are in numerical order in Section II and Section III.

(3) ITEM column. The item number identifies the item associated with the figure listed in the adjacent FIG. column. This item is also identified by the NSN listed on the same line.

b. PART NUMBER INDEX. Part numbers in this index are listed by part number in ascending alphanumeric sequence (i.e., vertical arrangement of letter and number combination which places the first letter or digit of each group In order A through Z, followed by the numbers 0 through 9 and each following letter or digit in like order).

(1) CAGEC column. The Contractor and Government Entity Code (CAGEC) is a 5-digit numeric code used to identify the manufacturer, distributor, or Government agency, etc., that supplies the item.

(2) PART NUMBER column. Indicates the primary number used by the manufacturer (individual, firm, corporation, or Government activity), which controls the design and characteristics of the item by means of Its engineering drawings, specifications standards, and inspection requirements to identify an Item or range of Items.

(3) STOCK NUMBER column. This column lists the NSN for the associated part number and manufacturer identified In the PART NUMBER and CAGEC columns to the left.

(4) FIG. column. This column lists the number of the figure where the item is identified/located in Sections II and III.

(5) ITEM column. The item number is that number assigned to the item as it appears in the figure referenced in the adjacent figure number column.

C-5. SPECIAL INFORMATION.

a. ASSEMBLY INSTRUCTION. Detailed assembly instructions for items source coded to be assembled from component spare/repair parts are found in Chapter 3. Items that make up the assembly are listed immediately following the assembly item entry or reference is made to an applicable figure.

b. ASSOCIATED PUBLICATIONS. The publications listed below pertain to the M16A2 rifle, M4/M4A1 carbine, and its components:

Publication	Short Title
TM 9-1005-319-10	M16A2 Rifle and M4/M4A1 Carbine

c. USABLE ON CODE. The usable on code appears in the lower left corner of the DESCRIPTION column heading, Usable on codes are shown as "UOC..." in DESCRIPTION column (justified left) on the first line applicable item description/nomenclature. Uncoded items are applicable to all models. Identification of the usable on codes used in the RPSTL are:

Code	Used On
AR8	M16A2 Rifle
AS1	M4 Carbine
AY6	M4A1 Carbine

C-6. HOW TO LOCATE REPAIR PARTS.

a. When National Stock Number or Part Number is Not Known.

(1) First. Using the table of contents, determine the assembly group or subassembly group to which the item belongs. This is necessary since figures are prepared for assembly groups and subassembly groups, and listings are divided into the same groups.

C-6. HOW TO LOCATE REPAIR PARTS (CONT).

(2) **Second.** Find the figure covering the assembly group or subassembly group to which the item belongs.

(3) **Third**. Identify the item on the figure and use the Figure and Item Number Index to find the NSN.

b. When National Stock Number or Part Number is Known.

(1) **First.** Using the National Stock Number or the Part Number Index, find the pertinent National Stock Number or Part Number. The NSN index is in National Item Identification Number (NIIN) sequence. (See C4.a.(1).) The pat numbers in the Part Number Index are listed in ascending alpha numeric sequence. (See C4.b.) Both indexes cross-reference you to the illustration/figure and item number of the item you are looking for.

(2) **Second.** Turn to the figure and item number, verify that the item is the one you're looking for, then locate the item number in the repair parts list for the figure.

C-7. ABBREVIATIONS.

Not applicable.

(1) ITEM NO QTY	(2) SMR CODE	(3) NSN	(4) CAGEC	(5) PART NUMBER	(6) DESCRIPTION AND USABLE	(7) ON CODE (UOC)
					GROUP 00 (M16A2) RIFLE 5.56MM 9349000,(M4) CARBINE 5.56MM 9390000, AND (M4A1) CARBINE 5.56MM 12972700 FIG. C-1. (M16A2) RIFLE, (M4) AND CM4A1) CARBINES	
1	PAOOO	1005000179546	19204	8448517	HANDLE ASSEMBLY,CHA CHARGING..............................1	
2	AFFFF		19204	8448501	BOLT CARRIER ASSEMBLY (M16A2)..................................1 UOC:AR8	
2A	AFFFF	19200	12011849	BOLT	AND BOLT CARRI ASSEMBLY (M4.....................................1 AND M4A1)... UOC:AS1,AY6	
3	AFFFF	1005011879716	19200	9349050	UPPER RECEIVER AND BARREL ASSEMBLY ....1 (M16A2)... UOC:AR8	
3A	AFFFF	19200	12972680	UPPER	RECEIVER AND BARREL ASSEMBLY.................................1 (M4 AND M4A1).. UOC:AS1,AY6	
4	PACZZ	1005009215004	19200	8448670	MAGAZINE,CARTRIDGE (30 ROUND)................................1	
5	PACZZ	1005012164510	19204	12624561	SLING,SMALL ARMS (M16A2)..1 UOC:AR8	
5A	PACZZ	1005013689852	19200	12011996	SLING,SMALL ARMS (M4 AND M4A1)..............................1 UOC:AS1,AY6	
6	XAFFA	19200	9349100	LOWER	RECEIVER AND BUTTSTOCK ....1 ASSEMBLY (M16A2).. UOC:AR8	
6A	XAFFA		19200	9390011	LOWER RECEIVER AND BUTTSTOCK ....1 ASSEMBLY (4)... UOC:AS1	
6B	XAFFA		19200	12972690	LOWER RECEIVER AND BUTTSTOCK ....1 ASSEMBLY (M4A1)... UOC:AY6	
7	AFFFF		19200	12951011	CARRYING HANDLE ASS (M4 AND M4A1).........................1 UOC:AS1,AY6	

END OF FIGURE

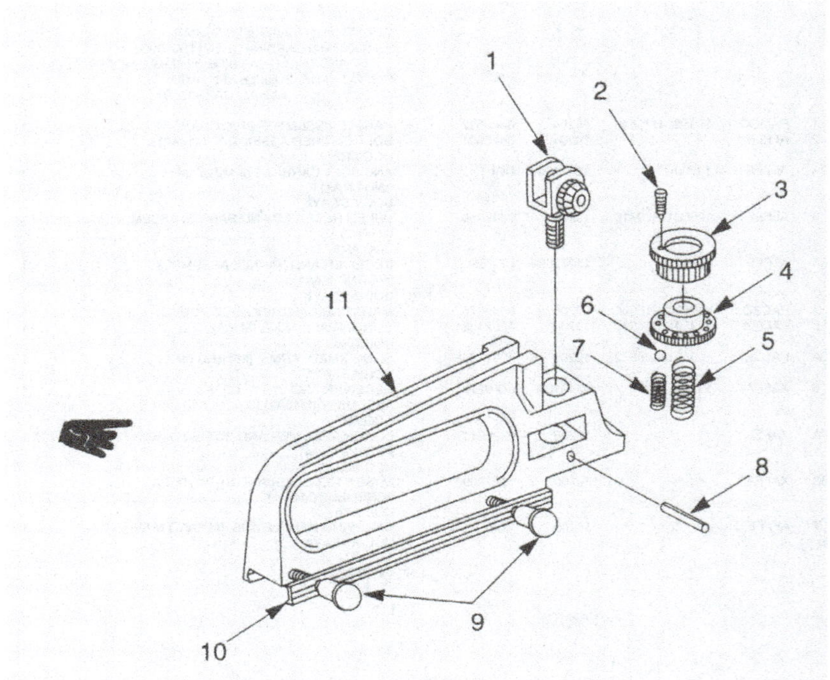

Figure C-1A. (M4 and M4A1) Carrying Handle Assembly 12951011.

(1) ITEM NO	(2) SMR CODE	(3) NSN	(4) CAGEC	(5) PART NUMBER	(6) DESCRIPTION AND USABLE ON CODE (UOC)	(7) QTY
					GROUP 00.1 (M4 AND M4A1) CARRYING HANDLE ASSEMBLY 12951011 FIG. C-1A. (M4 AND M4A1) CARRYING HANDLE ASSEMBLY	
1	AFFFF		19200	12951026	REAR SIGHT ASSY UOC:AS1,AY6	1
2	PAFZZ	5305011343622	19200	9349065	SCREW,INDEX	1
3	PAFZZ	1005013827089	19200	12951018	ELEVATING MECHANISM UOC:AS1 ,AY6	1
4	PAFZZ	5355013826801	19200	12951019	KNOB UOC:AS1,AY6	1
5	PAFZZ	5360011343710	19200	9349070	SPRING ,HELICAL,COMP	1
6	PAFZZ	3110001839175	96906	MS19060-4808	BALL,BEARING UOC:AS1,AY6	1
7	PAFZZ	5360013826802	19200	12951020	SPRING,HELICAL,COMP UOC:AS1,AY6	1
8	PAFZZ	5315008403812	96906	MS16562-121	PIN,SPRING	1
9	PAOZZ	5310013826793	19200	12951023	NUT,PLAIN,KNURLED UOC:AS1,AY6	2
10	PAOZZ	5340013823201	19200	12951017	BRACKET,MOUNTING UOC:AS1,AY6	1
11	PAFZZ	1005013827083	19200	12951021	HANDLE,GUN CARRYING UOC:AS1,AY6	1

END OF FIGURE

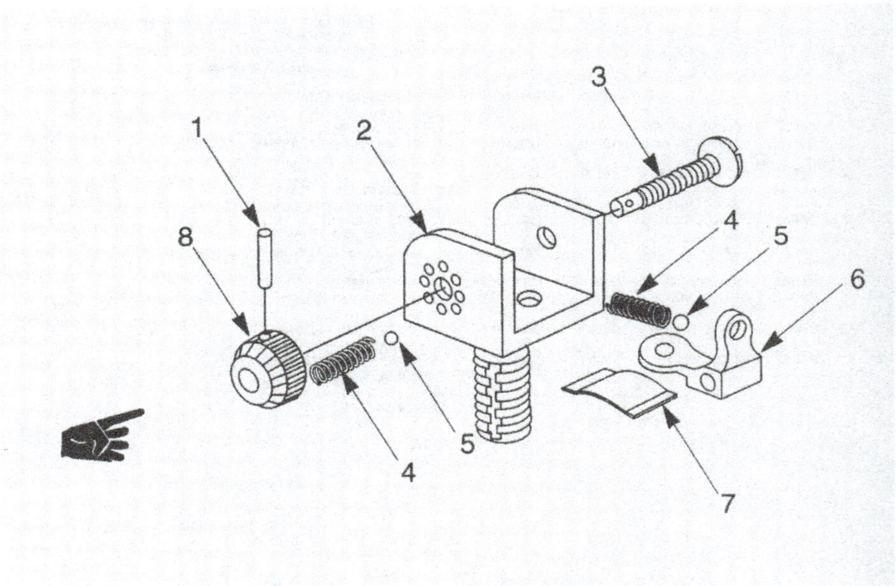

Figure C-1B. (M4 and M4A1) Rear Sight Assembly 12951026.

(1) ITEM NO	(2) SMR CODE	(3) NSN	(4) CAGEC	(5) PART NUMBER	(6) DESCRIPTION AND USABLE ON CODE (UOC)	(7) QTY
					GROUP 00.101 (M4 AND M4A1) REAR SIGHT ASSEMBLY 12951026 FIG. C-18. (M4 AND M4A1) REAR SIGHT ASSEMBLY	
1	PAFZZ	5315000586678	96906	MS16562-103	PIN,SPRING	1
2	PAFZZ	1005013827086	19200	12951028	BASE,REAR SIGHT UOC:AS1,AY6	1
3	PAFZZ	5305011441490	19200	9349076	SCREW,EXTERNALLY RE	1
4	PAFZZ	5360011481751	19200	9349069	SPRING,HELICAL,COMP UOC:AS1,AY6	2
5	PAFZZ	3110001839175	96906	MS19060-4808	BALL,BEARING UOC:AS1,AY6	2
6	PAFZZ	1005011353697	19200	9349075	APERTURE,SIGHT	1
7	PAFZZ	5360013816183	19200	12011987	SPRING,FLAT	1
8	PAFZZ	5355011343627	19200	9349077	KNOB,WINDAGE	1

END OF FIGURE

SECTION II

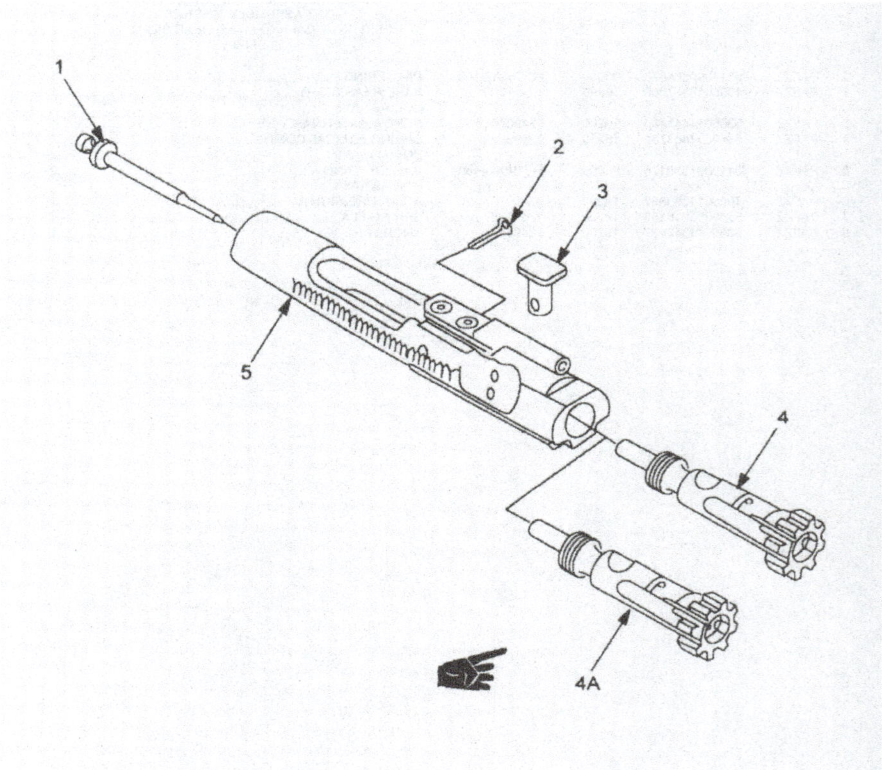

Figure C-2. Bolt Carrier Assembly (M16A2) 8448501 and (M4 and M4A1) 12011849.

(1) ITEM NO	(2) SMR CODE	(3) NSN	(4) CAGEC	(5) PART NUMBER	(6) DESCRIPTION AND USABLE ON CODE (UOC)	(7) QTY
					GROUP 01 BOLT CARRIER ASSEMBLY (M16A2) 8448501 AND (M4 AND M4A1) 12011849 FIG. C-2. BOLT CARRIER ASSEMBLY	
1	PAFZZ	1005000179547	19204	8448503	PIN,FIRING	1
2	PAOZZ	1005009991509	19204	8448504	PIN,FIRING PIN RETA RETAINING	1
3	PAOZZ	5315009927294	19204	8448502	PIN,GROOVED,HEADED BOLT CAM	1
4	PAFFF	1005009927285	19200	8448509	BOLT ASSEMBLY (M16A2) UOC:AR8	1
4A	PAFFF	1005014223770	19200	12972691	BOLT,BREECH ASSEMBLY (M4 AND M4A1) UOC:AS1,AY6	1
5	AFFFF		19204	8448505	KEY AND BOLT	1

END OF FIGURE

SECTION II

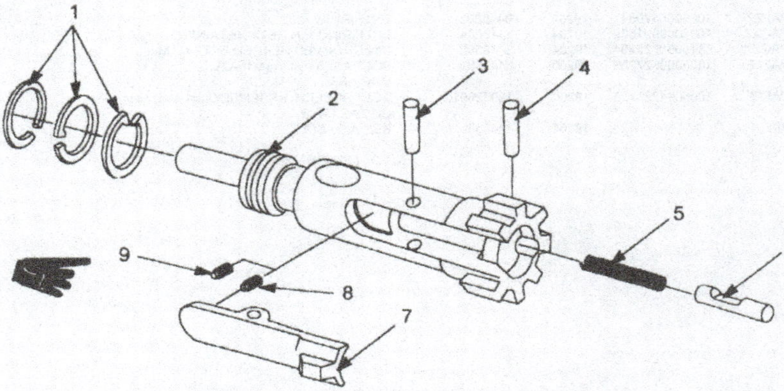

Figure C-3. Bolt Assembly (M16A2) 8448509 and (M4 and M4A1) 12972691.

(1) ITEM NO	(2) SMR CODE	(3) NSN	(4) CAGEC	(5) PART NUMBER	(6) DESCRIPTION AND USABLE ON CODE (UOC)	(7) QTY
					GROUP 0101 BOLT ASSEMBLY (M16A2) 8448509 AND (M4 AND M4A1) 12972691 FIG. C-3. BOLT ASSEMBLY	
1	PAFZZ	1005009927287	19204	8448511	RING,BOLT	3
2	XAFZZ		19204	8448510	BOLT	1
3	PAOZZ	1005009927290	19204	8448513	PIN,EXTRACTOR	1
4	PAOZZ	5315005975086	96906	MS16562-9B	PIN,SPRING EJECTOR	1
5	PAOZZ	5360009927292	19204	8448516	SPRING,HELICAL,COMP COMPRESSION, EJECTOR	1
6	PAOZZ	1005009927291	19204	8448515	EJECTOR,CARTRIDGE	1
7	PAOZZ	1005009927288	19204	8448512	EXTRACTOR,CARTRIDGE	1
8	PAOZZ	1005007603768	19200	8448755	SPRING ASSEMBLY,EXT EXTRACTOR (M16A2) UOC:AR8	1
9	PAOZZ	1005014245899	19200	12972692	SPRING ASSEMBLY,EXT EXTRACTOR (M4 AND M4A1)(BLACK) UOC:AS1 ,AY6	1

END OF FIGURE

SECTION II

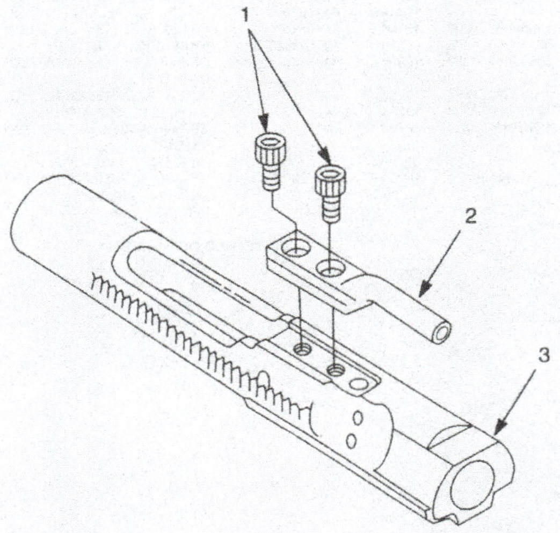

Figure C-4. Key and Bolt Carrier Assembly 8448505.

(1) ITEM NO	(2) SMR CODE	(3) NSN	(4) CAGEC	(5) PART NUMBER	(6) DESCRIPTION AND USABLE ON CODE (UOC)	(7) QTY
					GROUP 0102 KEY AND BOLT CARRIER ASSEMBLY 8448505 FIG. C-4. KEY AND BOLT CARRIER ASSEMBLY	
1	PAFZZ	5305009927284	19204	8448508	SCREW,CARRIER AND K KEY	2
2	PAFZZ	1005009927283	19200	8448506	KEY,BOLT CARRIER	1
3	PAFZZ		19200	8448507	CARRIER,BOLT	1

END OF FIGURE

SECTION II

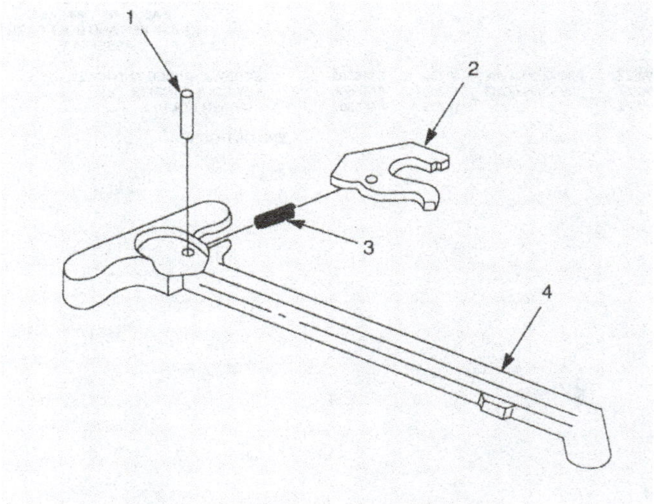

Figure C-5. Handle Assembly 8448517.

(1) ITEM NO	(2) SMR CODE	(3) NSN	(4) CAGEC	(5) PART NUMBER	(6) DESCRIPTION AND USABLE ON CODE (UOC)	(7) QTY
					GROUP 02 HANDLE ASSEMBLY 8448517 FIG. C-5. HANDLE ASSEMBLY	
1	PAOZZ	5315010489372	19204	8448521-2	PIN,SPRING CHARGING HANDLE	1
2	PAOZZ	5340009990405	19200	8448519	LATCH,CHARGING HAND HANDLE	1
3	PAOZZ	5360009990404	19204	8448520	SPRING,HELICAL,COMP COMPRESSION, CHARGING HANDLE	1
4	XAOZZ		19204	8448518	HANDLE ,CHARGING	1

END OF FIGURE

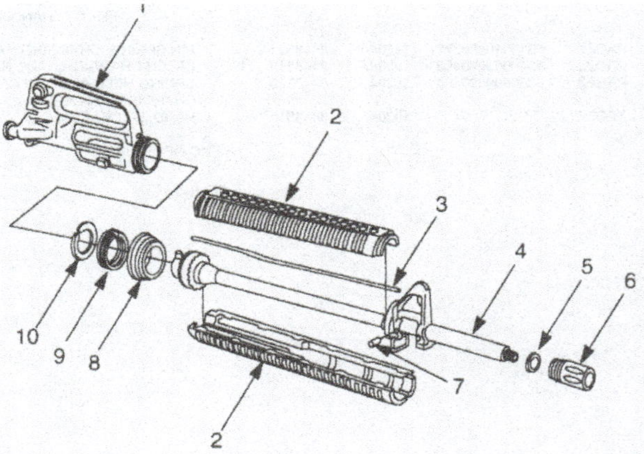

M16A2 Rifle

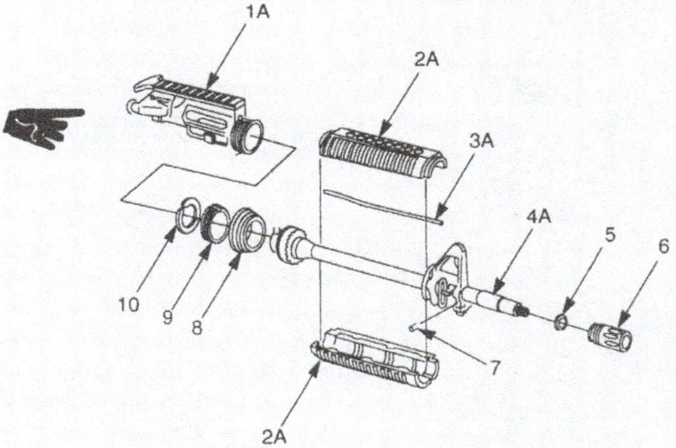

M4/M4A1 Carbine

Figure C-6. Upper Receiver and Barrel Assembly (M16A2) 9349050, and (M4 and M4A1) 12972680.

(1) ITEM NO	(2) SMR CODE	(3) NSN	(4) CAGEC	(5) PART NUMBER	(6) DESCRIPTION AND USABLE ON CODE (UOC)	(7) QTY
					GROUP 03 UPPER RECEIVER AND BARREL ASSEMBLY (M16A2) 9349050, (M4 AND M4A1] 12972680 FIG. C-6. UPPER RECEIVER AND BARREL ASSEMBLY	
1	AFFFF	19200	9349062	UPPER	RECEIVER ASSEMBLY (M16A2).. UOC:AR8	1
1A	AFFFF	19200	12972675	UPPER	RECEIVER ASSY (M4 AND M4A1)...................................... UOC:AS1,AY6	1
2	PAOZZ	1005011343629	19200	9349059	HANDGUARD ASSEMBLY (M16A2)................................... UOC:AR8	2
2A	PAOZZ	1005012342297	19200	9390003	GUARD,HAND,GUN (M4 AND M4A1)................................ UOC:AS1,AY6	2
3	PAFZZ	4710009781038	19200	8448567	TUBE,BENT,METALLIC (M16A2).. UOC:AR8	1
3A	PAFZZ	4710012338637	19200	9390016	TUBE,BENT,METALLIC (M4 AND M4A1) UOC:AS1,AY6	1
4	PAFFF	1005011467684	19200	9349124	BARREL ASSEMBLY (M16A2).. UOC:AR8	1
4A	PAFFF	1005012338529	19200	9390007	BARREL AND FRONT SI SIGHT ASSEMBLY, ... REPLACEMENT (M4 AND M4A1).. UOC:AS1,AY6	1
5	PAFZZ	5365011441496	19200	9349052	SHIM COMPENSATOR...	1
6	PAFZZ	1005011343633	19200	9349051	COMPENSATOR..	1
7	PAFZZ	5315000586044	96906	MS16562-106	PIN,SPRING ,GAS TUBE	1
8	PAFZZ	1005000878998	19204	8448712	RING,SLIP,HAND GUAR GUARD	1
9	PAFZZ	5360009781036	19204	8448555	SPRING,SLIP RING,HA HANDGUARD................................ UPPER RECEIVER	1
10	PAFZZ	5325009990863	96906	M5S16626-3137	RING,RETAINING .... END OF FIGURE	1

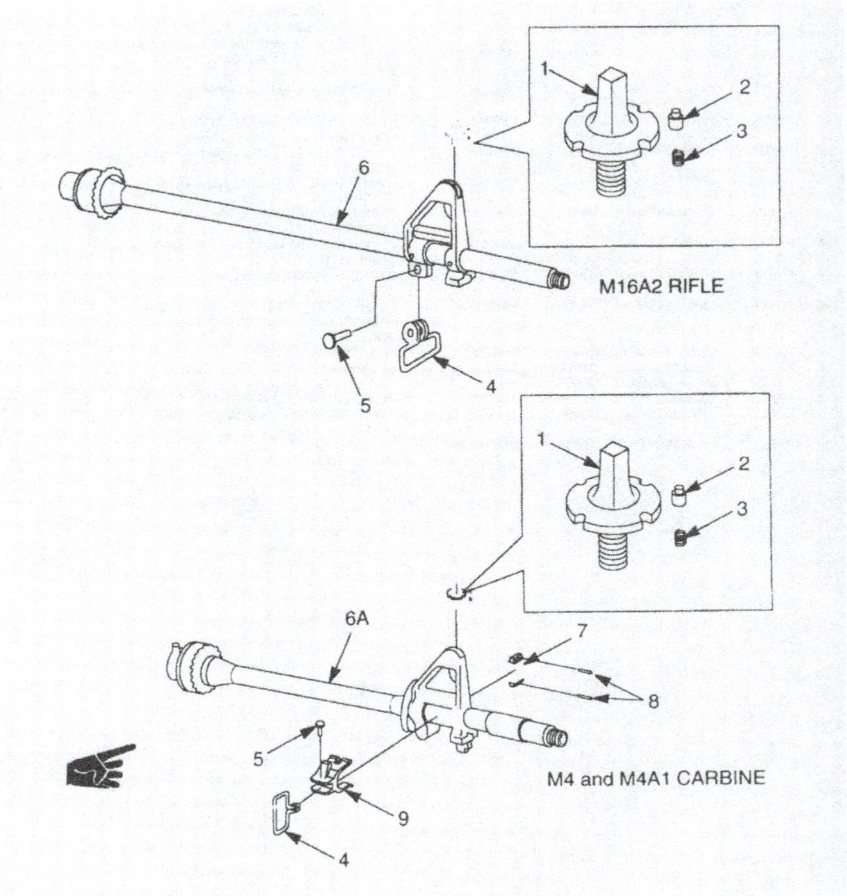

Figure C-7. (M16A2) Barrel Assembly 9349124 and (M4 and M4A1) Replacement Barrel and Front Sight Assembly 9390007.

(1) ITEM NO	(2) SMR CODE	(3) NSN	(4) CAGEC	(5) PART NUMBER	(6) DESCRIPTION AND USABLE ON CODE (UOC)	(7) QTY
					GROUP 0301 (M16A2) BARREL ASSEMBLY 9349124, AND (M4 AND M4A1) REPLACEMENT BARREL AND FRONT SIGHT ASSEMBLY 9390007 FIG. C-7. (M16A2) BARREL ASSEMBLY, AND (M4 AND M4A1) REPLACEMENT BARREL AND FRONT SIGHT ASSEMBLY	
1	PAOZZ	1005011343625	19200	9349056	POST,FRONT SIGHT	1
2	PAOZZ	5315009793930	19204	8448573	PIN,SHOULDER,HEADLE HEADLESS (DETENT, FRONT SIGHT)	1
3	PAOZZ	5360009793931	19204	8448574	SPRING,HELICAL,COMP ,COMPRESSION, FRONT SIGHT	1
4	PAOZZ	1005000179543	19204	8448571	SWIVEL,SLING,SMALL	1
5	PAOZZ	5320010637635	19204	8448697	RIVET,TUBULAR	1
6	XAFZZ		19200	9349054	BARREL AND BARREL E EXTENSION ASSEMBLY (M16A2) UOC:AR8	1
6A	XAFZZ		19200	9390009	BARREL AND BARREL EXTENSION ASSEMBLY (M4 AND M4A1) UOC:AS1,AY6	1
7	PAOZZ	5340012646530	19200	12598618	CLAMP,RIM CLENCHING (M4 AND M4A1) UOC:AS1,AY6	1
8	PAOZZ	5315006900544	96906	MS39086-93	PIN,SPRING (M4 AND M4A1) UOC:AS1,AY6	2
9	PAOZZ	1010012646517	19200	12598617	MOUNT,SWIVEL (M4 AND M4A1) UOC:AS1,AY6	1

END OF FIGURE

SECTION II

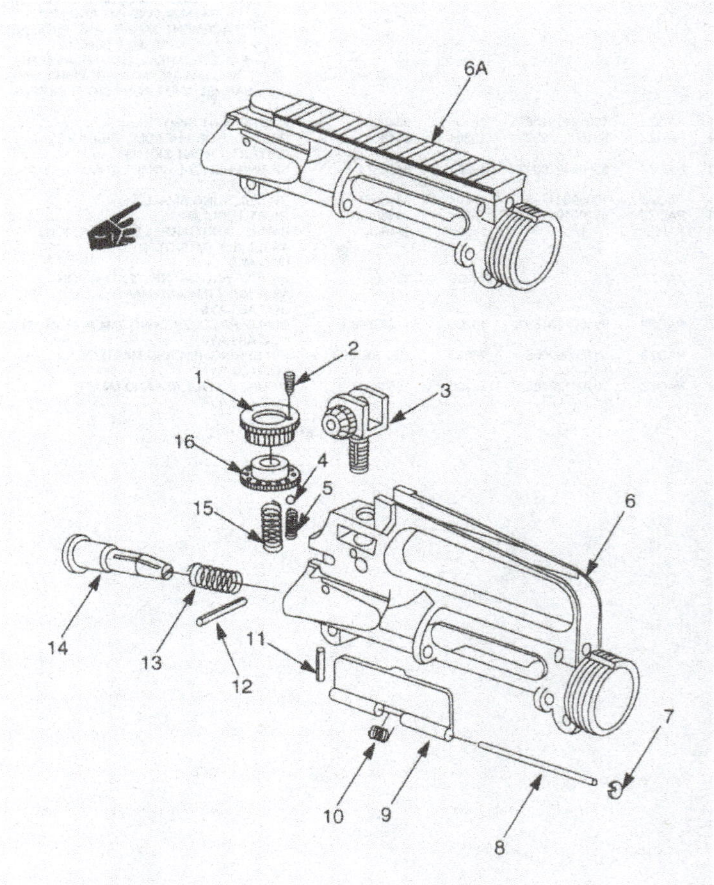

Figure C-8. Upper Receiver Assembly (M16A2) 9349062, and (M4 and M4A1) 12972675.

(1) ITEM NO	(2) SMR CODE	(3) NSN	(4) CAGEC	(5) PART NUMBER	(6) DESCRIPTION AND USABLE ON CODE (UOC)	(7) QTY
					GROUP 0302 UPPER RECEIVER ASSEMBLY (M16A2) 9349062, (M4 AND M4A1) 12972675 FIG. C-8. UPPER RECEIVER ASSEMBLY	
1	PAFZZ	1005011343621	19200	9349066	INDEX,ELEVATION (M16A2) UOC:AR8	1
2	PAFZZ	5305011343622	19200	9349065	SCREW,INDEX (M16A2)	1
3	AFFFF	19200	9349072	REAR	SIGHT ASSEMBLY (M16A2) UOC:AR8	1
4	PAFZZ	3110001839175	96906	MS19060-4808	BALL,BEARING (M16A2) UOC:AR8	1
5	PAFZZ	5360011481751	19200	9349069	SPRING,HELICAL,COMP COMPRESSION, ... INDEX (M16A2) UOC:AR8	1
6	PAFZZ	1005011343701	19200	9349063	RECEIVER,UPPER (M16A2) UOC:AR8	1
6A	PAFZZ	1005013826795	19200	12972670	RECEIVER,CARTRIDGE (M4 AND M4A1) UOC:AS1,AY6	1
7	PAOZZ	5325009990864	96906	MS16632-3012	RING,RETAINING ,COVER	1
8	PAOZZ	5315009781023	19204	8448533	PIN,GROOVED,HEADLES COVER	1
9	PAOZZ	1005009781022	19204	8448525	COVER,EJECTION PORT	1
10	PAOZZ	5360009781025	19204	8448532	SPRING,HELICAL,TORS ,COVER	1
11	PAFZZ	5315008403812	96906	M5S16562-121	PIN,SPRING FORWARD ASSIST	1
12	PAFZZ	5315008403812	96906	M5S16562-121	PIN,SPRING (M16A2)	1
13	PAFZZ	5360000179541	19200	8448540	SPRING,HELICAL,COMP COMPRESSION .... FORWARD ASSIST	1
14	AFFFF		19204	9349086	FORWARD ASSIST ASSY ASSEMBLY	1
15	PAFZZ	5360011343710	19200	9349070	SPRING,HELICAL,COMP COMPRESSION, ... ELEVATION (M16A2)	1
16	PAFZZ	5355011354972	19200	9349067	KNOB,ELEVATION (M16A2) UOC:AR8	1

END OF FIGURE

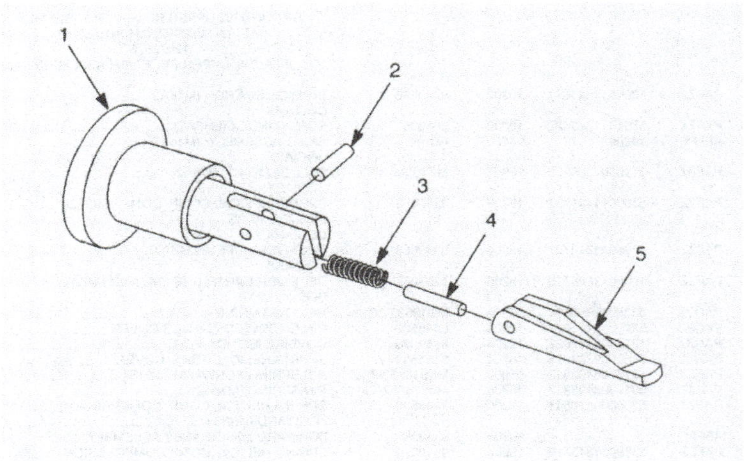

Figure C-9. Forward Assist Assembly 9349086.

(1) ITEM NO	(2) SMR CODE	(3) NSN	(4) CAGEC	(5) PART NUMBER	(6) DESCRIPTION AND USABLE ON CODE (UOC)	(7) QTY
					GROUP 030201 FORWARD ASSIST ASSEMBLY 9349086	
					FIG. C-9. FORWARD ASSIST ASSEMBLY	
1	PAFZZ	1005011441468	19200	9349085	PLUNGER ASSEMBLY	1
2	PAFZZ	5315010489372	19204	8448521-2	PIN,SPRING ,PAWL	1
3	PAFZZ	5360005238084	19200	8448542	SPRING,HELICAL,COMP COMPRESSION PAWL	1
4	PAFZZ	1005000179540	19204	8448544	DETENT,PAWL	1
5	PAFZZ	3040000179539	19204	8448543	PAWL,FORWARD ASSIST	1

END OF FIGURE

SECTION II

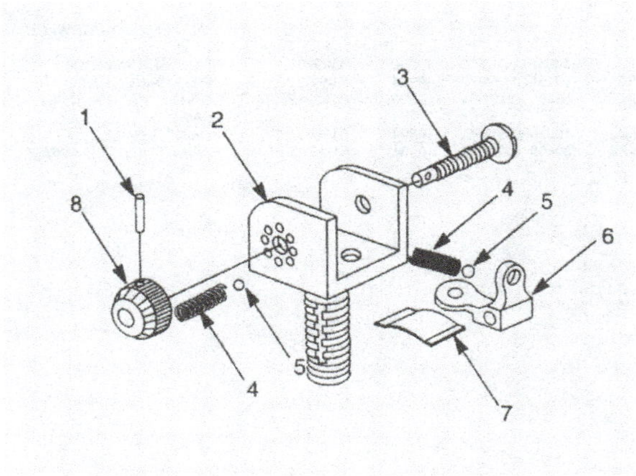

Figure C-10. (M16A2) Rear Sight Assembly 9349072.

(1) ITEM NO	(2) SMR CODE	(3) NSN	(4) CAGEC	(5) PART NUMBER	(6) DESCRIPTION AND USABLE ON CODE (UOC)	(7) QTY
					GROUP 030202 (M16A2) REAR SIGHT ASSEMBLY 9349072 FIG. C-10. REAR SIGHT ASSEMBLY	
1	PAFZZ	5315000586678	96906	MS16562-103	PIN,SPRING ,WINDAGE	1
2	PAFZZ	1005011343631	19200	9349074	BASE,REAR SIGHT .... UOC:ARB	1
3	PAFZZ	5305011441490	19200	9349076	SCREW,EXTERNALLY RE	1
4	PAFZZ	5360011481751	19200	9349069	SPRING,HELICAL,COMP ,REAR SIGHT UOC:AR8	2
5	PAFZZ	3110001839175	96906	M5S19060-4808	BALL,BEARING.. UOC:AR8	2
6	PAFZZ	1005011353697	19200	9349075	APERTURE,SIGHT	1
7	PAFZZ	5360013816183	19200	12011987	SPRING,FLAT REAR SIGHT..	1
8	PAFZZ	5355011343627	19200	9349077	KNOB,WINDAGE	1

END OF FIGURE

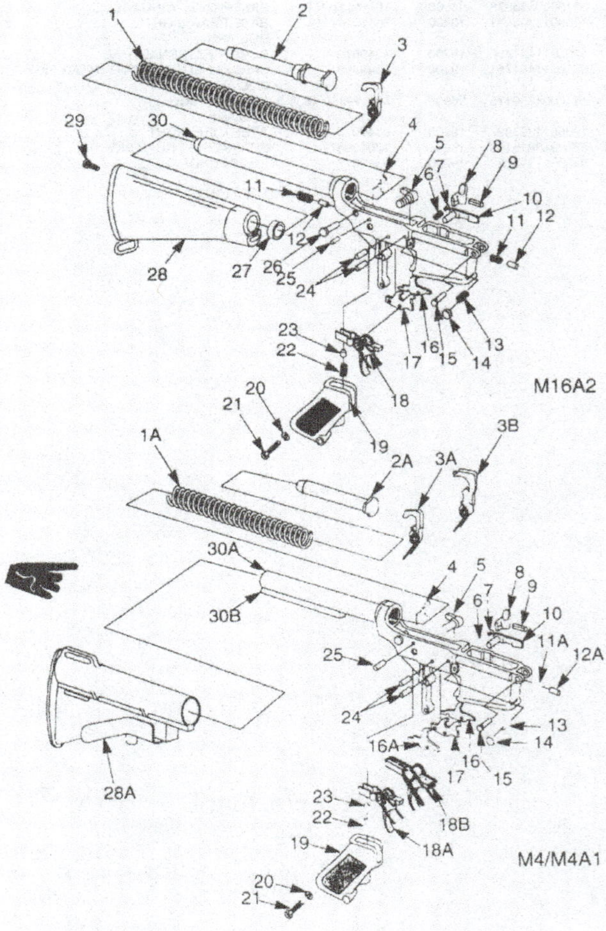

Figure C-11. Lower Receiver and Buttstock Assembly (MI 6A2) 9349100, (M4) 9390011 and (M4A1) 12972690.

(1) ITEM NO	(2) SMR CODE	(3) NSN	(4) CAGEC	(5) PART NUMBER	(6) DESCRIPTION AND USABLE ON CODE (UOC)	(7) QTY
					GROUP 04 LOWER RECEIVER AND BUTTSTOCK ASSEMBLY (M16A2) 9349100, (M4) 9390011, AND (M4A1) 12972690 FIG C-11 LOWER RECEIVER AND BUTTSTOCK ASSEMBLY	
1	PAOZZ	5360009926665	19204	8448629	SPRING, HELICAL, COMP ACTION (M16A2) UOC:AR8	1
1A	PAOZZ	5360012338617	19200	9390022	SPRING, HELICAL, COMP COMPRESSION (M4 AND M4A1) UOC:AS1, AY6	1
2	PAOZZ	1005009373078	19200	8448615	BUFFER ASSEMBLY (MI6A2) UOC:AR8	1
2A	PAOZZ	1005012313138	19200	9390023	BUFFER ASSEMBLY (M4 AND M4AI) UOC:AS1, AY6	1
3	AFFFF		19200	9349106	HAMMER ASSEMBLY (M16A2) UOC:AR8	1
3A	AFFFD		19200	9390032	HAMMER ASSEMBLY (M4) UOC:AS1	1
3B	AFFFF		19204	8448610	HAMMER ASSY (M4A1) UOC:AY6	1
4	PAFZZ	1005009926649	19200	8448595	SEAR	1
5	PAFZZ	5340012258339	19200	9381367	SELECTOR, FIRE CONTROL	1
6	PAFZZ	5360000562246	19204	8448633	SPRING, HELICAL, COMP COMPRESSION, BOLT CATCH	1
7	PAFZZ	1005000562247	19204	8448634	PLUNGER, BOLT CATCH	1
8	PAFZZ	1005000179548	19200	8448628	CATCH, BOLT	1
9	PAFZZ	5315008123312	96906119	MS16562-119	PIN, SPRING , BOLT CATCH	1
10	PAFZZ	1005000562201	19204	8448638	CATCH, MAGAZINE	1
11	PAOZZ	5360009926655	19204	8448586	SPRING, HELICAL, COMP COMPRESSION, TAKE DOWN/PIVOT PIN (M16A2) UOC:AR8	2
11A	PAOZZ	5360009926655	19204	8448586	SPRING, HELICAL, COMP COMPRESSION (M4 AND M4AI) UOC:AS1	1
12	PAOZZ	5315009926654	19204	8448585	PIN, STRAIGHT, HEADLE DETENT, TAKEDOWN PIN (M16A2) UOC:ARS	2
12A	PAOZZ	5315009926654	19204	8448585	PIN, STRAIGHT, HEADLE HEADLESS (M4 AND M4A1) UOC:AS1, AY6	1
13	PAFZZ	5360009927301	19204	8448637	SPRING, HELICAL, COMP COMPRESSION, MAGAZINE CATCH	1
14	PAFZZ	1005009927302	19204	8448636	BUTTON, MAGAZINE CAT CATCH	1
15	PAOZZ	5315000179537	19204	8448621	PIN, GROOVED, HEADED (PIVOT PIN)	1
16	PAFZZ	5340011457910	19200	9349114	LEVER, LOCK-RELEASE SEMI (M16A2 AND M4) UOC:AR8, AS1	1
16A	PAFZZ	1005009990406	19200	8448635	DISCONNECTOR (M4A1) UOC:AY6	1
17	PAFZZ	5340011441499	19200	9349113	LEVER, LOCK-RELEASE , BURST (MI6A2 AND M4)	1

(1) ITEM NO	(2) SMR CODE	(3) NSN	(4) CAGEC	(5) PART NUMBER	(6) DESCRIPTION AND USABLE ON CODE (UOC)	(7) QTY
					UOC:AR8, AS1	
18	AFFFF		19200	9349115	TRIGGER ASSEMBLY (M16A2)	1
					UOC:AR8	
18A	AFFFF		19200	12972697	TRIGGER ASSEMBLY (M4)	1
					UOC:AS1	
18B	AFFFF		19200	12972698	TRIGGER ASSEMBLY (M4A1)	1
					UOC:AY6	
19	PAOZZ	1005011484805	19200	9349127	GRIP, RIFLE PLASTIC, BLACK	1
20	PAOZZ	5310005273634	96906	MS35335-61	WASHER, LOCK , RIFLE GRIP	1
21	PAOZZ	5305012681191	88044	AN501D416-18	SCREW, MACHINE , RIFLE GRIP	1
22	PAOZZ	5360009927292	19204	8448516	SPRING, HELICAL, COMP SAFETY	1
23	PAOZZ	1005009926667	19204	8448631	DETENT, SAFETY	1
24	PAFZZ	5315009927309	19204	8448609	PIN, GROOVED, HEADLES TRIGGER AND HAMMER	2
25	PAFZZ	5315009926650	19204	8448599	PIN, GROOVED, HEADLES , AUTOMATIC SEAR	1
26	PAFZZ	5315009926653	19204	8448584	PIN, GROOVED, HEADED (M16A2)	1
27	PAOZZ	5365012672169	19200	12597640	SPACER, STEPPED (M16A2)	1
					UOC:AR8	
28	PAOOO	1005011354973	19200	9349119	BUTTSTOCK ASSEMBLY, (M16A2)	1
					UOC:AR8	
28A	AOOOO		19200	9390012	BUTTSTOCK ASSY (M4 AND M4A1)	1
					UOC:AS1, AY6	
29	PAOZZ	5305011478585	19200	9349128	SCREW, MACHINE , BUTTCAP (M16A2)	1
					UOC:AR8	
30	XAFFA		19200	9349101	LOWER RECEIVER AND RECEIVER EXTENSION ASSEMBLY (M16A2)	1
					UOC:AR8	
30A	XAFFA		99999	NPN	LOWER RECEIVER ASSY ASSEMBLY (M4)	1
					UOC:AS1	
30B	XAFFA		99999	NPN	LOWER RECEIVER ASSY (M4AI)	1
					UOC:AY6	
					END OF FIGURE	

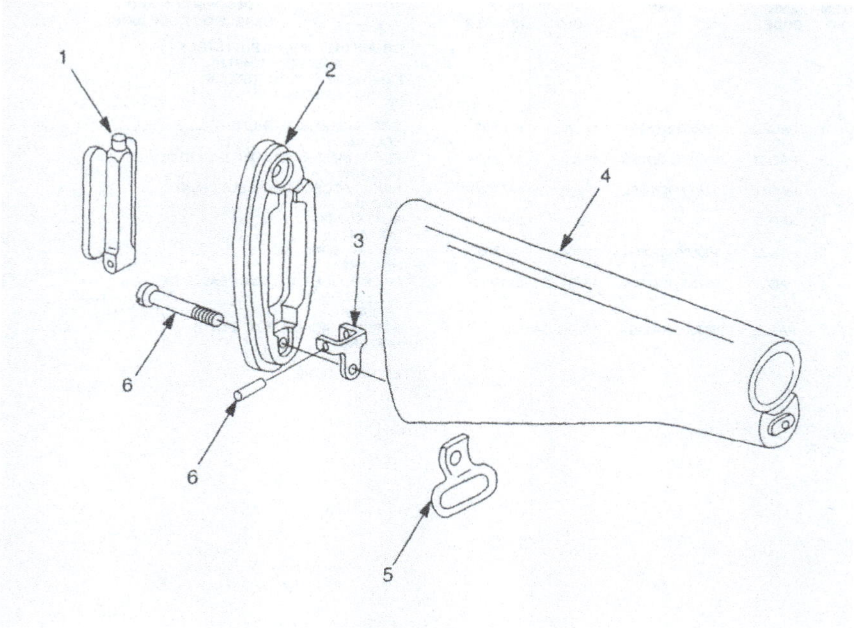

Figure C-12. (M16A2) Buttstock Assembly 9349119.

(1) ITEM NO	(2) SMR CODE	(3) NSN	(4) CAGEC	(5) PART NUMBER	(6) DESCRIPTION AND USABLE ON CODE (UOC)	(7) QTY
					GROUP 0401 (M16A2) BUTTSTOCK ASSEMBLY 9349119 FIG C-12 (M16A2) BUTTSTOCK ASSEMBLY	
1	PAOZZ	1005012288504	19200	9381380	DOOR ASSEMBLY, THUMB UOC:AR8	1
2	PAOZZ	1005011467685	19200	9349130	PLATE, BUTT, SHOULDER GUN STOCK UOC:AR8	1
3	PAOZZ	5340004633892	19200	8448653	HINGE, ACCESS DOOR BUTT PLATE.......... UOC:AR8	1
4	XAOZZ		19200	9349121	BUTTSTOCK.......... UOC:AR8	1
5	PAOZZ	1005004030964	19204	8448652	SWIVEL, SLING, SMALL.......... UOC:AR8	1
6	PAOZZ	5315004633894	19204	8448655	PIN, STRAIGHT, HEADLE HEADLESS, ACCESS DOOR UOC:AR8	1
7	PAOZZ	5305011441494	19200	9349120	SCREW, MACHINE BUTT PLATE UOC:AR8	1
					END OF FIGURE	

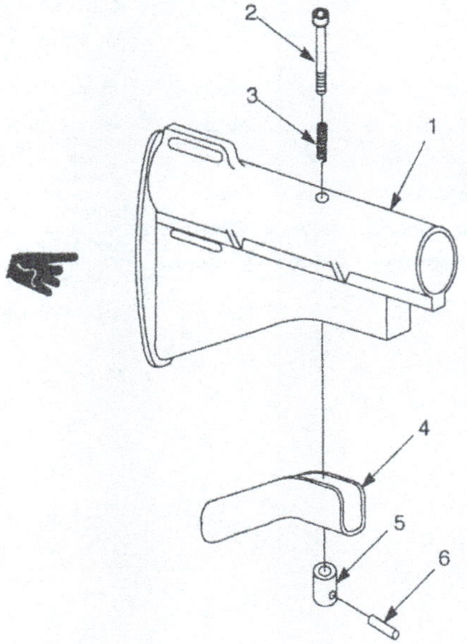

Figure C12A. (M4 and M4A1) Buttstock Assembly 9390012.

(1) ITEM NO	(2) SMR CODE	(3) NSN	(4) CAGEC	(5) PART NUMBER	(6) DESCRIPTION AND USABLE ON CODE (UOC)	(7) QTY
					GROUP 0401A (M4 AND M4A1) 9390012 BUTTSTOCK ASSEMBLY FIG C12A (M4 AND M4A1) BUTTSTOCK ASSEMBLY	
1	PAOZZ	1005012338636	19200	9390013	STOCK, GUN, SHOULDER UOC:AS1, AY6	1
2	PAOZZ	5315012338608	19200	9390025	PIN, SHOULDER, HEADLE UOC:AS1, AY6	1
3	PAOZZ	5360012338616	19200	9390027	SPRING, HELICAL, COMP COMPRESSION UOC:AS1, AY6	1
4	PAOZZ	5340012338638	19200	9390014	LEVER, LOCK-RELEASE UOC:AS1, AY6	1
5	PAOZZ	5310012338626	19200	9390026	NUT, SELF-LOCKING, RO UOC:AS1, AY6	1
6	PAOZZ	5315008439487	96906	MS16562-202	PIN, SPRING UOC:AS1, AY6	1

END OF FIGURE

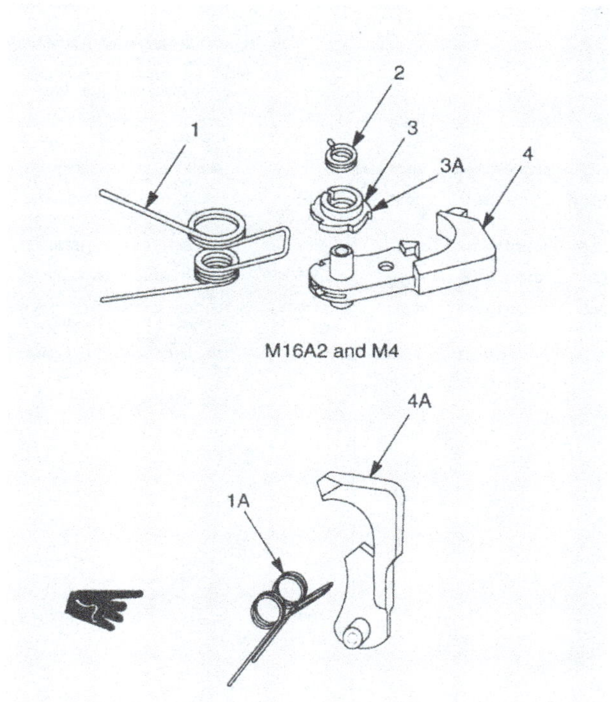

Figure C-13. Hammer Assembly(M16A2) 9349106, (M4) 9390032, and (M4A1) 8448610.

(1) ITEM NO	(2) SMR CODE	(3) NSN	(4) CAGEC	(5) PART NUMBER	(6) DESCRIPTION AND USABLE ON CODE (UOC)	(7) QTY
					GROUP 0402 HAMMER ASSEMBLY (M16A2) 9349106, (M4) 9390032, AND (M4A1) 8448610 FIG C-13 HAMMER ASSEMBLY	
1	PAFZZ	5360011441492	19200	9349107	SPRING, HELICAL, TORS HAMMER (MI6A2 AND M4) UOC:AR8, AS1	1
1A	PAFZZ	5360009926648	19204	8448611	SPRING, HELICAL, TORS (M4A1) UOC:AY6	1
2	PAFZZ	5360011365471	19200	9349109	SPRING, HELICAL, TORS BURST CAM (M16A2 AND M4) UOC:AR8, AS1	1
3	PAFZZ	1005011480172	19200	9349108	CAM, BURST (M16A2) (BLACK) UOC:AR8	1
3A	PAFZZ	3040012477969	19200	9390031	CAM, CONTROL (M4) (NICKEL/SHINY) UOC:AS1	1
4	PAFZZ	1005011343630	19200	9349110	HAMMER AND HAMMER P PIN RETAINER ASSEMBLY (MI6A2 AND M4) UOC:AR8, AS1	1
4A	PAFZZ	1005000179551	19200	8448612	HAMMER, FIRING, SMALL (M4A1) UOC:AY6	1

END OF FIGURE

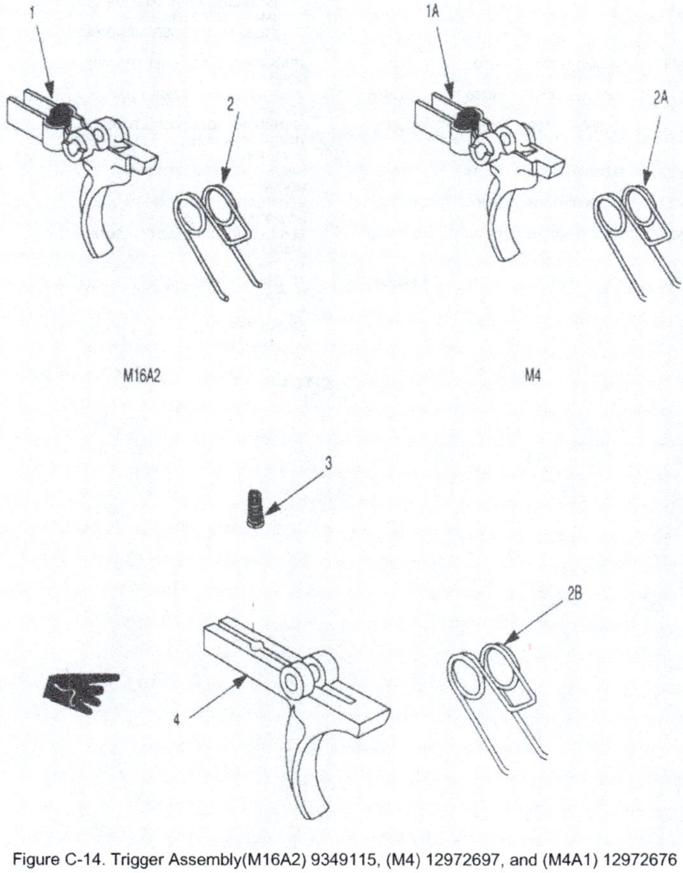

Figure C-14. Trigger Assembly(M16A2) 9349115, (M4) 12972697, and (M4A1) 12972676

(1) ITEM NO	(2) SMR CODE	(3) NSN	(4) CAGEC	(5) PART NUMBER	(6) DESCRIPTION AND USABLE ON CODE (UOC)	(7) QTY
					GROUP 0403 TRIGGER ASSEMBLY (M16A2) 9349115, (M4) 12972697, AND (M4A1) 12972676 FIG C-14 TRIGGER ASSEMBLY	
1	PAFFF	1005012192402	19200	9392518	TRIGGER SUB ASSEMBL (M16A2)...... UOC:AR8	1
1A	PAFFF	1005013954257	19200	12972696	TRIGGER SUBASSEMBLY (M4)...... UOC:AS1	1
2	PAFZZ	5360009927308	19204	8448593	SPRING, HELICAL, TORS TORSION, ... TRIGGER (MI6A2)...... UOC:AR8	1
2A	PAFZZ	5360009927308	19204	8448593	SPRING, HELICAL, TORS (M4)...... UOC:ASI	1
2B	PAFZZ	5360009927308	19204	8448593	SPRING, HELICAL, TORS (M4A1)...... UOC:AY6	1
3	PAFZZ	5360013960256	19200	12972695	SPRING, HELICAL, COMP DISCONNECT ...... (M4A1)...... UOC:AY6	1
4	PAFZZ	1005009927307	19204	8448592	TRIGGER (M4A1)...... UOC:AY6	1

END OF FIGURE

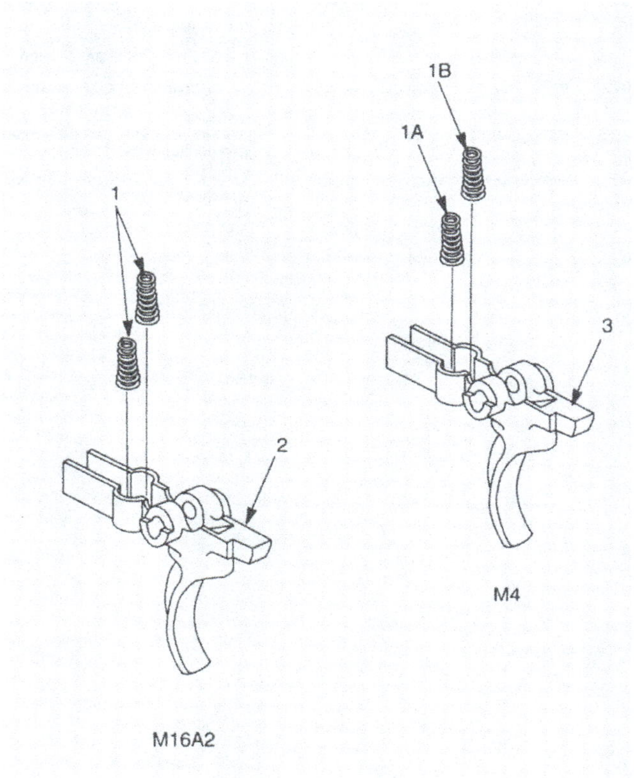

Figure C-15. Trigger Subassembly(M16A2) 9392518, (M4) 12972696.

(1) ITEM NO	(2) SMR CODE	(3) NSN	(4) CAGEC	(5) PART NUMBER	(6) DESCRIPTION AND USABLE ON CODE (UOC)	(7) QTY
					GROUP 040301 TRIGGER SUBASSEMBLY (M6A2) 9392518 AND (M4) 12972696 FIG C-15 TRIGGER SUBASSEMBLY	
1	PAFZZ	5360011350353	19200	9349116	SPRING, HELICAL, COMP DISCONNECT (M16A2) UOC:AR8	2
1A	PAFZZ	5360011350353	19200	9349116	SPRING, HELICAL, COMP DISCONNECT (M4) (NICKEL/SHINY) UOC:AS1	1
1B	PAFZZ	5360013960256	19200	12972695	SPRING, HELICAL, COMP DISCONNECT (M4) (BLACK) UOC:AS1	1
2	XAFZZ		19200	9390736	TRIGGER (M16A2) UOC:AR8	1
3	XAFZZ		19200	9390736	TRIGGER (M4) UOC:AS1	1
					END OF FIGURE	

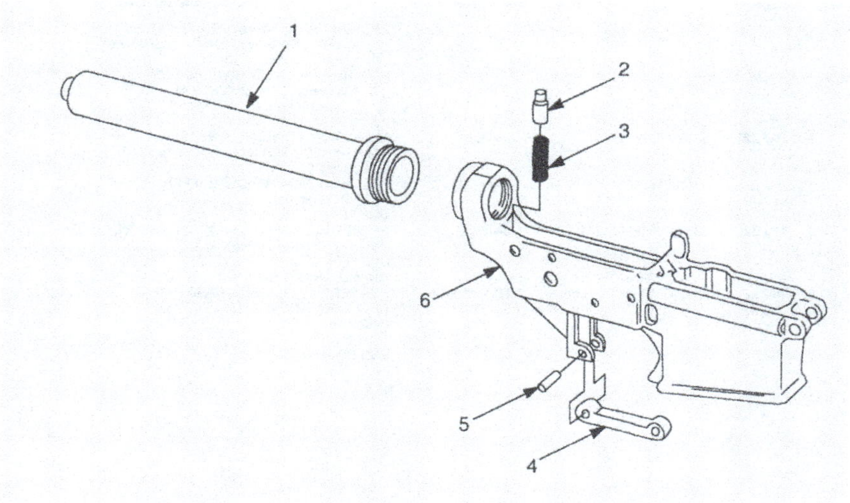

Figure C-16. (M16A2) Lower Receiver and Receiver extension Assembly 9349101.

(1) ITEM NO	(2) SMR CODE	(3) NSN	(4) CAGEC	(5) PART NUMBER	(6) DESCRIPTION AND USABLE ON CODE (UOC)	(7) QTY
					GROUP 0404 (M16A2) LOWER RECEIVER AND RECEIVER EXTENSION ASSEMBLY 9349101	
					FIG C-16 (M16A2) LOWER RECEIVER AND RECEIVER EXTENSION ASSEMBLY	
1	PAFZZ	5340009927297	19200	8448581	EXTENSION, LOWER REC RECEIVER .... (LOWER RECEIVER EXTENSION) (M1SA2)........................ UOC:AR8	1
2	PAFZZ	5315009926651	19204	8448582	PIN, SHOULDER, HEADLE BUFFER.................................... RETAINER (M16A2)... UOC:AR8	1
3	PAFZZ	5360009926652	19200	8448583	SPRING, HELICAL, COMP COMPRESSION, ... BUFFER RETAINER (NI6A2) UOC:AR8	1
4	PAFZZ	1005009927299	19204	8448587	GUARD, TRIGGER (M16A2)...	1
5	PAFZZ	5315000586081	96906	MS16562-129	PIN, SPRING TRIGGER GUARD (M16A2)............................	1
6	XAFDA		19200	9349102	RECEIVER, (M16A2).. UOC:AR8	1

END OF FIGURE

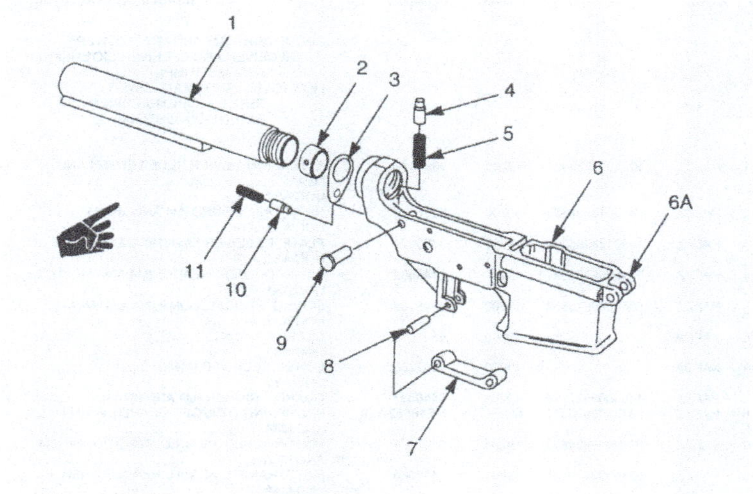

Figure C16A. (M4 and M4A1) Lower Receiver and Receiver extension Assembly, (NPN)

(1) ITEM NO	(2) SMR CODE	(3) NSN	(4) CAGEC	(5) PART NUMBER	(6) DESCRIPTION AND USABLE ON CODE (UOC)	(7) QTY
					GROUP 0404A (M4 AND M4A1) LOWER RECEIVER AND RECEIVER EXTENSION ASSEMBLY (NPN) FIG C16A (M4 AND M4A1) LOWER RECEIVER AND RECEIVER EXTENSION ASSEMBLY	
1	PAFZZ	1005012338531	19200	9390019	EXTENSION, LOWER RECEIVER (M4 AND M4A1) UOC:AS1, AY6	1
2	PAFZZ	5310012338625	19200	9390020	NUT, PLAIN, ROUND (M4 AND M4A1) UOC:AS1, AY6	1
3	PAFZZ	105012338530	19200	9390021	PLATE, RECEIVER END (M4 AND M4A1) UOC:AS1, AY6	1
4	PAFZZ	5315009926651	19204	8448582	PIN, SHOULDER, HEADLE (M4 AND M4A1) UOC:AS1, AY6	1
5	PAFZZ	5360009926652	19200	8448583	SPRING, HELICAL, COMP (M4 AND M4A1) UOC:AS1, AY6	1
6	XAFDA		19200	9390015	RECEIVER, (M4) UOC:AS1	1
6A	XAFDA		19200	12972652	LOWER RECEIVER (M4A1) UOC:AY6	1
7	PAFZZ	1005009927299	19204	8448587	GUARD, TRIGGER (M4 AND M4A1)	1
8	PAFZZ	5315000586081	96906	MS16562-129	PIN, SPRING (TRIGGER GUARD PIN) (M4 AND M4A1)	1
9	PAFZZ	5315009926653	19204	8448584	PIN, GROOVED, HEADED (TAKE DOWN) (M4 AND M4A1)	1
10	PAFZZ	5315009926654	19204	8448585	PIN, STRAIGHT, HEADLE HEADLESS (M4 AND M4A1) UOC:AS1, AY6	1
11	PAFZZ	5360009926655	19204	8448586	SPRING, HELICAL, COMP COMPRESSION (M4 AND M4A1) UOC:AS1, AY6	1
					END OF FIGURE	

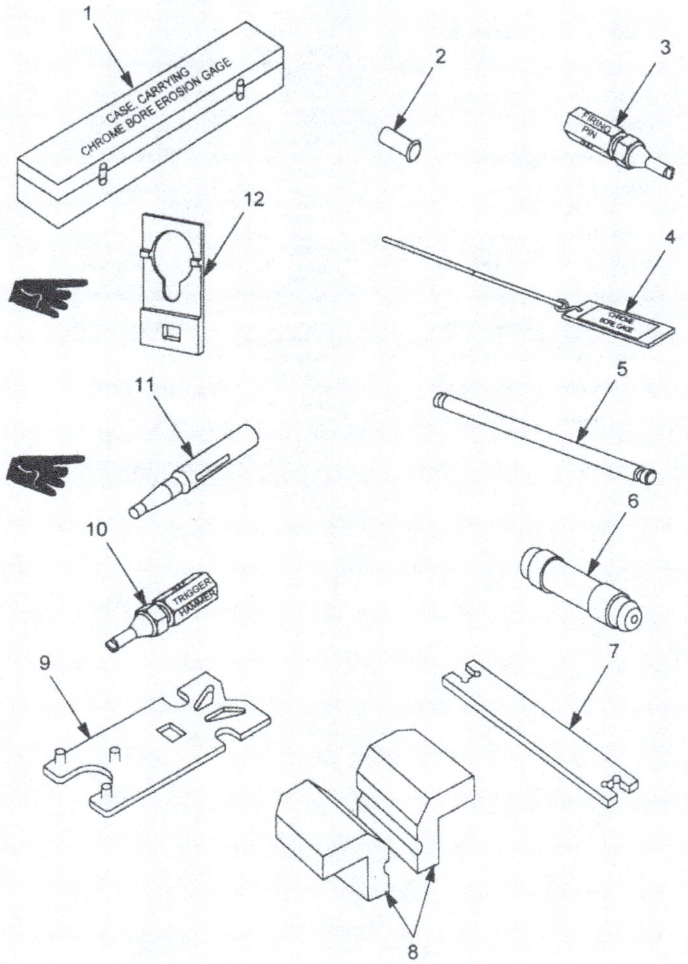

Figure C-17. Special Tools

(1) ITEM NO	(2) SMR CODE	(3) NSN	(4) CAGEC	(5) PART NUMBER	(6) DESCRIPTION AND USABLE ON CODE (UOC)	(7) QTY
					GROUP 9500 SPECIAL TOOLS FIG C-17 SPECIAL TOOLS	
	PEFZZ	1005000567106	19204	8426685	MAINTENANCE KIT, GUN DS/GS SUPPORT MAINTENANCE FOR 5.S6MM RIFLE, M16 RIFLE SERIES BOI: 2 PER SUPPORTING DSU/GSU	
1	PAFZZ	4933010355607	19204	12006359	CASE, BORE GAGE PART OF KIT P/N 8426685	
2	PAFZZ	4933008007508	19204	8448201	..REFLECTOR TOOL, CHAMBER, PART OF KIT P/N 8426685	
3	PAFZZ	5220010755004	19200	12620101	..GAGE, PLUG, PLAIN PART OF KIT P/N 8426685	
4	PAFZZ	5220010148183	19204	8448496	..GAGE, BARREL, EROSION BARREL EROSION PART OF KIT P/N 8426685(CHROME BARREL)	
5	PAFZZ	5220002219391	19204	8448202	..GAGE, STRAIGHTNESS PART OF KIT P/N 8426685	
6	PAFZZ	5220000707814	19204	7799734	..GAGE, HEADSPACE PART OF KIT P/N 8426685	
7	PAFZZ	6695000707815	19204	7799735	..GAGE, FIRING PIN PROTRUSION PART OF KIT P/N 8426685	
8	PAFZZ	4933000709151	19204	11010032	..FIXTURE, BARREL REMOVAL PART OF KIT P/N 8426685	
9	PAFZZ	5120000709152	19204	11010033	..WRENCH, COMBINATION PART OF KIT P/N 8426685	
10	PAFZZ	5220010439473	19204	12006472	GAGE, PLUG, TAPER CYLINDER PART OF KIT P/N 8426685	
11	PAOZZ	5315013100370	19200	12926769	KEY, MACHINE KEY, 2 PER ORGANIZATIONAL OR DIRECT SUPPORT SHOP	
12	PAFZZ	5120013246631	19200	9390035	WRENCH, SPANNER 2 PER DS/GS SUPPORT SHOP A/R FOR M4/M4A1 CARBINE UOC:AS1, AY6	
					END OF FIGURE	

CROSS-REFERENCE INDEXES

NATIONAL STOCK NUMBER INDEX

STOCK NUMBER	FIG	ITEM	STOCK NUMBER	FIG	ITEM
5315-00-017-9537	C-11	15	5315-00-979-3930	C-7	2
3040-00-017-9539	C-9	5	5360-00-979-3931	C-7	3
1005-00-017-9540	C-9	4	5360-00-992-6648	C-13	1A
5360-00-017-9541	C-8	13	1005-00-992-6649	C-11	4
1005-00-017-9543	C-7	4	5315-00-992-6650	C-11	25
1005-00-017-9546	C-1	1	5315-00-992-6651	C-16	2
1005-00-017-9547	C-2	1		C16A	4
1005-00-017-9548	C-11	8	5360-00-992-6652	C-16	3
1005-00-017-9551	C-13	4A		C16A	5
1005-00-056-2201	C-11	10	5315-00-992-6653	C-11	26
5360-00-056-2246	C-11	6		C16A	9
1005-00-056-2247	C-11	7	5315-00-992-6654	C-11	12
1005-00-056-7106	C-17			C-11	12A
5315-00-058-6044	C-6	7		C16A	10
5315-00-058-6081	C-16	5	5360-00-992-6655	C-11	11
	C16A	8		C-11	11A
5315-00-058-6678	C-1B	1		C16A	11
	C-10	1	5360-00-992-6665	C-11	1
5220-00-070-7814	C-17	6	1005-00-992-6667	C-11	23
6695-00-070-7815	C-17	7	1005-00-992-7283	C-4	2
4933-00-070-9151	C-17	8	5305-00-992-7284	C-4	1
5120-00-070-9152	C-17	9	1005-00-992-7285	C-2	4
1005-00-087-8998	C-6	8	1005-00-992-7287	C-3	1
3110-00-183-9175	C-8	4	1005-00-992-7288	C-3	7
	C-1A	6	1005-00-992-7290	C-3	3
	C-1B	5	1005-00-992-7291	C-3	6
	C-10	5	5360-00-992-7292	C-3	5
5220-00-221-9391	C-17	5		C-11	22
1005-00-403-0964	C-12	5	5315-00-992-7294	C-2	3
5340-00-463-3892	C-12	3	5340-00-992-7297	C-16	1
5315-00-463-3894	C-12	6	1005-00-992-7299	C-16	4
5360-00-523-8084	C-9	3		C16A	7
5310-00-527-3634	C-11	20	5360-00-992-7301	C-11	13
5315-00-597-5086	C-3	4	1005-00-992-7302	C-11	14
5315-00-690-0544	C-7	8	1005-00-992-7307	C-14	4
1005-00-738-6213	C-4	3	5360-00-992-7308	C-14	2
1005-00-760-3768	C-3	8		C-14	2A
4933-00-800-7508	C-17	2		C-14	2B
5315-00-812-3312	C-11	9	5315-00-992-7309	C-11	24
5315-00-840-3812	C-8	11	5360-00-999-0404	C-5	3
	C-8	12	5340-00-999-0405	C-5	2
	C-1A	8	1005-00-999-0406	C-11	16A
5315-00-843-9487	C12A	6	5325-00-999-0863	C-6	10
1005-00-921-5004	C-1	4	5325-00-999-0864	C-8	7
1005-00-937-3078	C-11	2	1005-00-999-1509	C-2	2
1005-00-978-1022	C-8	9	5220-01-014-8183	C-17	4
5315-00-978-1023	C-8	8	4933-01-035-5607	C-17	1
5360-00-978-1025	C-8	10	5220-01-043-9473	C-17	10
5360-00-978-1036	C-6	9	5315-01-048-9372	C-5	1
4710-00-978-1038	C-6	3		C-9	2

CROSS-REFERENCE INDEXES

NATIONAL STOCK NUMBER INDEX

STOCK NUMBER	FIG	ITEM	STOCK NUMBER	FIG	ITEM
5320-01-063-7635	C-7	5	5310-01-233-8626	C12A	5
5220-01-075-5004	C-17	3	1005-01-233-8636	C12A	1
1005-01-134-3621	C-8	1	4710-01-233-8637	C-6	3A
5305-01-134-3622	C-8	2	5340-01-233-8638	C12A	4
	C-1A	2	1005-01-234-2297	C-6	2A
1005-01-134-3625	C-7	1	3040-01-247-7969	C-13	3A
5355-01-134-3627	C-1B	8	1010-01-264-6517	C-7	9
	C-10	8	5340-01-264-6530	C-7	7
1005-01-134-3629	C-6	2	5365-01-267-2169	C-11	27
1005-01-134-3630	C-13	4	5305-01-268-1191	C-11	21
1005-01-134-3631	C-10	2	5315-01-310-0370	C-17	11
1005-01-134-3633	C-6	6	5120-01-324-6631	C-17	12
1005-01-134-3701	C-8	6	1005-01-368-9852	C-1	5A
5360-01-134-3710	C-8	15	5360-01-381-6183	C-1B	7
	C-1A	5		C-10	7
5360-01-135-0353	C-15	1	5340-01-382-3201	C-1A	10
	C-15	1A	5310-01-382-6793	C-1A	9
1005-01-135-3697	C-1B	6	1005-01-382-6795	C-8	6A
	C-10	6	5355-01-382-6801	C-1A	4
5355-01-135-4972	C-8	16	5360-01-382-6802	C-1A	7
1005-01-135-4973	C-11	28	1005-01-382-7083	C-1A	11
5360-01-136-5471	C-13	2	1005-01-382-7086	C-1B	2
1005-01-144-1468	C-9	1	1005-01-382-7089	C-1A	3
5305-01-144-1490	C-1B	3	1005-01-395-4257	C-14	1A
	C-10	3	5360-01-396-0256	C-14	3
5360-01-144-1492	C-13	1		C-15	1B
5305-01-144-1494	C-12	7	1005-01-422-3770	C-2	4A
5365-01-144-1496	C-6	5	1005-01-424-5899	C-3	9
5340-01-144-1499	C-11	17			
5340-01-145-7910	C-11	16			
1005-01-146-7684	C-6	4			
1005-01-146-7685	C-12	2			
5305-01-147-8585	C-11	29			
1005-01-148-0172	C-13	3			
5360-01-148-1751	C-8	5			
	C-1B	4			
	C-10	4			
1005-01-148-4805	C-11	19			
1005-01-216-4510	C-1	5			
1005-01-219-2402	C-14	1			
5340-01-225-8339	C-11	5			
1005-01-228-8504	C-12	1			
1005-01-231-3138	C-11	2A			
1005-01-233-8529	C-6	4A			
1005-01-233-8530	C16A	3			
1005-01-233-8531	C16A	1			
5315-01-233-8608	C12A	2			
5360-01-233-8616	C12A	3			
5360-01-233-8617	C-11	1A			
5310-01-233-8625	C16A	2			

CROSS-REFERENCE INDEXES

NATIONAL STOCK NUMBER INDEX

STOCK NUMBER	FIG	ITEM	STOCK NUMBER	FIG	ITEM
88044	AN5010416-18		5305-01-268-1191	C-11	21
96906	MS16562-103		5315-00-058-6678	C-1B	1
				C-10	1
96906	MS16562-106		5315-00-058-6044	C-6	7
96906	MS16562-119		5315-00-812-3312	C-11	9
96906	MS16562-121		5315-00-840-3812	C-8	11
				C-8	12
				C-1A	8
96906	MS16562-129		5315-00-058-6081	C-16	5
				C16A	8
96906	MS16562-202		5315-00-843-9487	C12A	6
96906	MS16562-98		5315-00-597-5086	C-3	4
96906	MS16626-3137		5325-00-999-0863	C-6	10
96906	MS16632-3012		5325-00-999-0864	C-8	7
96906	MS19060-4808		3110-00-183-9175	C-8	4
				C-1A	6
				C-1B	5
				C-10	5
96906	MS35335-61		5310-00-527-3634	C-11	20
96906	MS39086-93		5315-00-690-0544	C-7	8
99999	NPN			C-11	30A
				C-11	30B
19204	11010032		4933-00-070-9151	C-17	8
19204	11010033		5120-00-070-9152	C-17	9
19204	12006359		4933-01-035-5607	C-17	1
19204	12006472		5220-01-043-9473	C-17	10
19200	12011849			C-1	2A
19200	12011987		5360-01-381-6183	C-1B	7
				C-10	7
19200	12011996		1005-01-368-9852	C-1	5A
19200	12597640		5365-01-267-2169	C-11	27
19200	12598617		1010-01-264-6517	C-7	9
19200	12598618		5340-01-264-6530	C-7	7
19200	12620101		5220-01-075-5004	C-17	3
19204	12624561		1005-01-216-4510	C-1	5
19200	12926769		5315-01-310-0370	C-17	11
19200	12951011			C-1	7
19200	12951017		5340-01-382-3201	C-1A	10
19200	12951018		1005-01-382-7089	C-1A	3
19200	12951019		5355-01-382-6801	C-1A	4
19200	12951020		5360-01-382-6802	C-1A	7
19200	12951021		1005-01-382-7083	C-1A	11
19200	12951023		5310-01-382-6793	C-1A	9
19200	12951026			C-1A	1
19200	12951028		1005-01-382-7086	C-1B	2
19200	12972652			C16A	6A
19200	12972670		1005-01-382-6795	C-8	6A
19200	12972675			C-6	1A
19200	12972680			C-1	3A
19200	12972690			C-1	6B
19200	12972691		1005-01-422-3770	C-2	4A

CROSS-REFERENCE INDEXES

PART NUMBER INDEX

CAGEC	PART NUMBER	STOCK NUMBER	FIG	ITEM
19200	12972692	1005-01-424-5899	C-3	9
19200	12972695	5360-01-396-0256	C-14	3
			C-15	1B
19200	12972696	1005-01-395-4257	C-14	1A
19200	12972697		C-11	18A
19200	12972698		C-11	18B
19204	7799734	5220-00-070-7814	C-17	6
19204	7799735	6695-00-070-7815	C-17	7
19204	8426685	1005-00-056-7106	C-17	
19204	8448201	4933-00-800-7508	C-17	2
19204	8448202	5220-00-221-9391	C-17	5
19204	8448496	5220-01-014-8183	C-17	4
19204	8448501		C-1	2
19204	8448502	5315-00-992-7294	C-2	3
19204	8448503	1005-00-017-9547	C-2	1
19204	8448504	1005-00-999-1509	C-2	2
19204	8448505		C-2	5
19200	8448506	1005-00-992-7283	C-4	2
19200	8448507	1005-00-738-6213	C-4	3
19204	8448508	5305-00-992-7284	C-4	1
19200	8448509	1005-00-992-7285	C-2	4
19204	8448510		C-3	2
19204	8448511	1005-00-992-7287	C-3	1
19204	8448512	1005-00-992-7288	C-3	7
19204	8448513	1005-00-992-7290	C-3	3
19204	8448515	1005-00-992-7291	C-3	6
19204	8448516	5360-00-992-7292	C-3	5
			C-11	22
19204	8448517	1005-00-017-9546	C-1	1
19204	8448518		C-5	4
19200	8448519	5340-00-999-0405	C-5	2
19204	8448520	5360-00-999-0404	C-5	3
19204	8448521-2	5315-01-048-9372	C-5	1
			C-9	2
19204	8448525	1005-00-978-1022	C-8	9
19204	8448532	5360-00-978-1025	C-8	10
19204	8448533	5315-00-978-1023	C-8	8
19200	8448540	5360-00-017-9541	C-8	13
19200	8448542	5360-00-523-8084	C-9	3
19204	8448543	3040-00-017-9539	C-9	5
19204	8448544	1005-00-017-9540	C-9	4
19204	8448555	5360-00-978-1036	C-6	9
19200	8448567	4710-00-978-1038	C-6	3
19204	8448571	1005-00-017-9543	C-7	4
19204	8448573	5315-00-979-3930	C-7	2
19204	8448574	5360-00-979-3931	C-7	3
19200	8448581	5340-00-992-7297	C-16	1
19204	8448582	5315-00-992-6651	C-16	2
			C16A	4
19200	8448583	5360-00-992-6652	C-16	3
			C16A	5

CROSS-REFERENCE INDEXES

PART NUMBER INDEX.

CAGEC	PART NUMBER	STOCK NUMBER	FIG	ITEM
19204	8448584	5315-00-992-6653	C-11	26
			C16A	9
19204	8448585	5315-00-992-6654	C-11	12
			C-11	12A
			C16A	10
19204	8448586	5360-00-992-6655	C-11	11
			C-11	11A
			C16A	11
19204	8448587	1005-00-992-7299	C-16	4
			C16A	7
19204	8448592	1005-00-992-7307	C-14	4
19204	8448593	5360-00-992-7308	C-14	2
			C-14	2A
			C-14	2B
19200	8448595	1005-00-992-6649	C-11	4
19204	8448599	5315-00-992-6650	C-11	25
19204	8448609	5315-00-992-7309	C-11	24
19204	8448610		C-11	3B
19204	8448611	5360-00-992-6648	C-13	1A
19200	8448612	1005-00-017-9551	C-13	4A
19200	8448615	1005-00-937-3078	C-11	2
19204	8448621	5315-00-017-9537	C-11	15
19200	8448628	1005-00-017-9548	C-11	8
19204	8448629	5360-00-992-6665	C-11	1
19204	8448631	1005-00-992-6667	C-11	23
19204	8448633	5360-00-056-2246	C-11	6
19204	8448634	1005-00-056-2247	C-11	7
19200	8448635	1005-00-999-0406	C-1l	11A
19204	8448636	1005-00-992-7302	C-11	14
19204	8448637	5360-00-992-7301	C-11	13
19204	8448638	1005-00-056-2201	C-11	10
19204	8448652	1005-00-403-0964	C-12	5
19200	8448653	5340-00-463-3892	C-12	3
19204	8448655	5315-00-463-3894	C-12	6
19200	8448670	1005-00-921-5004	C-1	4
19204	8448697	5320-01-063-7635	C-7	5
19204	8448712	1005-00-087-8998	C-6	8
19200	8448755	1005-00-760-3768	C-3	8
19200	9349050		C-1	3
19200	9349051	1005-01-134-3633	C-6	6
19200	9349052	5365-01-144-1496	C-6	5
19200	9349054		C-7	6
19200	9349056	1005-01-134-3625	C-7	1
19200	9349059	1005-01-134-3629	C-6	2
19200	9349062		C-6	1
19200	9349063	1005-01-134-3701	C-8	6
19200	9349065	5305-01-134-3622	C-8	2
			C-1A	2
19200	9349066	1005-01-134-3621	C-8	1
19200	9349067	5355-01-135-4972	C-8	16
19200	9349069	5360-01-148-1751	C-8	5

CROSS-REFERENCE INDEXES

PART NUMBER INDEX

CAGEC	PART NUMBER	STOCK NUMBER	FIG	ITEM
19200	9349069	5360-01-148-1751	C-1B	4
			C-10	4
19200	9349070	5360-01-134-3710	C-8	15
			C-1A	5
19200	9349072		C-8	3
19200	9349074	1005-01-134-3631	C-10	2
19200	9349075	1005-01-135-3697	C-1B	6
			C-10	6
19200	9349076	5305-01-144-1490	C-1B	3
			C-10	3
19200	9349077	5355-01-134-3627	C-1B	8
			C-10	8
19200	9349085	1005-01-144-1468	C-9	1
19204	9349086		C-8	14
19200	9349100		C-1	6
19200	9349101		C-11	30
19200	9349102		C-16	6
19200	9349106		C-11	3
19200	9349107	5360-01-144-1492	C-13	1
19200	9349108	1005-01-148-0172	C-13	3
19200	9349109	5360-01-136-5471	C-13	2
19200	9349110	1005-01-134-3630	C-13	4
19200	9349113	5340-01-144-1499	C-11	17
19200	9349114	5340-01-145-7910	C-11	16
19200	9349115		C-11	18
19200	9349116	5360-01-135-0353	C-15	1
			C-15	1A
19200	9349119	1005-01-135-4973	C-11	28
19200	9349120	5305-01-144-1494	C-12	7
19200	9349121		C-12	4
19200	9349124	1005-01-146-7684	C-6	4
19200	9349127	1005-01-148-4805	C-11	19
19200	9349128	5305-01-147-8585	C-11	29
19200	9349130	1005-01-146-7685	C-12	2
19200	9381367	5340-01-225-8339	C-11	5
19200	9381380	1005-01-228-8504	C-12	1
19200	9390003	1005-01-234-2297	C-6	2A
19200	9390007	1005-01-233-8529	C-6	4A
19200	9390009		C-7	6A
19200	9390011		C-1	6A
19200	9390012		C-11	28A
19200	9390013	1005-01-233-8636	C12A	1
19200	9390014	5340-01-233-8638	C12A	4
19200	9390015		C16A	6
19200	9390016	4710-01-233-8637	C-6	3A
19200	9390019	1005-01-233-8531	C16A	1
19200	9390020	5310-01-233-8625	C16A	2
19200	9390021	1005-01-233-8530	C16A	3
19200	9390022	5360-01-233-8617	C-11	1A
19200	9390023	1005-01-231-3138	C-11	2A
19200	9390025	5315-01-233-8608	C12A	2

CROSS-REFERENCE INDEXES

PART NUMBER INDEX

CAGEC	PART NUMBER	STOCK NUMBER	FIG	ITEM
19200	9390026	5310-01-233-8626	C12A	5
19200	9390027	5360-01-233-8616	C12A	3
19200	9390031	3040-01-247-7969	C-13	3A
19200	9390032		C-11	3A
19200	9390035	5120-01-324-6631	C-17	12
19200	9390736		C-15	2
			C-15	3
19200	9392518	1005-01-219-2402	C-14	1

APPENDIX D
EXPENDABLE/DURABLE SUPPLIES AND MATERIALS LIST

Section I. INTRODUCTION

D-1. SCOPE. This appendix lists expendable durable supplies and materials you will need to operate and maintain the M16A2 Rifle This listing is for informational purposes only and Is not authority to requisition the listed Items. These items are authorized to you by CTA 50-970, Expendable, Durable Items (Except Medical, Class V, Repair Parts, and Heraldic Items), or CTA 8-100, Army Medical Department Expendable/Durable Items.

D-2. EXPLANATION OF COLUMNS.

a. Column (1)-Item Number. This number is assigned to the entry in the listing and is referenced In the narrative instructions to identify the material (e.g., "Use cloth, abrasive, crocus, Item 12, app D").

b. Column (2)-Level. This column identifies the lowest level of maintenance that requires the listed Item.

C - Operator/Crew Maintenance
O - Unit Maintenance
F - Direct Support Maintenance

c. Column (3)-National Stock Number. This is the National stock number assigned to the Item, use It to request or requisition the Item.

d. Column (4)-Description. Indicates the Federal Item name and, if required, a description to identify the item. The last line for each Item indicates the Commercial and Government Entity Code (CAGEC) in parentheses followed by the part number.

e. Column (5)-Unit of Measure (U/M). Indicates the measure used in performing the actual maintenance function. This measure is expressed by a two-character alphabetical abbreviation (e g., ea, In , pr). If the unit of measure differs from the unit of issue, requisition the lowest unit of issue that will satisfy your requirements.

Section II. EXPENDABLE/DURABLE SUPPLIES AND MATERIALS LIST

(1) ITEM NUMBER	(2) LEVEL	(3) NATIONAL STOCK NUMBER	(4) DESCRIPTION PART NUMBER AND CAGE	(5) UNIT OF MEAS
1	F	8040 00-944 7292	ADHESIVE, I<IT, (81348) MMM A 1754	KT
2	0	8020 00 244 0153	BRUSH, ARTIST'S metal ferrule, flat chisel ed(i(e 7 16 w, 1 1/8, exposed bristle (813481 H B 241	EA
3	F	1005-00-716 2702	BRUSH, CLEANING, SMALL ARMS (192051 7162702	EA
4	C	1005-00-903 1296	BRUSH, CLEANING, SMALL ARMS bore (19204) 11686340	EA
5	C	1005-00 999-1435	BRUSH, CLEANING, SMALL ARMS chamber (19204) 8432358	EA
6	C	1005 00-444-6602	BRUSH, CLEANING, SMALL ARMS tooth 1192041 8448462	EA
7	O	7920-00 205-2401	BRUSHCLEANING, TOOLS AND PARTS (813491 MILS43871	EA
8	O	6850-00-965-2332	CARBON REMOVING COMPOUND (813481 P-C-111	GL
9			CLEANER, LUBRICANT AND PRESERVATIVE (27412)	
	O	9150-01-079-6124	CLP- 4 oz (118 30 ml) bottle	EA
	O	9150-01-054 6453	CLP- 1 pt (0 47 1l bottle	EA
	O	9150-01 053-6688	CLP 7 gal (26 50 11 bottle	EA
10	C	9150 01-102 1473	CLEANER, LUBRICANT AND PRESERVATIVE (81349i MIL L 63460 1/2 oz (14 79 ml) bottle	EA
11	C	9920-00-292 9946	CLEANER, TOBACCO PIPE cotton turf, wire core (89855) DILLSPIPE cleaner (6 per pkg)	
			CLEANING COMPOUND, RIFLE BORE small arms bore cleaner solution (RBC) 1813491 MIL C 372	
	C	6850-00-224-6656	2 oz 159.15 mil) bottle	OZ
	O	6850-00-224-6657	8 oz (236 59 ml) can	CN
	O	6850-00-224-6663	1 gal (3.79 l) can	CN

EXPENDABLE/DURABLE SUPPLIES AND MATERIALS LIST (CONT)

(1) ITEM NUMBER	(2) LEVEL	(3) NATIONAL STOCK NUMBER	(4) DESCRIPTION PART NUMBER AND CAGE	(5) UNIT OF MEAS
13	0	5350-00-221-0872	CLOTH, ABRASIVE: (58536) A-A-1206	SH
- 14			Deleted See Item 23.2	
15	F		DICHLOROMETHANE, TECHNICAL: (81346) ASTM D 4701	
		6810-00-244-0290	5 gal (18.93 1) pail	CN
		6810-00-616-9188	600 lb (272.16 kg) drum	DR
16	0	6850-00-281-1985	DRY CLEANING SOLVENT: (58536) A-A-711 1 gal(3.79 I1) can	GL
17	0	8010-00-297-0560	ENAMEL: olive drab no 3407 (81348) TT-E-527 1 gal (3.791) can	GL
18	0		GLOVES, CHEMICAL AND OIL PROTECTIVE: (81348) ZZ-G-381	
		8415-00-823-7455	Size 9	PR
		8415-00-823-7456	Size10	PR
		8415-00-823-7457	Size 11	PR
19	F	9150-00-754-2595	GREASE, MOLYBDENUM DISULFIDE: (81349) MIL-G-21164	LB
20	C	1005-01-113-0321	HANDLE SECTION, CLEANING ROD, SMALL ARMS: (19204)8436776	EA
21	0	9150-01-260-2534	LUBRICANT, SOLID FILM: (81349) MIL-L-23398 16 OZ (473.18 ml) spray can	OZ
22	C	9150-00-292-9689	LUBRICATING OIL, WEAPONS: (LAW) (81349) MIL-L-14107 1 qt (0.95 1) can	QT

EXPENDABLE/DURABLE SUPPLIES AND MATERIALS LIST (CONT)

(1) ITEM NUMBER	(2) LEVEL	(3) NATIONAL STOCK NUMBER	(4) DESCRIPTION PART NUMBER AND CAGE	(5) UNIT OF MEAS
23			LUBRICATING OIL, WEAPONS: (LSA), semifluid (81349) MIL-L-46000	
	C	9150-00-935-6597	2 oz (59.15 ml) plastic btl	OZ
	C	9150-00-889-3522	4 oz (118.30 ml) plastic btl	OZ
	O	9150-00-687-4241	1 qt (0.95 l1) can	CN
	O	9150-00-753-4686	1 gal(3.79 l1) can	CN
23.1	0	5340-01-230-3181	MOUNTING BRACKET (M4/M4A1 ONLY) (19200)12556036	EA
23.2	0	8010-00-087-0102	PAINT, ENAMEL, SEMIGLOSS: paint for blank firing attachment (M1 5A2) (81348) TT-E-529 1 qt can (RED - M16A2)	EA
23.3	0	8010-01-031-1274	PAINT, ENAMEL, SEMIGLOSS: paint for blank firing attachment (M23) (81348) TT-E-529 1 pt can (YELLOW - M4/M4A1)	EA
24	0	4940-00-795-3595	PAN, WASH: (94453) 1211	EA
25	F	6850-00-826-0981	PENETRANT KIT: (81349) MIL-I-25135	KT
26	F	8135-01-019-1691	POLYETHYLENE SHEET: (84744) PE88-80-2	EA
26.1	0	1005-01-394-7677	PROTECTOR, RAIL (M4/M4A1 ONLY) (19200)12972676	EA
27	C	7920-00-205-1711	RAG, WIPING: (58536) A-A-531 50 lb (22.68 kg) bdl	LB
28	C	1005-00-050-6357	ROD SECTION, CLEANING SMALL ARMS: (19204) 8436775 (3 REQUIRED)	EA
29	F	8030-00-670-8553	SEALING COMPOUND (16059) DEVCONF	KT

EXPENDABLE/DURABLE SUPPLIES AND MATERIALS LIST (CONT)

(1) ITEM NUMBER	(2) LEVEL	(3) NATIONAL STOCK NUMBER	(4) DESCRIPTION PART NUMBER AND CAGE	(5) UNIT OF MEAS
30	C	1005-00-937-2250	SWAB HOLDER SECTION, CLEANING ROD, SMALL ARMS: (19204) 11686327	EA
31	C	1005-00-912-4248	SWAB, SMALL ARMS: (19204)11686408	SH
32	C	6920-01-395-2949	TARGET, 25 METER ZEROING, M16A2: (19200)12012024	BX

APPENDIX E

ILLUSTRATED LIST OF MANUFACTURED ITEMS

E-1. INTRODUCTION.

a. This appendix includes complete instructions for making items authorized to be manufactured or fabricated at unit or direct support maintenance.

b. A part number in alphanumeric order is provided for cross-referencing the part number of the item to be manufactured to the figure which covers fabrication criteria.

c. All bulk matenals needed for manufacture of an item are listed by part number or specification number In a tabular list on the illustration.

E-2. MANUFACTURED ITEMS PART NUMBER INDEX.

INDEX

Item	Figure Number
Front sight detent depressor	E-1
Front sight post removal and installation tool	E-2
Pivot pin removal tool	E-3
Pivot pin installation tool	E-4
Slave pin	E-5
Adapter bar for M12 arms rack	E-6
Modified needle nose pliers	E-7

E-3. MANUFACTURED ITEMS ILLUSTRATIONS - FORMAT AND CONTENT.

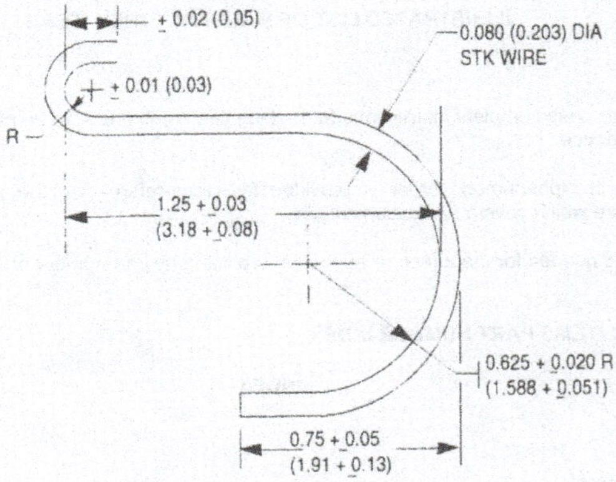

FABRICATE FROM 0.08 IN. MUSIC WIRE OR EQUIVALENT.
FINISH: NO. 5.3.1.2 OR 5.3.2.2 OF MIL-STD 17.

NOTE: ALL DIMENSIONS SHOWN ARE IN INCHES WITH THE METRIC CONVERSION TO CENTIMETERS IN PARENTHESIS.

Figure E-1. Front sight detent depressor.

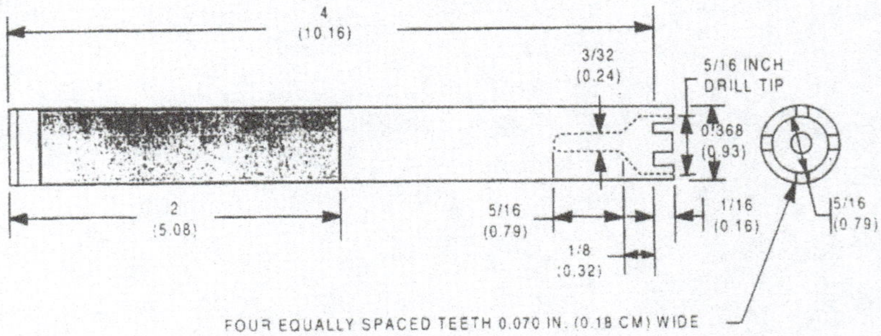

FABRICATE FROM 0.375 INCH ROUND METAL BAR,
ASTM A686. FSCM 81346, GRADE C. CLASS W2-09.
NSN 9510-00-640-4407 OR EQUIVALENT.

NOTES: 1. ALL DIMENSIONS SHOWN ARE IN INCHES WITH THE METRIC CONVERSION TO CENTIMETERS IN PARENTHESIS.
2 TEETH MUST BE HAND FILED TO FIT FRONT SIGHT POST.

Figure E-2. Front sight post removal and Installation tool.

FABRICATE FROM 1'16 IN. SOCKET HEAD SCREW KEY
NSN 5120-00-198-5398 OR EQUIVALENT.

NOTE: ALL DIMENSIONS SHOWN ARE IN INCHES WITH THE METRIC CONVERSION TO CENTIMETERS IN PARENTHESES

Figure E-3. Pivot pin removal tool.

E-3. MANUFACTURED ITEMS ILLUSTRATIONS-FORMAT AND CONTENT (CONT).

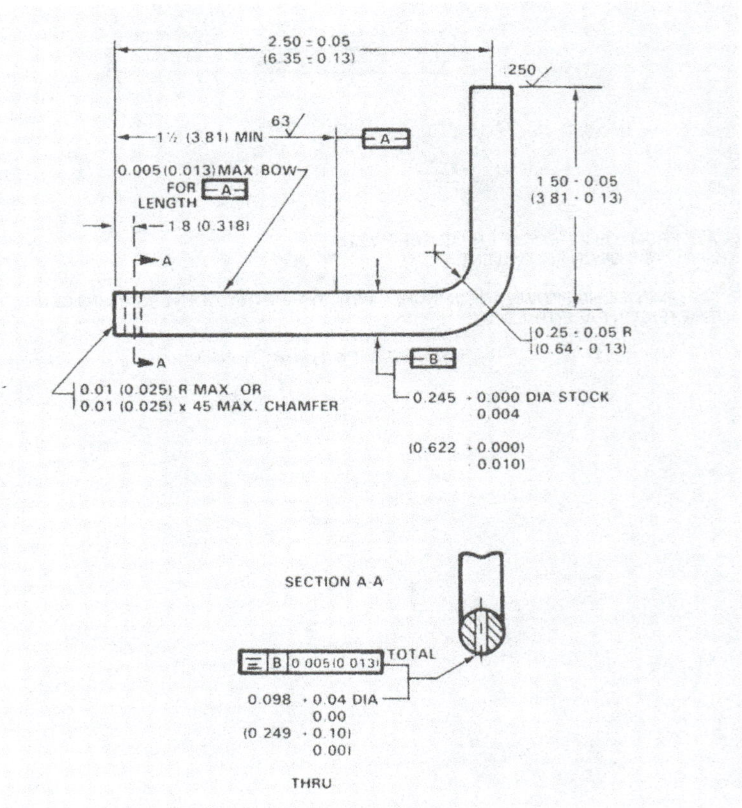

FABRICATE FROM 0 245 IN STEEL AISI 1095 OR EQUIVALENT
HARDEN AND TEMPER TO RC-57-61 FOR LENGTH -A-
FINISH 5 3 1.2 OR 5 3.2.2 OF MIL-STD-171

NOTE: ALL DIMENSIONS SHOWN ARE IN INCHES WITH THE METRIC CONVERSION TO CENTIMETERS IN PARENTHESES

Figure E-4. Pivot pin installation tool.

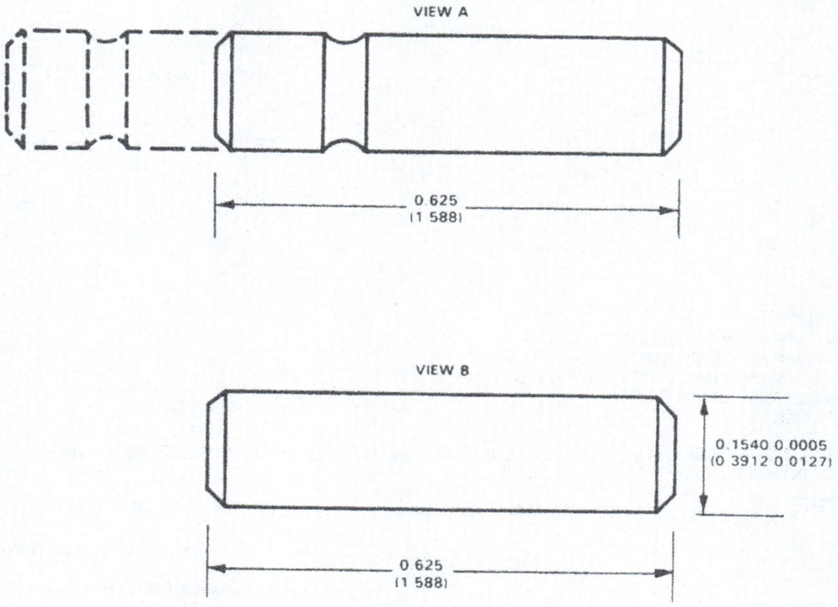

FABRICATE FROM OLD TRIGGER PIN (VIEW AI P,/N 8448609 OR FABRICATE SLAVE PIN (VIEW B) FROM MATERIAL BLOCK, WIRE, STEEL ALLOY, GRADE 4140, ASTM-A547 OR EQUIVALENT

NOTE: ALL DIMENSIONS SHOWN ARE IN INCHES WITH THE METRIC CONVERSION TO CENTIMETERS IN PARENTHESES

Figure E-5. Slave pin.

E-3. MANUFACTURED ITEMS ILLUSTRATIONS -- FORMAT AND CONTENT (CONT).

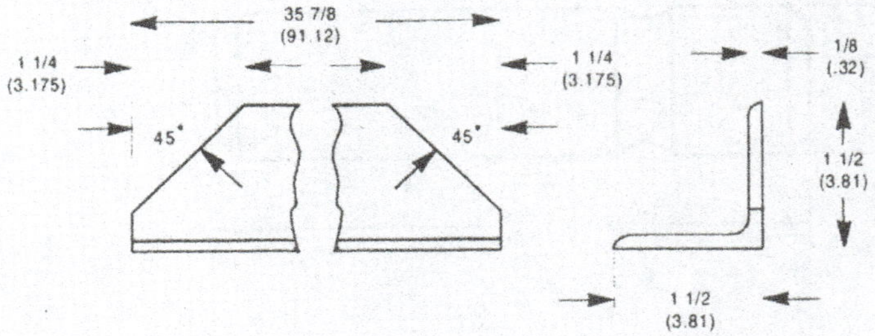

ADAPTER
FABRICATE FROM 1 1/2 INCH BY 1 1/2 INCH
BY 1/8 INCH ANGLE IRON NSN 9520-00-277-4902
OR EQUIVALENT. PAINT WITH OLIVE DRAB
ENAMEL PAINT, NSN 8010-01-350-5249, OR
EQUIVALENT.

NOTE: ALL DIMENSIONS SHOWN ARE IN INCHES WITH THE METRIC CONVERSION TO CENTIMETERS IN PARENTHESES.

Figure E-6. Adapter bar for M12 arms rack.

PIN: 060514-004

***U S Government Printing Office 1995- 648-378**

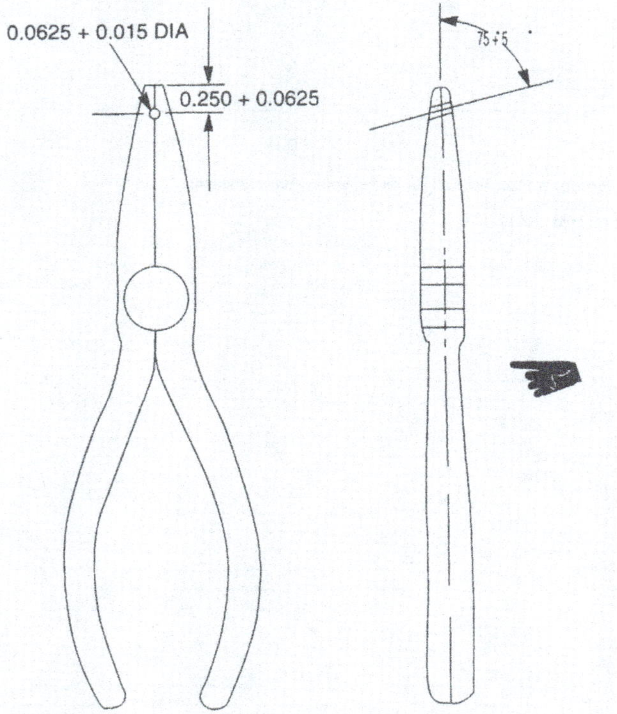

FABRICATE FROM NEEDLE NOSE PLIERS,
NSN 5120-00-268-3579 OR EQUIVALENT.

NOTE: ALL DIMENSIONS ARE IN INCHES.

Figure E-7. Modified needle nose pliers.

***U. S. GOVERNMENT PRINTING OFFICE 1997 - 545-039/4 077**

ALPHABETICAL INDEX

ALPHABETICAL INDEX (CONT)

ALPHABETICAL INDEX (CONT)

ALPHABETICAL INDEX (CONT)

Subject	Page

Made in the USA
Monee, IL
05 November 2026

67813163R00292